VOLUME ONE HUNDRED AND SIXTY ONE

CURRENT TOPICS IN DEVELOPMENTAL BIOLOGY

Retinoids in Development and Disease

CURRENT TOPICS IN DEVELOPMENTAL BIOLOGY

"A meeting-ground for critical review and discussion of developmental processes"
A.A. Moscona and Alberto Monroy (Volume 1, 1966)

SERIES EDITOR
Paul M. Wassarman
Department of Cell, Developmental and Regenerative Biology
Icahn School of Medicine at Mount Sinai
New York, NY, USA

CURRENT ADVISORY BOARD

Blanche Capel
Denis Duboule
Anne Ephrussi
Susan Mango

Philippe Soriano
Claudio Stern
Cliff Tabin
Magdalena Zernicka-Goetz

FOUNDING EDITORS
A.A. Moscona and Alberto Monroy

FOUNDING ADVISORY BOARD

Vincent G. Allfrey
Jean Brachet
Seymour S. Cohen
Bernard D. Davis
James D. Ebert
Mac V. Edds, Jr.

Dame Honor B. Fell
John C. Kendrew
S. Spiegelman
Hewson W. Swift
E.N. Willmer
Etienne Wolff

VOLUME ONE HUNDRED AND SIXTY ONE

Current Topics in
DEVELOPMENTAL BIOLOGY
Retinoids in Development and Disease

Edited by

GREGG DUESTER
*Development, Aging, and Regeneration Program,
Sanford Burnham Prebys Medical Discovery Institute,
La Jolla, CA, United States*

NORBERT B. GHYSELINCK
*Université de Strasbourg, IGBMC UMR 7104; CNRS,
UMR 7104; Inserm, UMR-S 1258; IGBMC, Institut
de Génétique et de Biologie Moléculaire et Cellulaire, Illkirch, France*

Academic Press is an imprint of Elsevier
125 London Wall, London, EC2Y 5AS, United Kingdom
50 Hampshire Street, 5th Floor, Cambridge, MA 02139, United States
525 B Street, Suite 1650, San Diego, CA 92101, United States

First edition 2025

Copyright © 2025 Elsevier Inc. All rights are reserved, including those for text and data mining, AI training, and similar technologies.

Publisher's note: Elsevier takes a neutral position with respect to territorial disputes or jurisdictional claims in its published content, including in maps and institutional affiliations.

No part of this publication may be reproduced or transmitted in any form or by any means, electronic or mechanical, including photocopying, recording, or any information storage and retrieval system, without permission in writing from the publisher. Details on how to seek permission, further information about the Publisher's permissions policies and our arrangements with organizations such as the Copyright Clearance Center and the Copyright Licensing Agency, can be found at our website: www.elsevier.com/permissions.

This book and the individual contributions contained in it are protected under copyright by the Publisher (other than as may be noted herein).

Notices
Knowledge and best practice in this field are constantly changing. As new research and experience broaden our understanding, changes in research methods, professional practices, or medical treatment may become necessary.

Practitioners and researchers must always rely on their own experience and knowledge in evaluating and using any information, methods, compounds, or experiments described herein. In using such information or methods they should be mindful of their own safety and the safety of others, including parties for whom they have a professional responsibility.

To the fullest extent of the law, neither the Publisher nor the authors, contributors, or editors, assume any liability for any injury and/or damage to persons or property as a matter of products liability, negligence or otherwise, or from any use or operation of any methods, products, instructions, or ideas contained in the material herein.

ISBN: 978-0-323-91700-1
ISSN: 0070-2153

For information on all Academic Press publications
visit our website at https://www.elsevier.com/books-and-journals

Publisher: Zoe Kruze
Editorial Project Manager: Devwart Chauhan
Production Project Manager: Maria Shalini
Cover Designer: Arumugam Kothandan

Typeset by MPS Limited, India

Contents

Contributors xi
Preface xiii

1. Early retinoic acid signaling organizes the body axis and defines domains for the forelimb and eye   1
Gregg Duester

1. Introduction 2
   1.1 ATRA signaling begins in mouse at E7.5 when retinol is first converted to ATRA 4
   1.2 ATRA generated in somites controls early posterior body axis formation and somitogenesis by repressing *Fgf8* 6
   1.3 ATRA repression of *Fgf8* regulates early heart anteroposterior patterning 9
2. Hindbrain anteroposterior patterning requires ATRA signaling 9
3. ATRA controls dorsoventral patterning of the spinal cord 10
4. What is the role of ATRA signaling during limb development? 12
   4.1 ATRA is not required for limb patterning (either anteroposterior or proximodistal) 12
   4.2 ATRA is required during forelimb initiation 17
   4.3 How is *Tbx5* activated in the forelimb field? 19
5. Control of optic cup formation and eye morphogenesis by ATRA signaling 20
   5.1 ATRA is required for optic cup formation 21
   5.2 ATRA is required for eye morphogenesis after optic cup formation 22
6. Conclusions 24
Acknowledgments 25
Funding 25
References 25

2. Multiple roles for retinoid signaling in craniofacial development   33
Masahiro Nakamura and Lisa L. Sandell

1. Spatio-temporal regulation of RA signaling 35
2. Retinoid transport 35
3. RA generating enzymes 37

4. RA signal transduction through nuclear receptors	38
5. RA degrading enzymes	39
6. RA role in axial patterning of hindbrain and NCC	39
6.1 Midface	42
7. RA role in early eye development	44
8. RA role in nasal airway and upper lip morphogenesis	45
9. Salivary gland morphogenesis requires RA	47
10. Secondary palate	48
11. Bone and sutures	50
12. Stage-specific RA roles	51
References	52

3. Meiosis and retinoic acid in the mouse fetal gonads: An unforeseen twist   59

Giulia Perrotta, Diana Condrea, and Norbert B. Ghyselinck

1. Introduction	60
2. ATRA and CYP26B1 as meiotic inducing and preventing substances	62
3. Mutant mouse models disqualify ATRA as the MIS	63
4. How to reconcile seemingly irreconcilable observations	67
5. What else, if not ATRA?	69
6. Mutant mouse models support the idea that CYP26B1 acts as an MPS	70
7. The evidence that *Cyp26b1* ablation increases testis ATRA is all refutable	71
8. Meiosis initiates in CYP26B1-deficient testes lacking ALDH1A1	73
9. If it does not degrade ATRA, what does CYP26B1 do?	76
10. What about ATRA in the human fetal gonad?	79
11. Concluding remark	80
Acknowledgments	81
Competing interests statement	81
Funding	81
Data availability statement	81
References	81

4. Retinoids and retinoid-binding proteins: Unexpected roles in metabolic disease   89

William S. Blaner, Jisun Paik, Pierre-Jacques Brun, and Marcin Golczak

1. Introduction	90
2. Recent advances in understanding the biochemistry of retinoid-binding proteins (RBPs)	91

3. The unexpected role of RBP2 in regulating both retinoid and neutral lipid signaling and its potential consequences of this for metabolic disease development	97
4. Diverse and unanticipated actions of retinol-binding protein 4 (RBP4) in retinoid signaling, metabolism and metabolic disease	104
Acknowledgments	108
References	108

5. Rethinking retinoic acid self-regulation: A signaling robustness network approach — 113
Abraham Fainsod and Rajanikanth Vadigepalli

1. Retinoic acid is a major regulator of developmental processes and tissue homeostasis	114
2. Environmental dependence of ATRA biosynthesis	116
3. Polymorphisms in ATRA network components and the induction of disease	119
4. Robustness of the ATRA metabolic and signaling network	120
5. The ATRA robustness response exhibits Pareto optimality	122
Acknowledgments	128
Funding	128
References	129

6. The action of retinoic acid on spermatogonia in the testis — 143
Shelby L. Havel and Michael D. Griswold

1. Male germ cell development in the mouse	144
2. The essential role of sertoli cells in spermatogenesis	146
3. The cycle of the seminiferous epithelium	147
4. Retinoic acid synthesis in the testis	148
5. Elucidating testicular *at*RA synthesis activity through genetic knockout studies	151
5.1 Rdh10 expression by sertoli cells is required for juvenile spermatogenesis	151
5.2 RAL dehydrogenase activity in Sertoli cells, but not germ cells, is essential for the initial A to A1 transition	152
5.3 Loss of Cyp26b1, but not Cyp26a1, affects spermatogenesis	154
5.4 Gene expression changes induced by atRA support spermatogonial differentiation	154
6. Synchronizing spermatogenesis using the drug WIN 18,446	156
7. Conclusions	159
References	160

7. The interplay between retinoic acid binding proteins and retinoic acid degrading enzymes in modulating retinoic acid concentrations — **167**
Nina Isoherranen and Yue Winnie Wen

1. Introduction — 168
2. Tissue and cell partitioning and protein binding of *at*RA — 172
   2.1 Binding of *at*RA in plasma and passive distribution to tissues — 172
   2.2 *at*RA binding to CRABPs — 176
3. Pathways of retinoic acid clearance and patterns of enzyme expression — 181
   3.1 Enzymes that metabolize *at*RA — 181
   3.2 Role of CYP26 enzymes in *at*RA metabolism — 183
4. Interplay of CYP26 enzymes and CRABPs in regulating tissue *at*RA concentrations — 188
5. Conclusions — 192
References — 193

8. Retinoic acid homeostasis and disease — **201**
Maureen A. Kane

1. Introduction — 202
   1.1 ATRA biosynthetic pathway and homeostatic mechanisms — 202
   1.2 Binding proteins in ATRA homeostasis — 205
2. ATRA homeostasis and disease — 207
   2.1 CYP26B1 — 209
   2.2 POR — 209
   2.3 DHRS3 — 209
   2.4 Cellular ROL-binding proteins in disease — 210
   2.5 Cellular ATRA binding proteins — 215
   2.6 Potential for gene-environment interaction — 215
3. Other considerations — 217
   3.1 Direct quantification of ATRA is essential to understanding ATRA homeostasis — 217
   3.2 Reporter assays — 222
   3.3 Surrogates for ATRA and indirect measurement — 222
4. Conclusions and future opportunities — 224
Acknowledgments — 224
References — 224
Further reading — 233

9. The multifaceted roles of retinoids in eye development, vision, and retinal degenerative diseases   235
Zachary J. Engfer and Krzysztof Palczewski

1. Introduction   236
2. Vitamin A: The root of all retinoids   238
3. Retinoic acid signaling in the developing and mature retina   246
4. Retinaldehydes, retinyl esters, and retinols in vision   256
5. Conclusion   277
Acknowledgments   278
Author Contributions   278
Competing Interest Statement   278
References   278

10. Retinoid signaling in pancreas development, islet function, and disease   297
Manuj Bandral, Lori Sussel, and David S. Lorberbaum

1. Introduction   297
2. The ATRA signaling pathway   298
3. Pancreas development   300
4. hPSC differentiations and therapeutic interventions   306
5. Maintaining islet function   308
6. ATRA signaling in disease   311
7. Conclusion   313
Acknowledgments   313
References   313

11. Vitamin A supply in the eye and establishment of the visual cycle   319
Sepalika Bandara and Johannes von Lintig

1. Introduction   320
2. Classes of retinoid metabolizing enzymes   322
3. Unlocking retinoids: the role of carotenoid cleavage dioxygenases in vitamin A conversion   322
4. Retinol and retinal dehydrogenases interconvert the oxidation states of vitamin A   324
5. Mastering retinoic acid levels: the role of cytochrome P450 enzymes   327
6. Lecithin: retinol acyltransferase: beyond vitamin A storage   328

7. Retinoid couriers: binding proteins navigating vitamin A across the body — 329
8. Regulating vitamin A production: unveiling gut control mechanisms — 330
9. Vitamin A express: delivering essential nutrients to the eyes — 331
10. The visual cycle: the art of synthesis and recycling — 334
11. Seeing in color: unveiling the cone visual pigment regeneration mechanism(s) — 336

References — 339

Contributors

Sepalika Bandara
Department of Pharmacology, School of Medicine, Case Western Reserve University, Cleveland, OH, United States

Manuj Bandral
University of Michigan, Department of Pharmacology, Caswell Diabetes Institute, Ann Arbor, MI, United States

William S. Blaner
Department of Medicine, College of Physicians and Surgeons, Columbia University, New York, NY, United States

Pierre-Jacques Brun
Department of Medicine, College of Physicians and Surgeons, Columbia University, New York, NY, United States

Diana Condrea
Université de Strasbourg, IGBMC UMR 7104; CNRS, UMR 7104; Inserm, UMR-S 1258; IGBMC, Institut de Génétique et de Biologie Moléculaire et Cellulaire, Illkirch, France

Gregg Duester
Development, Aging, and Regeneration Program, Sanford Burnham Prebys Medical Discovery Institute, La Jolla, CA, United States

Zachary J. Engfer
Center for Translational Vision Research, Department of Ophthalmology, Gavin Herbert Eye Institute; Department of Physiology and Biophysics, University of California, Irvine, Irvine, CA, United States

Abraham Fainsod
Department of Developmental Biology and Cancer Research, Institute for Medical Research Israel-Canada, Faculty of Medicine, The Hebrew University of Jerusalem, Jerusalem, Israel

Norbert B. Ghyselinck
Université de Strasbourg, IGBMC UMR 7104; CNRS, UMR 7104; Inserm, UMR-S 1258; IGBMC, Institut de Génétique et de Biologie Moléculaire et Cellulaire, Illkirch, France

Marcin Golczak
Department of Pharmacology and Cleveland Center for Membrane and Structural Biology, Case Western Reserve University, Cleveland, OH, United States

Michael D. Griswold
School of Molecular Biosciences, Washington State University, Pullman, Washington, United States

Shelby L. Havel
School of Molecular Biosciences, Washington State University, Pullman, Washington, United States

Nina Isoherranen
Department of Pharmaceutics, School of Pharmacy, University of Washington

Maureen A. Kane
Department of Pharmaceutical Sciences, University of Maryland School of Pharmacy, Baltimore, MD, United States

David S. Lorberbaum
University of Michigan, Department of Pharmacology, Caswell Diabetes Institute, Ann Arbor, MI, United States

Masahiro Nakamura
Department of Oral Immunology and Infectious Diseases, University of Louisville School of Dentistry, Louisville, KY, United States

Jisun Paik
Department of Comparative Medicine, University of Washington, Seattle, WA, United States

Krzysztof Palczewski
Center for Translational Vision Research, Department of Ophthalmology, Gavin Herbert Eye Institute; Department of Physiology and Biophysics; Department of Chemistry; Department of Molecular Biology and Biochemistry, University of California, Irvine, Irvine, CA, United States

Giulia Perrotta
Université de Strasbourg, IGBMC UMR 7104; CNRS, UMR 7104; Inserm, UMR-S 1258; IGBMC, Institut de Génétique et de Biologie Moléculaire et Cellulaire, Illkirch, France

Lisa L. Sandell
Department of Oral Immunology and Infectious Diseases, University of Louisville School of Dentistry, Louisville, KY, United States

Lori Sussel
University of Colorado Denver Anschutz Medical Campus, Barbara Davis Center for Diabetes, Aurora, CO, United States

Rajanikanth Vadigepalli
Daniel Baugh Institute for Functional Genomics and Computational Biology, Department of Pathology and Genomic Medicine, Sidney Kimmel Medical College, Thomas Jefferson University, Philadelphia, PA, United States

Yue Winnie Wen
Department of Pharmaceutics, School of Pharmacy, University of Washington

Johannes von Lintig
Department of Pharmacology, School of Medicine, Case Western Reserve University, Cleveland, OH, United States

Preface

The importance of retinoids in biology began first in the 1930s and 1940s with the observation that its precursor retinol (vitamin A) is required for embryonic development and vision as shown by vitamin A deficiency studies (Hale, 1935; Wilson & Warkany, 1947). Subsequently, various naturally occurring and synthetic retinoids have been described including all-*trans* retinoic acid (ATRA), the predominant natural retinoid generated from all-*trans* retinol that controls embryonic development (Kastner, Mark, & Chambon, 1995), as well as 11-*cis* retinaldehyde for vision (Wald & Brown, 1958). 11-*cis* retinaldehyde binds to opsin in the retina to form rhodopsin that enables vision by converting visible light into signals to the brain. When 11-*cis* retinaldehyde absorbs light it is converted to all-*trans* retinaldehyde, which then stimulates a conformational shift in rhodopsin that produces a signaling event (Saari, 2012). Dietary sources of natural retinoids include animal products (dairy, eggs, liver) that contain all-*trans* retinol and retinyl esters, and plant products (fruits and vegetables) containing β-carotene that can be metabolized to all-*trans* retinaldehyde and all-*trans* retinol.

Early studies showed that treatment of vertebrate embryos with ATRA resulted in various teratogenic processes, thus demonstrating that ATRA has biological actions (Kochhar, 1967; Shenefelt, 1972). Key to understanding how ATRA acts was the discovery of dedicated nuclear receptors (RARs) that control transcription of key genes when activated by ATRA as a ligand (Kastner et al., 1995). In this regard, ATRA follows the examples of cholesterol-derived hormones such as testosterone, estradiol, or cortisol. Its action is mediated through modulating uptake, distribution, and storage of a precursor (retinol), synthesis and genomic activity of the "hormone-like" active compound (ATRA) through nuclear receptors, and finally degradation by dedicated enzymes. This sophisticated system relies on ATRA synthesizing and degrading enzymes, the expression domains of which are often non-overlapping or even mutually exclusive (Rhinn & Dollé, 2012). In the end, ATRA exerts its biological effects through binding to and activating nuclear receptors (RARA, RARB and RARG) that function as ligand-dependent transcriptional regulators bound to regulatory regions located in ATRA-regulated genes (Petkovich & Chambon, 2022). RARs were found to bind DNA as heterodimers with related nuclear receptors known as RXRs (Kastner et al., 1995; Mangelsdorf et al., 1995). RAR/RXR binding occurs at DNA sequences referred to as the ATRA response element (RARE) comprised of two 6 bp repeated DNA sequences. When a RAR/RXR heterodimer is bound to a RARE, this complex can then recruit either nuclear receptor coactivators (NCOA) that activate transcription or nuclear receptor corepressors (NCOR) that repress transcription, with these events being regulated by binding of ATRA to RAR. Thus, control of gene expression by endogenous ATRA is a natural process known as ATRA signaling.

Proteins other than RARs and opsin that regulate ATRA signaling and vision include enzymes that catalyze conversion of β-carotene to all-*trans* retinaldehyde (BCO1, BCO2), conversion of all-*trans* retinaldehyde to all-*trans* retinol (DHRS3), conversion of all-*trans* retinol to retinyl esters for retinoid storage (LRAT), conversion of all-*trans* retinol to all-*trans* retinaldehyde (RDH10), isomerization of all-*trans* retinaldehyde to 11-*cis* retinaldehyde (RPE65), conversion of all-*trans* retinaldehyde to ATRA (ALDH1A1, ALDH1A2, ALDH1A3), and catabolic metabolism of ATRA to hydroxylated retinoids that are destined for excretion (CYP26A1, CYP26B1, CYP26C1). Other proteins include retinol-binding proteins (RBP1, RBP2, RBP4), retinaldehyde-binding protein (RLBP1), and ATRA-binding proteins (CRABP1, CRABP2) that function to transport retinol, retinaldehyde, or ATRA. The functions of these and other retinoid-related enzymes and proteins have been described in part, but they are still being examined for their roles in vivo.

This volume of *Current Topics in Developmental Biology* spotlights the role of natural retinoids during development and disease. Gene knockout studies of the various retinoid-related enzymes and proteins mentioned above have provided a wealth of information to reveal retinoid functions. Topics covered herein include the role of retinoids in control of several developmental processes including embryonic organogenesis (somites/vertebrae, limbs, hindbrain/spinal cord, eye, craniofacial tissue, pancreas) and germ cell differentiation/meiosis initiation. Also covered is the role of various retinoids in adult processes including the visual cycle, metabolism/obesity, and retinoid homeostasis (balance of synthesis/degradation). These topics will provide readers a sense of the current state-of-the art in retinoids that will also inspire further research to understand the normal functions of retinoids.

GREGG DUESTER
Development, Aging, and Regeneration Program,
Sanford Burnham Prebys Medical Discovery Institute,
La Jolla, CA, USA

NORBERT B. GHYSELINCK
Université De Strasbourg, IGBMC UMR 7104,
Illkirch, France
CNRS, UMR 7104, Illkirch, France
INSERM, UMR-S 1258, Illkirch, France
IGBMC, Institut de Génétique et de Biologie
Moléculaire et Cellulaire, Illkirch, France

References

Hale, F. (1935). The Relation of Vitamin a to Anophthalmos in Pigs. *American Journal of Ophthalmology, 18*, 1087–1093.

Kastner, P., Mark, M., & Chambon, P. (1995). Nonsteroid nuclear receptors: what are genetic studies telling us about their role in real life? *Cell, 83*, 859–869. PMID:8521510.

Kochhar, D. M. (1967). Teratogenic activity of retinoic acid. *Acta Pathologica et Microbiologica Scandinavica, 70*, 398–404. PMID:4867280.

Mangelsdorf, D. J., Thummel, C., Beato, M., Herrlich, P., Schütz, G., Umesono, K., ... Evans, R. M. (1995). The nuclear receptor superfamily: the second decade. *Cell, 83*, 835–839. PMID:8521507.

Petkovich, M., & Chambon, P. (2022). Retinoic acid receptors at 35 years. *Journal of Molecular Endocrinology, 69*, T13–T24. PMID:36149754.

Rhinn, M., & Dollé, P. (2012). Retinoic acid signalling during development. *Development (Cambridge, England), 139*, 843–858. PMID:22318625.

Saari, J. C. (2012). Vitamin A metabolism in rod and cone visual cycles. *Annual Review of Nutrition, 32*, 125–145. PMID:22809103.

Shenefelt, R. E. (1972). Morphogenesis of malformations in hamsters caused by retinoic acid: relation to dose and stage at treatment. *Teratology, 5*, 103–118. PMID:5014447.

Wald, G., & Brown, P. K. (1958). Human rhodopsin. *Science (New York, N.Y.), 127*, 222–226. PMID:13495499.

Wilson, J. G., & Warkany, J. (1947). Anomalies of the genito-urinary tract induced by maternal vitamin A deficiency in fetal rats. *The Anatomical Record, 97*, 376. PMID:20341872.

CHAPTER ONE

Early retinoic acid signaling organizes the body axis and defines domains for the forelimb and eye

Gregg Duester[*]

Development, Aging, and Regeneration Program, Sanford Burnham Prebys Medical Discovery Institute, La Jolla, CA, United States
*Corresponding author. e-mail address: duester@SBPdiscovery.org

Contents

1. Introduction	2
1.1 ATRA signaling begins in mouse at E7.5 when retinol is first converted to ATRA	4
1.2 ATRA generated in somites controls early posterior body axis formation and somitogenesis by repressing *Fgf8*	6
1.3 ATRA repression of *Fgf8* regulates early heart anteroposterior patterning	9
2. Hindbrain anteroposterior patterning requires ATRA signaling	9
3. ATRA controls dorsoventral patterning of the spinal cord	10
4. What is the role of ATRA signaling during limb development?	12
4.1 ATRA is not required for limb patterning (either anteroposterior or proximodistal)	12
4.2 ATRA is required during forelimb initiation	17
4.3 How is *Tbx5* activated in the forelimb field?	19
5. Control of optic cup formation and eye morphogenesis by ATRA signaling	20
5.1 ATRA is required for optic cup formation	21
5.2 ATRA is required for eye morphogenesis after optic cup formation	22
6. Conclusions	24
Acknowledgments	25
Funding	25
References	25

Abstract

All-*trans* RA (ATRA) is a small molecule derived from retinol (vitamin A) that directly controls gene expression at the transcriptional level by serving as a ligand for nuclear ATRA receptors. ATRA is produced by ATRA-generating enzymes that convert retinol to retinaldehyde (retinol dehydrogenase; RDH10) followed by conversion of retinaldehyde to ATRA (retinaldehyde dehydrogenase; ALDH1A1, ALDH1A2, or ALDH1A3). Determining what ATRA normally does during vertebrate development has been challenging as studies employing ATRA gain-of-function (RA treatment) often do not agree with genetic loss-of-function studies that remove ATRA via knockouts of ATRA-generating enzymes. In mouse embryos, ATRA is first generated at stage E7.5 by ATRA-generating enzymes whose genes are first expressed at that stage. This article focuses upon what ATRA normally does at early stages based upon these knockout studies. It has been observed that early-generated ATRA performs three essential functions: (1) activation of genes that control hindbrain and spinal cord patterning; (2) repression of *Fgf8* in the heart field and caudal progenitors to provide an FGF8-free region in the trunk essential for somitogenesis, heart morphogenesis, and initiation of forelimb fields; and (3) actions that stimulate invagination of the optic vesicle to form the optic cup.

1. Introduction

The role of all-*trans* RA (ATRA; also known simply as RA) in early mouse embryogenesis is the primary focus of this article, but many of the conclusions can be extrapolated to embryos from fish to human. The ATRA precursors retinol or retinaldehyde are made available to all cells of the embryo via maternal transfer of retinol for mammals (Horton & Maden, 1995) or incorporation of retinaldehyde into eggs for lower vertebrates (Costaridis, Horton, Zeitlinger, Holder, & Maden, 1996; Dong & Zile, 1995; Irie, Azuma, & Seki, 1991). ATRA production in mouse embryos is initiated by retinol dehydrogenase-10 (RDH10) that converts retinol to retinaldehyde beginning at embryonic day 7.5 (E7.5) (Cammas, Romand, Fraulob, Mura, & Dolle, 2007; Sandell et al., 2007). Subsequently, retinaldehyde is metabolized to ATRA by three retinaldehyde dehydrogenases; i.e. ALDH1A2 first expressed at E7.5 in trunk presomitic mesoderm then at E8.5 in the eye (Mic, Haselbeck, Cuenca, & Duester, 2002; Niederreither, Subbarayan, Dollé, & Chambon, 1999), ALDH1A3 first expressed at E8.5 in the eye (Dupé et al., 2003; Mic, Molotkov, Fan, Cuenca, & Duester, 2000), and ALDH1A1 first expressed at E9.5 in the eye (Fan et al., 2003; Molotkov, Molotkova, & Duester, 2006). Zebrafishes possess orthologs for ALDH1A2 and ALDH1A3 that generate ATRA in trunk mesoderm and eye, but they lack ALDH1A1 (Canestro, Catchen, Rodriguez-Mari, Yokoi, & Postlethwait, 2009). ATRA is degraded by P450 hydroxylase enzymes (CYP26A1, CYP26B1 and CYP26C1), resulting in a short ATRA half-life of about 1 h (Hernandez, Putzke, Myers, Margaretha, & Moens, 2007; Pennimpede et al., 2010).

ATRA regulates transcription during embryogenesis by functioning as a ligand for nuclear ATRA receptors (RARs) that exist as three isotypes in mouse, i.e. RARA, RARB, and RARG encoded by three different genes

(Ghyselinck et al., 1997; Lohnes et al., 1994; Mendelsohn et al., 1994). Zebrafishes possess two genes for RARA (*raraa* and *rarab*) and two genes for RARG (*rarga* and *rargb*), but they lack a gene for RARB (Linville, Radtke, Waxman, Yelon, & Schilling, 2009). RARs bind DNA as a heterodimer with retinoid X receptors (RXRA, RXRB, or RXRG) at a sequence known as the ATRA response element (RARE) (Kastner, Mark, & Chambon, 1995). Binding of ATRA to the RAR portion of RAR/RXR at a RARE stimulates recruitment of nuclear receptor coactivators (NCOA) or nuclear receptor corepressors (NCOR) that activate or repress transcription of a nearby gene (Cunningham & Duester, 2015). A ligand for RXR (such as 9-*cis* RA) is not required for ATRA signaling during embryogenesis (Mic, Molotkov, Benbrook, & Duester, 2003). Chromatin immunoprecipitation studies have identified 13,000–15,000 potential sites in the mouse genome where RARs bind RAREs (Chatagnon et al., 2015; Moutier et al., 2012). It is likely that most of these RAREs are not required for development due to weak binding affinity of RAR/RXR or due to being situated in genomic regions that cannot normally control nearby genes (Cunningham, Lancman, Berenguer, Dong, & Duester, 2018).

It has been hypothesized that ATRA signaling may also act by non-genomic events to control processes other than gene transcription in the nucleus. In particular, it has been proposed that the ATRA receptor RARA controls hippocampal cortical synaptic plasticity via a cytoplasmic non-genomic mechanism in which binding of RARA to the mRNA for GluA1, encoding a subunit of the alpha-amino-3-hydroxy-5-methyl-4-isoxazolepropionic acid (AMPA) receptor, inhibits translation, with such inhibition being reversed in the presence of ATRA (Aoto, Nam, Poon, Ting, & Chen, 2008). Support for this non-genomic mechanism has come from the RARA knockout mouse that exhibits defects in synaptic plasticity (Park, Tjia, Zuo, & Chen, 2018). However, a major shortcoming of the non-genomic hypothesis is a lack of mutational studies showing that specific amino acid residues of RARA can bind specific nucleotides in the GluA1 mRNA (Duester, 2023). Soon after ATRA receptors were originally discovered and proposed to bind specific DNA sequences (RAREs) to control gene transcription in the nucleus, mutational studies were performed that identified zinc fingers in ATRA receptors needed for DNA binding (Perlmann, Rangarajan, Umesono, & Evans, 1993) and mutational studies of RAREs were performed to determine the DNA sequence needed for ATRA receptor binding (Hoffmann et al., 1990; Vivanco Ruiz, Bugge, Hirschmann, & Stunnenberg, 1991). Thus, without clear evidence that RARA can bind GluA1 mRNA to function cytoplasmically, one should consider that the effect of RARA on synaptic plasticity may occur in the nucleus. Also, without a genetic loss-of-function model that removes the endogenous ligand ATRA in hippocampal cortical neurons, it is not possible to conclude that RARA and its ligand ATRA control synaptic plasticity in the cytoplasm through a non-genomic signaling mechanism (Duester, 2023).

During early development ATRA is produced in specific tissues, then diffuses to nearby tissues where it is either degraded by CYP26s in the cytoplasm or able to gain entry into the nucleus to control expression of genes required for development of specific tissues. Genetic loss-of-function studies of ATRA-generating enzymes have provided deep insight into the normal functions of ATRA during embryogenesis (Cunningham & Duester, 2015; Ghyselinck & Duester, 2019; Rhinn & Dolle, 2012). In particular, single knockouts of either *Aldh1a2* (Mic et al., 2002; Niederreither et al., 1999) or *Rdh10* (Cunningham et al., 2015; Sandell et al., 2007) are able to eliminate ATRA in all or most tissues of early embryos (E7.5-E8.5), thus avoiding redundancy with other ATRA-generating enzymes. Here, we will describe what is known about the early functions of ATRA signaling during development.

1.1 ATRA signaling begins in mouse at E7.5 when retinol is first converted to ATRA

Studies on wild-type embryos carrying the *RARE-lacZ* RA-reporter transgene (Tg(RARE-Hspa1b/lacZ)12Jrt) (Rossant, Zirngibl, Cado, Shago, & Giguère, 1991) have shown that ATRA signaling in mouse embryos is first observed at E7.5, a few hours before formation of the first somite, localized to a region spanning from the posterior hindbrain to the junction between the trunk and caudal epiblast that contains neuromesodermal progenitors (NMPs) needed for body axis extension (Sirbu, Gresh, Barra, & Duester, 2005). *Aldh1a2* knockouts carrying *RARE-lacZ* have no ATRA activity at E7.5-E8.0 (Sirbu et al., 2005). At E8.5, ATRA signaling is still absent in hindbrain/trunk/caudal epiblast of *Aldh1a2* knockouts, but ATRA activity is observed in the developing eye (optic vesicle) due to expression of *Aldh1a3* in the eye (Molotkov et al., 2006). Studies on *Rdh10* knockout embryos carrying *RARE-lacZ* also demonstrated that all ATRA signaling is lost at E7.5-E8.0, but ATRA activity was observed at E8.5 in caudal neural tissue (fated to become spinal cord) due to an unknown retinol-metabolizing enzyme (Cunningham et al., 2015). Unlike *Aldh1a2* knockouts, *Rdh10* knockouts lack ATRA activity in the eye at E8.5-E10.5, demonstrating that no redundant retinol-metabolizing enzyme exists that can replace RDH10 in the early eye field (Cunningham, Chatzi, Sandell, Trainor, & Duester, 2011; Sandell et al., 2007). Thus, during early embryogenesis ATRA signaling is initially confined to the posterior hindbrain, trunk (including somites, lateral plate mesoderm, endoderm, spinal cord), caudal epiblast, and eye field (Fig. 1).

As ATRA activity is first detected at E7.5 in mouse (late gastrulation stage), mouse embryonic stem cells (ESCs) which are derived from the E3.5 blastocyst are not normally exposed to endogenous ATRA during development. Although treatment of ESCs with a pharmacological dose of ATRA (1 μM) has been used to identify genes regulated by ATRA, a better approach is to first treat ESCs with Wnt and FGF to stimulate their differentiation to a stage similar to NMPs present at the junction of the

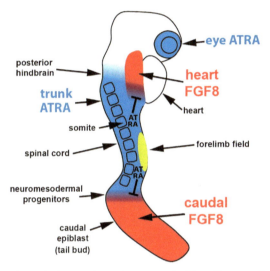

Fig. 1 ATRA signaling during early embryogenesis. The diagram shows a 9-somite (E8.5) mouse embryo with anterior at the top and posterior (caudal) at the bottom. At this stage, ATRA signaling activity (shown in blue) is limited to the developing trunk and the eye (optic vesicle). The anterior border of trunk ATRA activity occurs in the posterior portion of hindbrain, and the posterior border of ATRA activity occurs at the border between the trunk and caudal epiblast where neuromesodermal progenitors (NMPs) exist that differentiate into either spinal cord or somites as the body axis extends in the anterior to posterior direction. One major function of ATRA signaling is to repress FGF8 signaling (shown in red) produced by the caudal (A) epiblast and the anterior heart field (B). ATRA repression of the *Fgf8* gene ensures that the developing trunk (A) and posterior heart (B) are free of FGF8 signaling, which is essential for several processes including spinal cord neurogenesis, somitogenesis, heart patterning, and initiation of forelimb budding (C).

trunk and caudal epiblast at E7.5 where endogenous ATRA exists in the embryo (Gouti et al., 2014). Then treatment of these NMP-like cells with a physiological dose of ATRA (25 nM) can provide insight into what genes might normally be regulated by ATRA during early development to control developmental processes (Cunningham, Colas, & Duester, 2016).

A specific method of ATRA treatment of ESCs was reported to stimulate their reprogramming into a cell type similar to the 2-cell totipotent cells that exist at a very early stage of development (E0.5) just after fertilization (Iturbide et al., 2021). Furthermore, the authors reported that ATRA-generating enzymes encoded by *Rdh10*, *Aldh1a1*, *Aldh1a2*, *Aldh1a3* may be expressed at the 2-cell stage based on PCR analysis (though low expression was reported for all) and they proposed that endogenous ATRA generated at the 2-cell stage may normally mediate zygotic genome activation at the 2-cell stage to establish totipotency in embryos (Iturbide et al., 2021). However, those conclusions are not supported by previous studies described above showing that endogenous ATRA is not observed in mouse embryos until E7.5 (Duester, 2024). In addition, those conclusions can be challenged on the grounds that the available knockouts for *Rdh10*, *Aldh1a1*, *Aldh1a2*, and *Aldh1a3* do not display the expected arrest in early development if ATRA is required at the 2-cell stage. Importantly, redundancy among the ATRA-generating enzymes cannot explain the lack of a very early phenotype; the double knockout for *Aldh1a2/Aldh1a3* and the triple knockout for *Aldh1a1/Aldh1a2/Aldh1a3* both develop normally past the 2-cell stage and the blastocyst stage and do not display defects until E7.5 in the hindbrain and body axis similar to *Aldh1a2* single knockouts (Molotkov et al., 2006; Molotkova, Molotkov, & Duester, 2007). The *Aldh1a1/Aldh1a3* knockout shows that *Aldh1a1* and *Aldh1a3* are not required for ATRA synthesis until E8.5 as the first defects observed are in the eye field or frontonasal region (Matt et al., 2005; Molotkov et al., 2006). The studies reported by Iturbide et al. are also questionable as they rely on treatments with supraphysiological levels of ATRA or ATRA receptor antagonists that are known to stimulate abnormal processes (Kochhar, Jiang, Pennera, Beard, & Chandraratna, 1996; Lammer et al., 1985; Nau, 1995). In this case, the abnormal process is the ability to reprogram ESCs to a 2-cell-like stage (Duester, 2024).

1.2 ATRA generated in somites controls early posterior body axis formation and somitogenesis by repressing *Fgf8*

NMPs situated at the junction of the trunk and caudal epiblast express both *T* (*Brachyury*; *Tbxt*) (a marker of mesoderm) and *Sox2* (a marker of neuroectoderm). They can differentiate to either presomitic mesoderm that initiates expression of *Tbx6* or spinal cord neuroectoderm that initiates expression of *Sox1* during body axis extension (Henrique, Abranches, Verrier, & Storey, 2015). NMPs begin to contribute significantly to spinal cord and somites only posterior to the forelimb field (~10 somite stage). *T* activates *Wnt3a* expression which diffuses throughout the caudal epiblast and activates *Sox2*; *Wnt3a* also activates *T* in mutual feedback loop; NMP cells that ingress from the caudal epiblast to form mesoderm begin expressing *Tbx6* that represses *Wnt3a* to ensure that mesoderm does not express *Sox2*, whereas NMPs that remain in the caudal epiblast maintain *Sox2* expression and become neural plate then spinal cord (Kondoh & Takemoto, 2024). *Fgf8* expressed in the caudal epiblast is also needed to maintain *Wnt3a* expression and balanced NMP differentiation to neural and mesodermal fates (Henrique et al., 2015).

Loss of ATRA in *Aldh1a2* knockouts results in unbalanced NMP differentiation resulting in decreased *Sox1* expression that delays spinal cord formation and increased *Tbx6* expression that reduces the ability of presomitic mesoderm to coalesce into somites; somites are 50% their normal size showing that ATRA control of NMP differentiation is required for migratory presomitic mesoderm cells to condense and form epithelial somites (Cunningham et al., 2015).

Further studies identified a mechanism involving FGF8 through which ATRA controls balanced NMP differentiation. Genetic loss-of-function studies have shown that *Fgf8* and *Fgf4* expressed in mouse caudal epiblast function redundantly to control somitogenesis through their ability to maintain presomitic mesoderm progenitors caudally during body axis extension (Boulet & Capecchi, 2012; Naiche, Holder, & Lewandoski, 2011). Studies on vitamin A-deficient quail embryos, that have greatly reduced ATRA synthesis, exhibit small somites and ectopic expression of caudal *Fgf8* extending into the trunk (Diez del Corral et al., 2003). In mouse, loss of ATRA activity in either *Aldh1a2* or *Rdh10* knockouts results in ectopic *Fgf8* expression that now extends beyond the caudal epiblast into the posterior trunk where new somites are forming (Cunningham et al., 2015). When mouse *Aldh1a2* knockouts are treated with the FGF inhibitor SU5402, normal somite size is rescued (Cunningham et al., 2015). Thus, ATRA represses *Fgf8* activity at the border between the posterior trunk and caudal epiblast to allow formation of a region where balanced NMP differentiation can occur resulting in formation of normal somites and spinal cord (Fig. 1A).

Somites normally occur in bilateral left and right pairs of equal size on each side of the neural tube, however *Aldh1a2* knockouts exhibit left-right asymmetry of somite size associated with a corresponding left-right asymmetry of *Fgf8* expression at the border of the trunk and caudal epiblast (Vermot et al., 2005). *Fgf8* expressed in a structure called the node at this location regulates left-right asymmetry that normally occurs in some organs, however somites are not normally subjected to this left-right asymmetry mechanism (Meyers & Martin, 1999). Evidently, loss of ATRA disrupts the normal left-right asymmetry mechanism and now imparts this asymmetry on somites by allowing ectopic expansion of caudal *Fgf8* expression. *Aldh1a2* knockouts carrying *RARE-lacZ* that received a small maternal physiological dose of ATRA at E7.5 exhibited normal bilateral somites and normal caudal *Fgf8* expression that were rescued by ATRA activity located in the node ectoderm located at the junction of the posterior neuroectoderm and caudal epiblast ectoderm; thus ATRA did not need to act in the presomitic mesoderm itself but in node ectoderm to properly control caudal *Fgf8* expression so that it generates a bilateral diffusible FGF8 signal for presomitic mesoderm emerging from NMPs (Sirbu & Duester, 2006). These findings have shown that during somitogenesis ATRA acts in a paracrine fashion as it is normally generated in trunk presomitic mesoderm at the border with the caudal epiblast but diffuses to adjacent trunk neuroectoderm and node ectoderm at the border with the caudal epiblast to repress *Fgf8* and prevent its ectopic expression in trunk tissue.

The mechanism through which ATRA represses *Fgf8* in the caudal epiblast involves a RARE located 4.1 kb upstream of *Fgf8* that is required for repression at the border of the trunk and caudal epiblast (Kumar & Duester, 2014). The *Fgf8* RARE binds RAR/RXR and functions in an ATRA-dependent fashion to recruit nuclear receptor corepressors (NCOR1 and NCOR2) as well as Polycomb Repressive Complex 2 (PRC2) that together stimulate deposition of the repressive histone chromatin mark H3K27me3 near the *Fgf8* RARE (Kumar & Duester, 2014; Kumar, Cunningham, & Duester, 2016). Either a double knockout of *Ncor1/Ncor2* or a knockout of the *Fgf8* RARE resulted in an anterior extension of caudal *Fgf8* expression and a small somite defect at E8.5 (Kumar et al., 2016). These studies demonstrated for the first time that ATRA can function in vivo through NCOR to directly repress a gene through a RARE that functions as a silencer rather than an enhancer.

The mouse RARE upstream of *Fgf8* is conserved in human, other mammals, and chick, but it is not observed near the zebrafish *fgf8* gene

(Berenguer, Meyer, Yin, & Duester, 2020). Accordingly, zebrafish embryos do not require ATRA to control caudal *fgf8* expression, and loss of ATRA activity in zebrafish does result in ectopic caudal *fgf8* expression and no effects are observed on somitogenesis or body axis extension (Berenguer, Lancman, Cunningham, Dong, & Duester, 2018).

1.3 ATRA repression of *Fgf8* regulates early heart anteroposterior patterning

In mouse and zebrafish embryos, loss of ATRA in the *Aldh1a2* knockout results in defective heart tube looping that results in malformation of both ventricular (anterior) and atrial (posterior) heart chambers (Keegan, Feldman, Begemann, Ingham, & Yelon, 2005; Niederreither et al., 2001). Similar to the role of ATRA in somitogenesis, ATRA controls heart patterning in mouse embryos by repressing *Fgf8*, but in this case ATRA represses a domain of *Fgf8* expression in the heart to limit its expression to the anterior portion of the heart (Ryckebusch et al., 2008; Sirbu, Zhao, & Duester, 2008); see Fig. 1B. Interestingly, a similar domain of heart *fgf8* expression exists in zebrafish that is subject to ATRA repression even though the caudal *fgf8* domain in zebrafish is not repressed by ATRA as discussed above (Sorrell & Waxman, 2011). FGF8 signaling in the anterior heart field is required to establish heart ventricular identity, but ectopic FGF8 activity in the posterior heart field inhibits atrial identity (Pradhan et al., 2017). In mouse *Aldh1a2* knockouts at E8.0-E8.5, heart *Fgf8* expression extends more posteriorly than normal resulting in expansion of anterior ventricular progenitors into the posterior region where atria abnormally develop (Ryckebusch et al., 2008; Sirbu et al., 2008). These studies uncovered an essential RA-FGF8 interaction in heart conserved from fish to mammal.

2. Hindbrain anteroposterior patterning requires ATRA signaling

Examination of vitamin A-deficient quail embryos revealed defects in hindbrain anteroposterior patterning (Maden, Gale, Kostetskii, & Zile, 1996). Also, studies using mouse transgenes or knockouts identified RARE enhancers that activate *Hox* genes required for anteroposterior patterning of the hindbrain segments known as rhombomeres 1–8 (numbered from anterior to posterior) (Dupé et al., 1997; Marshall et al., 1994; Studer, Pöpperl, Marshall, Kuroiwa, & Krumlauf, 1994). These were the first studies that used methods other than

ATRA treatment to provide evidence that endogenous ATRA regulates hindbrain rhombomere patterning. ATRA generated early in development is derived only from *Rdh10* and *Aldh1a2* expressed in presomitic mesoderm, somites, and lateral plate mesoderm (Mic et al., 2002; Niederreither et al., 1999; Rhinn, Schuhbaur, Niederreither, & Dolle, 2011; Sandell et al., 2007). ATRA secreted from these mesodermal tissues diffuses into the adjacent neural tube including hindbrain as far anteriorly as rhombomere 3 (Sirbu et al., 2005). ATRA activates 3′-*Hox* genes (groups *Hox1–4*) that are known to be required for hindbrain rhombomere formation and identity (Begemann, Schilling, Rauch, Geisler, & Ingham, 2001; Krumlauf, 1993; Maden et al., 1996; Niederreither, Vermot, Schuhbaur, Chambon, & Dollé, 2000). ATRA arriving in the hindbrain directly activates *Hoxb1* expression via two RAREs, one that serves as an enhancer to activate expression in r4, and another that serves as a silencer to repress expression in r3 and r5 (Marshall et al., 1994; Studer et al., 1994). In this fashion, although ATRA is present in r3-r8, expression of *Hoxb1* can be limited to r4. In order to ensure that *Hoxb1* is not expressed further posterior in the hindbrain or spinal cord where ATRA is present, it was found that *Hnf1b* (*vhnf1*), encoding a repressor of *Hoxb1*, is activated by ATRA in r5-r8 and spinal cord; *Hnf1b* knockout in zebrafish and mouse results in ectopic expression of *Hoxb1* posterior of r4 (Hernandez, Rikhof, Bachmann, & Moens, 2004; Sirbu et al., 2005). These genetic studies and others focused on other *Hox* genes (Ahn, Mullan, & Krumlauf, 2014; Niederreither et al., 2000) have uncovered impressive mechanisms through which endogenous ATRA can stimulate expression of specific *Hox* genes in specific locations along the anteroposterior axis of the neural tube.

3. ATRA controls dorsoventral patterning of the spinal cord

After the hindbrain is formed, neural progenitors emerging from NMPs in the caudal epiblast differentiate into spinal cord neuroectoderm (Wilson, Olivera-Martinez, & Storey, 2009). Sonic hedgehog (SHH) generated in the floor plate cells of the ventral spinal cord is required for differentiation of ventral spinal cord cells to motor neurons that express *Isl1* (Ericson, Thor, Edlund, Jessell, & Yamada, 1992). Additional factors are also required for motor neuron differentiation including expression of *Pax6* (Ericson et al., 1997) and *Olig2* (Takebayashi, Nabeshima, Yoshida, Chisaka, & Ikenaka, 2002). Vitamin A-deficient quails were used to demonstrate that endogenous ATRA is required for expression of spinal cord *Pax6* and *Olig2* to generate dorsoventral patterning required for differentiation of motor neurons (Diez del Corral et al., (2003)). Mouse embryos carrying *RARE-lacZ* have shown that the entire spinal cord along both the anteroposterior and dorsoventral axes is exposed to ATRA derived from adjacent presomitic mesoderm and somites; all such ATRA activity is eliminated in *Aldh1a2* knockouts resulting in loss of *Pax6* and *Olig2* spinal cord expression (Molotkova, Molotkov, Sirbu, & Duester, 2005). Collectively, these findings demonstrated that ATRA activates *Pax6* expression in the spinal cord, and *Pax6* along with SHH function together to activate expression of *Olig2* in ventral spinal cord progenitors, with *Olig2* expression leading to a motor neuron fate marked by expression of *Isl1* (del Corral et al., 2003; England, Batista, Mich, Chen, & Lewis, 2011; Molotkova et al., 2005; Novitch, Wichterle, Jessell, & Sockanathan, 2003).

Olig2 expression is limited to a small region along the dorsoventral axis of the spinal cord where motor neurons arise, whereas *Pax6* is expressed across a greater portion of the dorsoventral axis, while ATRA activity is present all across the dorsoventral axis, with SHH activity being limited to the ventral spinal cord. Thus, differentiation of motor neurons in a specific region requires overlapping signals that activate *Olig2* in only a small domain. RNA-seq and ChIP-seq studies comparing wild-type and *Aldh1a2* knockouts, designed to identify ATRA-regulated genes with nearby alterations in epigenetic histone marks, demonstrated that ATRA activation of *Pax6* does regulate nearby histone marks, but control is indirect as it is not associated with a RARE; ATRA activation of *Pax6* appears to proceed through several indirect mechanisms including the ability of ATRA to directly activate RARE enhancers near *Sox2* and *Cdx1* that are both required for *Pax6* activation in spinal cord, as well as the ability of ATRA to directly repress caudal *Fgf8* via a RARE silencer that represses *Pax6* expression in the caudal epiblast (Berenguer et al., 2020). Other studies using *Aldh1a2* knockouts showed that ATRA repression of caudal *Fgf8* is required for decompaction of chromatin near *Pax6* to allow its expression in spinal cord neuroectoderm emerging from NMPs (Patel et al., 2013). A RARE has not been found near *Olig2*, suggesting ATRA does not directly activate *Olig2*. Thus, as spinal cord neural progenitors emerge from NMPs in the caudal epiblast, they move from a region of high FGF8 signaling to low, and they enter a region of high ATRA activity from all directions that activates *Pax6* in a more limited domain, with SHH signaling coming from only a ventral direction functioning in PAX6 + cells to activate *Olig2* in a small domain. This impressive signaling mechanism is designed to activate expression of *Olig2* in a limited number of ventral spinal cord cells that can then further differentiate into motor neurons.

4. What is the role of ATRA signaling during limb development?

4.1 ATRA is not required for limb patterning (either anteroposterior or proximodistal)

Early studies on treatment of embryos with pharmacological levels of ATRA described various teratogenic events including defects in limb anteroposterior patterning (thumb to little finger) resulting in limb duplications in chick embryos (Tickle, Alberts, Wolpert, & Lee, 1982) and other limb defects in mouse embryos (Kochhar, 1973). Treatment of regenerating adult axolotl limbs with ATRA results in a defect in limb proximodistal patterning (upper arm to hand) also resulting in limb duplication (Maden, 1982). These studies led to the belief that endogenous ATRA might normally function as a morphogen controlling limb patterning. However, subsequent studies proposed that endogenous ATRA is not required as a morphogen for normal anteroposterior patterning of limbs, but that pharmacological ATRA treatment can alter patterning (Noji et al., 1991; Wanek, Gardiner, Muneoka, & Bryant, 1991).

The original studies reporting a potential role for ATRA in proximodistal patterning of limbs were performed on regenerating axolotl limbs after amputation that exhibited altered proximodistal patterning when treated with ATRA; when the cut site is treated with ATRA an entirely new limb is regenerated instead of formation of just the region removed (Maden, 1982). Recent studies identified a gene encoding the protein Tig1 that is normally expressed differentially along the adult limb proximodistal axis (high proximal to low distal), but whose expression is greatly increased at the site of amputation if the cut site is treated with ATRA, thus stimulating regeneration of an entire limb from that cut site; Tig1 was proposed to control limb proximodistal patterning and is thus able to mediate the limb duplication effect of ATRA, however the hypothesis was also extended to propose that normal Tig1 proximodistal expression is regulated by endogenous ATRA (Oliveira et al., 2022). The conclusion that Tig1 controls proximodistal identity during normal limb regeneration to replace only the portion that was amputated appears valid. Also, the proposal that abnormal upregulation of Tig1 at the site of amputation by ATRA results in limb duplication also appears valid. However, no evidence was provided that endogenous ATRA is normally required to establish normal graded Tig1 expression along the limb proximodistal axis (Duester, 2022).

Mouse genetic loss-of-function studies have shown that limb proximodistal patterning is controlled by distal FGF8 signaling from the apical ectodermal ridge (Lewandoski, Sun, & Martin, 2000; Mariani, Ahn, & Martin, 2008). In order to investigate whether ATRA may also control limb proximodistal patterning, additional studies were performed on chick embryos treated with pharmacological levels of ATRA. In this case, ATRA treatment of distal forelimb or hindlimb bud tissue resulted in alteration of limb proximodistal patterning as measured by ectopic distal expression of homeobox genes *Meis1* and *Meis2* that are normally expressed in the proximal limb (upper arm) but not distal limb (lower arm and hand); it is important to note that *Meis1/Meis2* are also expressed throughout trunk mesoderm including trunk lateral plate mesoderm that expands into the limb as budding proceeds (Mercader et al., 2000). These studies also demonstrated that treatment of proximal limb tissue with FGF8 downregulates expression of *Meis1/Meis2*, leading to a two-signal hypothesis in which limb proximodistal patterning is controlled by mutual antagonism between ATRA activity in the proximal limb that activates *Meis1/Meis2*, and FGF8 activity in the distal limb that represses *Meis1/Meis2* (Mercader et al., 2000).

The above ATRA treatment studies are all ATRA gain-of-function studies. The availability of *Aldh1a2* knockouts allowed genetic loss-of-function studies to be performed to determine where endogenous ATRA activity is located and how loss of endogenous ATRA affects limb development. *Aldh1a2* knockouts carrying the RA-reporter *RARE-lacZ* demonstrated that endogenous ATRA produced in trunk mesoderm by ALDH1A2 diffuses into proximal limb tissue but does not enter the distal limb; ATRA is not generated in the proximal limb itself which does not express *Aldh1a2* (Mic et al., 2002; Niederreither et al., 1999). Additional studies showed that the ATRA-degrading enzyme encoded by *Cyp26b1* is expressed in distal limb tissue and the *Cyp26b1* knockout showed that CYP26B1 is required to eliminate ATRA in distal tissue to allow normal limb proximodistal patterning (Yashiro et al., 2004), Interestingly, *Fgf8* is required for *Cyp26b1* expression in distal limb (Probst et al., 2011). Compound knockouts of *Fgf8* and other redundant *Fgf* genes expressed distally in the apical ectodermal ridge exhibited ectopic distal expression of

Meis1/Meis2, showing that distal FGF activity normally represses *Meis1/Meis2* to establish its border in the proximal limb (Mariani et al., 2008). Further studies showed that the double knockout of *Meis1/Meis2* disrupts limb proximodistal patterning, and further evidence was reported showing that FGF8 represses *Meis1/Meis2* to limit its expression to proximal limb (Delgado et al., 2020). Together, these genetic loss-of-function studies in mouse were proposed to further support the two-signal model for limb proximodistal patterning in which a proximal-high gradient of ATRA signaling is required to activate *Meis1/Meis2* expression proximally, along with a distal-high gradient of FGF8 signaling that is required to repress *Meis1/Meis2* distally, thus creating a boundary in the middle of the limb that separates proximal and distal domains.

However, the proposal that ATRA activity in the proximal limb is required for proximodistal patterning was challenged by the mouse *Rdh10* knockout carrying *RARE-lacZ* that exhibits a complete loss of ATRA activity in limb buds, but normal limb proximodistal patterning in hindlimb buds; forelimbs exhibit a stunted phenotype which is discussed below (Sandell et al., 2007). Further studies on *Rdh10* knockouts and *Aldh1a2* conditional knockouts, that each completely eliminate ATRA activity in limb buds from their earliest stage of outgrowth, demonstrated that *Meis1/Meis2* are still expressed normally in the proximal limb, showing that ATRA activity is not required to activate *Meis1/Meis2* expression in the proximal limb; forelimbs were stunted but they still exhibited a domain of *Meis1/Meis2* expression limited to the proximal region, and a distal region lacking *Meis1/Meis2* expression that expresses *Fgf8* (Cunningham et al., 2011; Cunningham et al., 2013; Zhao et al., 2009). These knockouts also demonstrated that ATRA signaling is not required for limb anteroposterior patterning as posterior expression of *Shh* is observed in both stunted forelimbs and normal-sized hindlimbs (Cunningham et al., 2011). Instead of ATRA controlling limb patterning, the *Aldh1a2* and *Rdh10* knockouts demonstrated that ATRA signaling is required for initiation of forelimb (but not hindlimb) budding; further details on forelimb initiation are discussed below. Together, all of the studies described above demonstrate that the normal functions of endogenous ATRA cannot be determined by ATRA treatment (gain-of-function), stressing the importance of genetic loss-of-function studies to remove endogenous ATRA activity.

The opposing conclusions on limb patterning obtained with ATRA gain-of-function versus ATRA loss-of-function can be attributed to side effects of pharmacological levels of ATRA that ectopically activate *Meis1/Meis2* in distal limb. Based upon ATRA and *Fgf8* genetic loss-of-function studies a one-signal model for limb proximodistal patterning has been proposed in which distal FGF8 signaling controls genes needed for outgrowth of the limb in a proximal to distal direction, with FGF8 repressing *Meis1/Meis2* to generate a border that defines the proximal domain, and with FGF8 also activating expression of *Cyp26b1* distally to degrade ATRA that may diffuse from the trunk into the distal limb where it would disrupt digit patterning in the hand (Cunningham & Duester, 2015). Although proximal limb expression of *Meis1/Meis2* does not require ATRA, more recent *Aldh1a2* knockout studies have shown that endogenous ATRA is required for normal *Meis1/Meis2* expression in somites and trunk lateral plate mesoderm that later expands into limb buds during outgrowth; importantly, ChIP-seq studies comparing wild-type and *Aldh1a2* knockouts demonstrated that *Meis1/Meis2* both have nearby RAREs that function with endogenous ATRA to deposit the histone H3K27ac gene activation signal near *Meis1/Meis2* (Berenguer et al., 2020). Thus, a modified one-signal model for limb proximodistal patterning taking into account the results of Berenguer et al. (2020) goes as follows: (1) *Meis1/Meis2* expression is directly activated by ATRA in trunk lateral plate mesoderm as it emerges from the caudal progenitors (tail bud), thus significantly before initiation of limb budding; (2) as trunk lateral plate mesoderm moves away from the tail bud and then expands into limb buds after initiation of limb outgrowth, expression of *Meis1/Meis2* (now in limb mesoderm) continues to be expressed, however ATRA is no longer needed to maintain *Meis1/Meis2* expression and ATRA is thus not required to set its distal boundary of expression in the limb bud; (3) after the proximal region of the limb has formed (which still expresses *Meis1/Meis2*), *Fgf8* initiates expression at the distal tip of the limb resulting in diffusion of FGF8 that represses *Meis1/Meis2* in the distal domain as limb outgrowth continues, plus FGF8 activates *Cyp26b1* to eliminate any ATRA that may be able to diffuse from the trunk into the distal limb. Thus, distal FGF is the only diffusible signal needed to generate distinct proximal and distal domains by repressing *Meis1/Meis2* at a certain point in limb outgrowth (model is shown in Fig. 2).

Based on the observation that *Meis1/Meis2* genes have nearby RAREs that function along with endogenous ATRA to fully activate these genes in trunk mesoderm that emerges from the tail bud, it is possible that introduction of ATRA into a tissue that does not normally have endogenous ATRA (such as distal limb) can force ectopic expression of *Meis1/Meis2* via the RAREs they normally use for activation in another tissue (trunk mesoderm emerging from tail bud). Introduction of high pharmacological

RARE-lacZ - ATRA activity

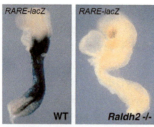

All ATRA activity is lost in *Raldh2* (*Aldh1a2*) knockout at E8.5: forelimbs/hindlimbs do not develop

Limb Development

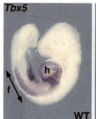 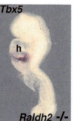

Forelimb initiation gene *Tbx5* is not expressed in forelimb (f) when all ATRA is lost but *Tbx5* is expressed in heart (h)

Most ATRA activity is lost in *Rdh10* knockout at E8.0, but some ATRA still in trunk tissue emerging from tail bud initiates normal *Meis2* in trunk but only small *Tbx5* forelimb domain

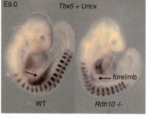

In *Rdh10* knockout, *Tbx5* has small forelimb expression domain due to small amount of ATRA still generated (*Uncx* co-staining of somites shows position along body axis)

At E10.5, the *Rdh10* knockout has no ATRA activity in trunk or limb mesoderm and displays a stunted forelimb but normal size hindlimb

At E10.5, the *Rdh10* knockout has normal *Meis2* expression along body axis and in proximal region of hindlimb without ATRA activity (ATRA initiates *Meis2* in body axis near tail bud but is not needed to maintain *Meis2* in upper trunk or initiate *Meis2* in limb outgrowth from trunk)

Fig. 2 ATRA role in limb development deciphered by analysis of *Aldh1a2* and *Rdh10* knockout mice. ATRA is required to initiate forelimb budding by providing a permissive environment for expression of the forelimb initiation gene *Tbx5* in a specific region of trunk lateral plate mesoderm. ATRA performs this function by repressing *Fgf8* in the forelimb field that lies between the heart and tail bud which allows forelimb *Tbx5* expression (also see Fig. 1). ATRA also functions instructively to initiate *Meis2* expression in trunk lateral plate mesoderm as it emerges from the tail bud progenitors during body

levels of ATRA (1 μM or higher) would be more likely than normal endogenous ATRA levels (~25 nM) to ectopically activate RAREs. Such a chain of ectopic activation events may also occur for genes that have a nearby RARE that is not normally used to control that gene in any tissue. Many of the 13,000–15,000 RAREs present in the mouse genome identified by RAR ChIP-seq (Moutier et al., 2012) may not bind RAR with high affinity, and thus may not normally control any gene in the presence of endogenous levels of ATRA. However, such RAREs may become active when exposed to pharmacological levels of ATRA that increase the chance that the RARE plus RAR and ATRA are present together to activate a nearby gene.

4.2 ATRA is required during forelimb initiation

Complete loss of ATRA activity in *Aldh1a2* knockouts results in failure to initiate forelimb development at E8.5, plus these mutants do not develop far enough to initiate hindlimb development at E9.5 (Mic et al., 2002; Niederreither et al., 1999). *Rdh10* knockouts lose most ATRA activity, but still retain ATRA activity in trunk tissue near the tail bud that allows formation of stunted forelimbs but normal hindlimbs (Sandell et al., 2007). *Rdh10* knockout forelimbs and hindlimbs still retain normal patterning (Fig. 2 shows hindlimb *Meis2* expression with normal proximodistal patterning). Early studies also demonstrated that ATRA is required to initiate forelimb buds, i.e. chick embryos treated with the ATRA synthesis inhibitor disulfiram (Stratford, Horton, & Maden, 1996) or vitamin-A deficient rat embryos (White et al., 1998). Mouse genetic loss-of-function studies later demonstrated that *Tbx5* is required for initiation of forelimb budding at E8.5 (Agarwal et al., 2003; Rallis et al., 2003). *Aldh1a2* knockouts fail to initiate *Tbx5* expression in the trunk lateral plate mesoderm region that normally becomes the forelimb field at E8.5 (Zhao et al., 2009). Although *Aldh1a2* knockouts normally die at E9.0 prior to hindlimb initiation which occurs at E9.5, maternal treatment of *Aldh1a2* knockouts (carrying *RARE-lacZ*) with a small physiological dose of ATRA only at E7.5 (referred to as

---

axis extension. However, ATRA is not required to maintain *Meis2* expression in lateral plate mesoderm as the body axis extends or to initiate *Meis2* expression in the proximal region of forelimbs and hindlimbs. Thus, although *Meis2* is required for limb proximodistal patterning, ATRA is not required for this downstream function of *Meis2*. Instead, studies on *Aldh1a2* and *Rdh10* knockout mice support a one-signal model for limb proximodistal patterning in which expression of *Fgf8* in the distal limb represses *Meis2* and limits its expression to a proximal limb domain.

conditionally ATRA-rescued) allows these mutants to develop to E10.5 at which point it is observed that forelimbs are stunted with a small domain of *Tbx5* expression, but hindlimbs are normal even though no ATRA activity is detectable in trunk lateral plate mesoderm, forelimbs, and hindlimbs from E8.5-E10.5 (Zhao et al., 2009); hindlimbs do not require *Tbx5* for initiation but instead require *Tbx4* and *Pitx1* (Lanctôt, Moreau, Chamberland, Tremblay, & Drouin, 1999; Logan & Tabin, 1999; Naiche & Papaioannou, 2003). Time-course studies showed that the embryonic ATRA activity obtained from maternal ATRA administration is completely cleared within 24 h (Mic et al., 2002). *Rdh10* knockouts mimic conditionally ATRA-rescued *Aldh1a2* knockouts in many ways including the formation of stunted forelimbs and normal hindlimbs at E10.5 with no ATRA activity detectable in trunk lateral plate mesoderm, forelimbs, and hindlimbs from E8.5-E10.5 (Cunningham et al., 2011; Sandell et al., 2007). Analysis of *Rdh10* knockouts (carrying *RARE-lacZ*) at E7.5-E8.5 demonstrated that ATRA activity is initially absent at E7.5, but a domain of ATRA activity is observed at E8.5 in the posterior neural tube down to the border with the caudal epiblast (Cunningham et al., 2015). This ATRA activity is evidently generated by a redundant retinol-metabolizing enzyme that begins generating retinaldehyde at E8.5 in *Rdh10* knockouts that along with *Aldh1a2* generates enough ATRA in the posterior neuroectoderm to repress caudal *Fgf8* and allow normal differentiation of NMPs as well as the ability to develop to at least E10.5 (Cunningham et al., 2015). Further studies on forelimb development in *Rdh10* knockouts revealed a smaller than normal forelimb field at E8.5 and stunted forelimbs at E9.5 that exhibit reduced *Tbx5* expression compared to wild-type (Cunningham et al., 2013). Thus, sufficient ATRA is generated in *Rdh10* knockouts and conditionally RA-rescued *Aldh1a2* knockouts to allow a small forelimb *Tbx5* domain to form, but not enough ATRA to generate forelimbs of a normal size.

Investigation of the mechanism underlying the requirement of ATRA for initiation of forelimb budding revealed that a failure in *Fgf8* repression in the posterior heart and trunk/caudal epiblast border is an important part of the mechanism. *Aldh1a2* knockouts fail to repress *Fgf8* at E8.5 in both the posterior heart field and trunk/caudal epiblast border which allows ectopic FGF8 signaling to enter the trunk from both anterior and posterior directions into the trunk lateral plate mesoderm where the forelimb fields normally arise (Fig. 1C). Conditionally ATRA-rescued *Aldh1a2* knockouts receive enough ATRA to prevent ectopic *Fgf8* expression at the trunk/caudal epiblast border, but ectopic

Fgf8 expression still extends from the posterior heart into the trunk forelimb field (Zhao et al., 2009). *Rdh10* knockouts at E8.5 appear similar to conditionally ATRA-rescued *Aldh1a2* knockouts with ATRA repression of caudal *Fgf8*, but a failure to repress *Fgf8* in the posterior heart field (Cunningham et al., 2013). As ATRA does repress caudal *Fgf8* in these mutants, less overall ectopic FGF8 activity from the posterior heart reaches the forelimb field, allowing a small domain of forelimb *Tbx5* expression to initiate. Wild-type mouse embryos grown in vitro and treated with FGF8 fail to activate forelimb *Tbx5* expression, thus providing good evidence that the forelimb initiation defect in *Aldh1a2* and *Rdh10* knockouts is due to ectopic FGF8 activity in the trunk resulting in disruption of forelimb *Tbx5* activation (Cunningham et al., 2013). In vivo support for this ATRA-FGF8 antagonism model was provided by zebrafish studies demonstrating that failure in forelimb bud (pectoral fin) initiation in *aldh1a2* mutants is also due to loss of *tbx5a* expression and ectopic *fgf8* expression from the posterior heart (Sorrell & Waxman, 2011). Importantly, loss of forelimbs in zebrafish *aldh1a2* mutants can be rescued by crossing to fish containing a transgene that expresses a heat-shock inducible dominant-negative FGF receptor (Cunningham et al., 2013). Thus, ATRA functions in a permissive manner to allow initiation of *Tbx5* expression in the forelimb field by repressing caudal and heart *Fgf8* expression that somehow prevents *Tbx5* activation if allowed to enter the trunk. ATRA-FGF8 antagonism is a fundamental ATRA control mechanism as it is conserved from fish to mammal. The ATRA-FGF8 antagonism mechanism described above for heart is also conserved from zebrafish to mammal (Sorrell & Waxman, 2011). However, the ATRA-FGF8 antagonism that occurs at the trunk/caudal epiblast of mouse and chick embryos to ensure normal NMP differentiation, somitogenesis, and spinal cord neurogenesis, does not occur in zebrafish that have a different type of NMP cell population in a caudal region that is different than the caudal epiblast of mouse and chick (Berenguer et al., 2018).

4.3 How is *Tbx5* activated in the forelimb field?

Although ATRA functions permissively to allow *Tbx5* expression in the forelimb field through *Fgf8* repression, the question remains of whether ATRA or any other factor may act instructively to activate forelimb *Tbx5* expression. Transgenic mouse studies identified a potential enhancer located within intron 2 that stimulates *Tbx5* expression in the forelimb field; this potential enhancer contains several HOX-binding sites and a RARE (Minguillon et al., 2012; Nishimoto, Minguillon, Wood, & Logan, 2014; Nishimoto, Wilde, Wood, & Logan, 2015). This potential *Tbx5* enhancer is conserved only in mammals, but other studies in zebrafish identified another forelimb enhancer located downstream of *tbx5a* that is conserved from mammals to fish; this enhancer does not contain a RARE (Adachi, Robinson, Goolsbee, & Shubin, 2016). Unfortunately, gene knockout studies in mouse designed to knockout these endogenous elements either singly or together in double knockouts, uncovered no defects in forelimb bud initiation or later skeletal development (Cunningham et al., 2018).

Other studies did uncover ATRA-activated genes that function upstream of forelimb *Tbx5*. As mentioned above, *Meis1/Meis2* both require ATRA for to initiate normal expression in trunk lateral plate mesoderm as it emerges from the tail bud prior to when the forelimb bud arises from an out-pocketing of this lateral plate mesoderm (Berenguer et al., 2020). Generation of *Meis1/Meis2* double knockout embryos revealed a loss of forelimb buds at E9.5, thus providing evidence that *Meis1/Meis2* are direct ATRA target genes that act instructively to activate *Tbx5* expression in that portion of the trunk lateral plate mesoderm that becomes the forelimb field (Berenguer et al., 2020). However, as *Meis1/Meis2* are expressed widely along the anteroposterior axis of trunk lateral plate mesoderm, additional control must exist to activate *Tbx5* only at the anteroposterior level of the forelimb field. Candidates include *Hoxa4* and *Hoxa5* that are expressed in trunk lateral plate mesoderm only at the anteroposterior level corresponding to the forelimb field (Minguillon et al., 2012). MEIS proteins have been shown to bind DNA control elements as heterodimers with HOX proteins (Penkov et al., 2013).

5. Control of optic cup formation and eye morphogenesis by ATRA signaling

A role for ATRA in controlling eye development is conserved from mammals to fish (Mory et al., 2014; Nedelec, Rozet, & Fares Taie, 2019; Plaisancie et al., 2016; Slavotinek, 2019; Williams & Bohnsack, 2019). The human developmental eye defects known as anophthalmia and microphthalmia can be caused by mutations in four components of the ATRA signaling pathway (*RBP4* that facilitates transport of retinol in the bloodstream, *STRA6* that facilitates transport of retinol into eye tissues, *ALDH1A3* encoding one of the three enzymes needed to generate ATRA, and *RARB* encoding one of the three ATRA receptors) as well as two genes encoding transcription factors that are upregulated by ATRA

(*PITX2, FOXC1*) (Mory et al., 2014; Nedelec et al., 2019; Plaisancie et al., 2016; Slavotinek, 2019; Weisschuh et al., 2006; Williams & Bohnsack, 2019). These defects are due to ATRA-regulated events that occur after optic cup formation. However, ATRA is also required for optic cup formation. Here, the roles of ATRA in optic cup formation as well as further morphogenesis of the optic cup to form the complete eye are discussed.

5.1 ATRA is required for optic cup formation

The first stage of eye development begins in mouse embryos at E8.5 when the optic vesicles are created as out-pocketings of the forebrain. Then, the portion of the optic vesicle closest to the surface ectoderm invaginates to form an optic cup as the lens forms by invagination of the surface ectoderm surrounded by the optic cup. A neural connection remains between the optic cup and brain that becomes the optic stalk and later the optic nerve. Vitamin A deficiency in rat embryos (Warkany & Schraffenberger, 1946) and double knockouts of two RARs in mouse embryos (Ghyselinck et al., 1997; Lohnes et al., 1994), demonstrated that although the optic cup is formed, ATRA is required for further morphogenesis of the optic cup and proper development of anterior eye structures including cornea and eyelids. Three RAR genes exist that are all expressed during early eye development, but triple RAR knockout studies have not been reported. However, ATRA function can also be examined using knockout studies targeting ATRA-generating enzymes.

At E9.5 in mouse, when the optic vesicle begins invagination to form the optic cup, all three ALDH1A ATRA-generating enzymes are expressed; *Aldh1a1* is expressed in the dorsal optic vesicle, *Aldh1a2* is expressed in the optic mesenchyme, and *Aldh1a3* is expressed in the ventral optic vesicle (Molotkov et al., 2006). Double knockouts between each of the three *Aldh1a* genes still generate optic vesicles and optic cups, but they display defects during later eye morphogenesis; however, as *Aldh1a1*−/− mice survive as adults and are fertile (Fan et al., 2003; Matt et al., 2005), adult mice could be generated that are *Aldh1a1*−/−;*Aldh1a2*+/−;*Aldh1a3*+/− to perform matings to generate *Aldh1a1*/*Aldh1a2*/*Aldh1a3* triple knockout embryos to study optic cup formation in the absence of ATRA synthesis (Molotkov et al., 2006). As optic cup formation occurs between E9.5-E10.5 and *Aldh1a2* knockouts do not survive this long, the generation of E10.5 *Aldh1a* triple knockouts required conditional ATRA-rescue with physiological ATRA as described above. At E10.5 it was observed that these conditional *Aldh1a* triple knockouts completely lacked ATRA activity in the eye region, and the optic cup did not form due to a defect in ventral invagination of the optic vesicle although dorsal invagination did occur; in contrast, relatively normal invagination of the surface ectoderm occurred to form the lens (Molotkov et al., 2006). These studies revealed that ATRA is required for proper invagination of the optic vesicle to form the optic cup.

Subsequent to these mouse *Aldh1a* triple knockout studies, RDH10 was found to function as the only retinol dehydrogenase that catalyzes the first step of ATRA synthesis in the eye; *Rdh10* knockout embryos were found to survive to E10.5 with no ATRA activity in the optic field and a failure in optic cup formation with loss of ventral invagination similar to *Aldh1a* triple knockouts (Sandell et al., 2007). Further studies on *Rdh10−/−* embryos will be useful to generate enough embryos with ATRA-deficient eyes to further examine how ATRA controls optic cup formation, particularly identification of target genes controlled by ATRA needed for this process.

5.2 ATRA is required for eye morphogenesis after optic cup formation

After its formation, the optic cup undergoes morphogenesis to form posterior eye structures including retina, retinal pigment epithelium, and optic nerve. ATRA continues to be generated by the optic cup after E10.5 and it diffuses to nearby tissues to control later stages of eye morphogenesis that generate anterior eye structures including cornea and eyelids (Matt et al., 2005; Molotkov et al., 2006). During these stages, which occur after E10.5 in mouse, ATRA is produced by *Aldh1a1* expressed in dorsal retina and *Aldh1a3* expressed in ventral retina, whereas *Aldh1a2* expression is no longer observed in the optic vesicle/cup after E9.5 (Molotkov et al., 2006). *Aldh1a1* knockouts revealed no clear disruption of embryonic eye development due to compensation by *Aldh1a3*; *Aldh1a3* knockouts exhibited mild eye defects due to partial compensation by *Aldh1a1*, whereas *Aldh1a1/Aldh1a3* double knockouts studied at E14.5 revealed major eye defects (Matt et al., 2005; Molotkov et al., 2006). Curiously, although *Aldh1a1* and *Aldh1a3* exhibit expression in dorsal and ventral optic vesicle/cup/retina, respectively, *Aldh1a1/Aldh1a3* double knockouts demonstrated that ATRA activity is not required for further development of the retina or retinal pigment epithelium after optic cup formation, nor for dorsoventral patterning of the retina which was previously proposed based on the dorsoventral expression patterns of *Aldh1a1* and *Aldh1a3* (Molotkov et al., 2006). Instead, *Aldh1a1/Aldh1a3* double knockouts have shown that

ATRA produced in the dorsal or ventral retina functions by diffusing to nearby neural crest-derived perioptic mesenchyme to control anterior eye formation (Matt et al., 2005; Molotkov et al., 2006). ATRA functions by limiting invasion of neural crest-derived perioptic mesenchyme into the anterior eye during formation of corneal mesenchyme and eyelids which, in the *Aldh1a1/Aldh1a3* double knockout, results in excessive corneal thickening and overgrowth of eyelids. This mesenchymal overgrowth also exerts a mechanical force resulting in deformation of optic cup morphology (Matt et al., 2005; Molotkov et al., 2006). These studies demonstrated that ATRA-generating enzymes function cell non-autonomously to generate paracrine ATRA signals in retina that guide eye morphogenetic movements in neighboring cells.

ATRA targets during anterior eye development have been identified. Mouse *Aldh1a1/Aldh1a3* double knockouts were found to have decreased expression of *Pitx2* and *Foxc1* in perioptic mesenchyme (Matt et al., 2005). Studies in zebrafish revealed that ATRA controls eye development similar to mouse, plus ATRA was found to activate *Pitx2* expression in neural crest-derived perioptic mesenchyme (Chawla, Schley, Williams, & Bohnsack, 2016). Mouse knockout studies demonstrated that *Pitx2* knockouts exhibit perioptic mesenchymal overgrowth, plus it was found that loss of *Pitx2* results in decreased expression of *Dkk2* (encoding a WNT antagonist) in perioptic mesenchyme. *Dkk2* knockout eyes were found to have a similar eye defect, showing that *Pitx2* normally activates *Dkk* in perioptic mesenchyme to reduce WNT signaling that stimulates perioptic mesenchyme growth (Evans & Gage, 2005; Gage, Qian, Wu, & Rosenberg, 2008). *Aldh1a1/Aldh1a3* double knockouts lacking ATRA activity in the perioptic mesenchyme exhibited loss of both *Pitx2* and *Dkk2* expression in perioptic mesenchyme, as well as increased expression of *Wnt5a* in perioptic mesenchyme (Kumar & Duester, 2010). Together, these genetic loss-of-function studies revealed that ATRA generated in the retina by *Aldh1a1* and *Aldh1a3* diffuses to the perioptic mesenchyme where it activates *Pitx2* that then activates *Dkk2* to suppress WNT signaling and prevent perioptic mesenchyme overgrowth.

Pitx2 was observed to have a nearby RARE that can recruit ATRA receptors in ChIP studies on mouse eye, providing evidence that *Pitx2* may be a direct ATRA target gene required for anterior eye morphogenesis (Kumar & Duester, 2010). However, it remains unclear if *Pitx2* and *Foxc1* are direct ATRA target genes for eye morphogenesis, plus no

ATRA-regulated genes have been identified for optic cup formation. Identification of direct ATRA target genes for eye morphogenesis may be made possible using genomic studies similar to those published for other embryonic tissues (Berenguer et al., 2020).

As the mouse *Aldh1a3* knockout exhibits perinatal lethality, no studies were reported in the adult eye (Dupé et al., 2003). The mouse *Aldh1a1* knockout survives to adulthood, but no eye defects were originally observed in adults perhaps due to compensating ATRA production by *Aldh1a3* in the ventral retina (Fan et al., 2003; Matt et al., 2005). However, more recent studies on the adult *Aldh1a1* knockout revealed a reduction in dorsal eye choroidal vascular development, while ventral choroidal development was normal likely due to expression of *Aldh1a3* in the ventral retina (Goto et al., 2018). Interestingly, studies on human eye development demonstrated that ATRA generated by ALDH1A1 and ALDH1A3 promotes M cone expression in the central retina during fetal development; then as ATRA signaling decreases later in eye development, L cones are expressed in peripheral retina resulting in higher M cone expression in the central retina and higher L cone expression in peripheral retina; this differential expression of M and L cones is important for proper human color vision (Hadyniak et al., 2024).

6. Conclusions

Use of genetic loss-of-function studies have been essential for identification of numerous developmental processes that require endogenous ATRA signaling. Although genetic loss-of-function studies can be difficult to perform, they are required to determine the normal functions for any gene, protein, or molecule (such as ATRA) as gain-of-function studies cannot provide this answer. Knockouts for *Aldh1a1*, *Aldh1a2*, *Aldh1a3*, and *Rdh10* have been particularly important for discovering the earliest embryonic events regulated by ATRA. These knockouts uncovered roles for ATRA in hindbrain and spinal cord patterning, proper development of NMPs, somites, heart, and forelimbs that all require ATRA repression of *Fgf8* signaling, and eye that requires ATRA for invagination of optic vesicles to form optic cups (Cunningham & Duester, 2015).

Now that many ATRA-regulated processes have been identified during early embryogenesis, organogenesis and at adulthood, a major focus needs to be placed on identifying the genes that are directly controlled by ATRA signaling. Genomic and epigenetic methods allow a global identification of RAREs and ATRA-regulated genes that mediate the downstream effects of ATRA signaling. Identification of ATRA-direct target genes can be accomplished by comparing wild-type and ATRA knockout embryos or tissues using RNA-seq and ChIP-seq to discover genes with significant decreases or increases in expression that also have nearby ATRA-regulated alterations in deposition of H3K27ac (a gene activation mark) or H3K27me3 (a gene repression mark) associated with a nearby RARE (Berenguer et al., 2020). Although many histone epigenetic marks exist, H3K27ac and H3K27me3 are particularly useful as they involve the same lysine residue on the same histone, which allows one to see if genes in knockouts exhibit an increase in one mark along with a decrease in the other mark at particular genomic sites including a nearby RARE (Chatagnon et al., 2015; Moutier et al., 2012). Also, although H3K4me1 is often used to detect enhancers, this epigenetic mark is globally associated with enhancers independent of activity (Heintzman et al., 2007), whereas H3K27ac is characteristic of active enhancers (Creyghton et al., 2010; Rada-Iglesias et al., 2011). Direct ATRA target genes discovered in this manner can be subjected to knockout studies to determine if they are required to mediate the developmental or physiological processes controlled by ATRA. Such basic knowledge will improve our view of how ATRA normally functions to regulate specific biological processes. Indeed, this methodology employing RNA-seq combined with H3K27ac/H3K27me3 ChIP-seq can be used to discovery direct target genes for any transcription factor for which a knockout is available, thus improving our knowledge of transcriptional gene regulation in general.

Acknowledgments
Special thanks to former and current members of my laboratory for spearheading genetic loss-of-function studies that resulted in new insights into how ATRA controls development.

Funding
This work was funded by the National Institutes of Health (National Eye Institute) grant R01 EY031745 (G.D.) and the National Institutes of Health (National Institute of Arthritis and Musculoskeletal and Skin Diseases) grant R56 AR067731 (G.D.).

References
Adachi, N., Robinson, M., Goolsbee, A., & Shubin, N. H. (2016). Regulatory evolution of Tbx5 and the origin of paired appendages. *Proceedings of the National Academy of Sciences of the United States of America, 113*, 10115–10120.

Agarwal, P., Wylie, J. N., Galceran, J., Arkhitko, O., Li, C., Deng, C., et al. (2003). *Tbx5* is essential for forelimb bud initiation following patterning of the limb field in the mouse embryo. *Development, 130*, 623–633.

Ahn, Y., Mullan, H. E., & Krumlauf, R. (2014). Long-range regulation by shared retinoic acid response elements modulates dynamic expression of posterior Hoxb genes in CNS development. *Developmental Biology, 388*, 134–144.

Aoto, J., Nam, C. I., Poon, M. M., Ting, P., & Chen, L. (2008). Synaptic signaling by all-trans retinoic acid in homeostatic synaptic plasticity. *Neuron, 60*, 308–320.

Begemann, G., Schilling, T. F., Rauch, G. J., Geisler, R., & Ingham, P. W. (2001). The zebrafish *neckless* mutation reveals a requirement for *raldh2* in mesodermal signals that pattern the hindbrain. *Development, 128*, 3081–3094.

Berenguer, M., Lancman, J. J., Cunningham, T. J., Dong, P. D. S., & Duester, G. (2018). Mouse but not zebrafish requires retinoic acid for control of neuromesodermal progenitors and body axis extension. *Developmental Biology, 441*, 127–131.

Berenguer, M., Meyer, K. F., Yin, J., & Duester, G. (2020). Discovery of genes required for body axis and limb formation by global identification of retinoic acid-regulated epigenetic marks. *PLoS Biology, 18*, e3000719.

Boulet, A. M., & Capecchi, M. R. (2012). Signaling by FGF4 and FGF8 is required for axial elongation of the mouse embryo. *Developmental Biology, 371*, 235–245.

Cammas, L., Romand, R., Fraulob, V., Mura, C., & Dolle, P. (2007). Expression of the murine retinol dehydrogenase 10 (Rdh10) gene correlates with many sites of retinoid signalling during embryogenesis and organ differentiation. *Developmental Dynamics, 236*, 2899–2908.

Canestro, C., Catchen, J. M., Rodriguez-Mari, A., Yokoi, H., & Postlethwait, J. H. (2009). Consequences of lineage-specific gene loss on functional evolution of surviving paralogs: ALDH1A and retinoic acid signaling in vertebrate genomes. *PLoS Genetics, 5*, e1000496.

Chatagnon, A., Veber, P., Morin, V., Bedo, J., Triqueneaux, G., Semon, M., et al. (2015). RAR/RXR binding dynamics distinguish pluripotency from differentiation associated cis-regulatory elements. *Nucleic Acids Research, 43*, 4833–4854.

Chawla, B., Schley, E., Williams, A. L., & Bohnsack, B. L. (2016). Retinoic acid and Pitx2 regulate early neural crest survival and migration in craniofacial and ocular development. *Birth Defects Research Part B, 107*, 126–135.

Costaridis, P., Horton, C., Zeitlinger, J., Holder, N., & Maden, M. (1996). Endogenous retinoids in the zebrafish embryo and adult. *Developmental Dynamics, 205*, 41–51.

Creyghton, M. P., Cheng, A. W., Welstead, G. G., Kooistra, T., Carey, B. W., Steine, E. J., et al. (2010). Histone H3K27ac separates active from poised enhancers and predicts developmental state. *Proceedings of the National Academy of Sciences of the United States of America, 107*, 21931–21936.

Cunningham, T. J., Brade, T., Sandell, L. L., Lewandoski, M., Trainor, P. A., Colas, A., et al. (2015). Retinoic acid activity in undifferentiated neural progenitors is sufficient to fulfill Its role in restricting Fgf8 expression for somitogenesis. *PLoS One, 10*, e0137894.

Cunningham, T. J., Chatzi, C., Sandell, L. L., Trainor, P. A., & Duester, G. (2011). *Rdh10* mutants deficient in limb field retinoic acid signaling exhibit normal limb patterning but display interdigital webbing. *Developmental Dynamics, 240*, 1142–1150.

Cunningham, T. J., Colas, A., & Duester, G. (2016). Early molecular events during retinoic acid induced differentiation of neuromesodermal progenitors. *Biology Open, 5*, 1821–1833.

Cunningham, T. J., & Duester, G. (2015). Mechanisms of retinoic acid signalling and its roles in organ and limb development. *Nature Reviews Molecular Cell Biology, 16*, 110–123.

Cunningham, T. J., Lancman, J. J., Berenguer, M., Dong, P. D. S., & Duester, G. (2018). Genomic knockout of two presumed forelimb Tbx5 enhancers reveals they are non-essential for limb development. *Cell Reports, 23*, 3146–3151.

Cunningham, T. J., Zhao, X., Sandell, L. L., Evans, S. M., Trainor, P. A., & Duester, G. (2013). Antagonism between retinoic acid and fibroblast growth factor signaling during limb development. *Cell Reports, 3*, 1503–1511.

Delgado, I., Lopez-Delgado, A. C., Rosello-Diez, A., Giovinazzo, G., Cadenas, V., Fernandez-de-Manuel, L., et al. (2020). Proximo-distal positional information encoded by an Fgf-regulated gradient of homeodomain transcription factors in the vertebrate limb. *Science Advances, 6*, eaaz0742.

Diez del Corral, R., Olivera-Martinez, I., Goriely, A., Gale, E., Maden, M., & Storey, K. (2003). Opposing FGF and retinoid pathways control ventral neural pattern, neuronal differentiation, and segmentation during body axis extension. *Neuron, 40*, 65–79.

Dong, D., & Zile, M. H. (1995). Endogenous retinoids in the early avian embryo. *Biochemical and Biophysical Research Communications, 217*, 1026–1031.

Duester, G. (2022). Pharmacological retinoic acid alters limb patterning during regeneration but endogenous retinoic acid is not required. *Regenerative Medicine, 17*, 705–707.

Duester, G. (2023). Insufficient support for retinoic acid receptor control of synaptic plasticity through a non-genomic mechanism. *Frontiers in Neuroendocrinology, 71*.

Duester, G. (2024). Requirement of retinoic acid to establish mammalian totipotency is questioned. Nature Structural & Molecular Biology. (in press).

Dupé, V., Davenne, M., Brocard, J., Dollé, P., Mark, M., Dierich, A., et al. (1997). In vivo functional analysis of the *Hoxa-1* 3′ retinoic acid response element (3′RARE). *Development, 124*, 399–410.

Dupé, V., Matt, N., Garnier, J.-M., Chambon, P., Mark, M., & Ghyselinck, N. B. (2003). A newborn lethal defect due to inactivation of retinaldehyde dehydrogenase type 3 is prevented by maternal retinoic acid treatment. *Proceedings of the National Academy of Sciences of the United States of America, 100*, 14036–14041.

England, S., Batista, M. F., Mich, J. K., Chen, J. K., & Lewis, K. E. (2011). Roles of Hedgehog pathway components and retinoic acid signalling in specifying zebrafish ventral spinal cord neurons. *Development, 138*, 5121–5134.

Ericson, J., Rashbass, P., Schedl, A., Brenner-Morton, S., Kawakami, A., van Heyningen, V., et al. (1997). Pax6 controls progenitor cell identity and neuronal fate in response to graded Shh signaling. *Cell, 90*, 169–180.

Ericson, J., Thor, S., Edlund, T., Jessell, T. M., & Yamada, T. (1992). Early stages of motor neuron differentiation revealed by expression of homeobox gene *Islet-1*. *Science, 256*, 1555–1560.

Evans, A. L., & Gage, P. J. (2005). Expression of the homeobox gene *Pitx2* in neural crest is required for optic stalk and ocular anterior segment development. *Human Molecular Genetics, 14*, 3347–3359.

Fan, X., Molotkov, A., Manabe, S.-I., Donmoyer, C. M., Deltour, L., Foglio, M. H., et al. (2003). Targeted disruption of *Aldh1a1* (*Raldh1*) provides evidence for a complex mechanism of retinoic acid synthesis in the developing retina. *Molecular and Cellular Biology, 23*, 4637–4648.

Gage, P. J., Qian, M., Wu, D., & Rosenberg, K. I. (2008). The canonical Wnt signaling antagonist DKK2 is an essential effector of PITX2 function during normal eye development. *Developmental Biology, 317*, 310–324.

Ghyselinck, N. B., & Duester, G. (2019). Retinoic acid signaling pathways. *Development, 146*, dev167502.

Ghyselinck, N. B., Dupé, V., Dierich, A., Messaddeq, N., Garnier, J.-M., Rochette-Egly, C., et al. (1997). Role of the retinoic acid receptor beta (RARb) during mouse development. *The International Journal of Developmental Biology, 41*, 425–447.

Goto, S., Onishi, A., Misaki, K., Yonemura, S., Sugita, S., Ito, H., et al. (2018). Neural retina-specific Aldh1a1 controls dorsal choroidal vascular development via Sox9 expression in retinal pigment epithelial cells. *eLife, 7*, 03.

Gouti, M., Tsakiridis, A., Wymeersch, F. J., Huang, Y., Kleinjung, J., Wilson, V., et al. (2014). In vitro generation of neuromesodermal progenitors reveals distinct roles for wnt signalling in the specification of spinal cord and paraxial mesoderm identity. *PLoS Biology, 12,* e1001937.

Hadyniak, S. E., Hagen, J. F. D., Eldred, K. C., Brenerman, B., Hussey, K. A., McCoy, R. C., et al. (2024). Retinoic acid signaling regulates spatiotemporal specification of human green and red cones. *PLoS Biology, 22,* e3002464.

Heintzman, N. D., Stuart, R. K., Hon, G., Fu, Y., Ching, C. W., Hawkins, R. D., et al. (2007). Distinct and predictive chromatin signatures of transcriptional promoters and enhancers in the human genome. *Nature Genetics, 39,* 311–318.

Henrique, D., Abranches, E., Verrier, L., & Storey, K. G. (2015). Neuromesodermal progenitors and the making of the spinal cord. *Development, 142,* 2864–2875.

Hernandez, R. E., Putzke, A. P., Myers, J. P., Margaretha, L., & Moens, C. B. (2007). Cyp26 enzymes generate the retinoic acid response pattern necessary for hindbrain development. *Development, 134,* 177–187.

Hernandez, R. E., Rikhof, H. A., Bachmann, R., & Moens, C. B. (2004). vhnf1 integrates global RA patterning and local FGF signals to direct posterior hindbrain development in zebrafish. *Development, 131,* 4511–4520.

Hoffmann, B., Lehmann, J. M., Zhang, X., Hermann, T., Husmann, M., Graupner, G., et al. (1990). A retinoic acid receptor-specific element controls the retinoic acid receptor-b promoter. *Molecular Endocrinology, 4,* 1727–1736.

Horton, C., & Maden, M. (1995). Endogenous distribution of retinoids during normal development and teratogenesis in the mouse embryo. *Developmental Dynamics, 202,* 312–323.

Irie, T., Azuma, M., & Seki, T. (1991). The retinal and 3-dehydroretinal in *Xenopus laevis* eggs are bound to lipovitellin 1 by a Schiff base linkage. *Zoological Science, 8,* 855–863.

Iturbide, A., Ruiz Tejeda Segura, M. L., Noll, C., Schorpp, K., Rothenaigner, I., Ruiz-Morales, E. R., et al. (2021). Retinoic acid signaling is critical during the totipotency window in early mammalian development. *Nature Structural & Molecular Biology, 28,* 521–532.

Kastner, P., Mark, M., & Chambon, P. (1995). Nonsteroid nuclear receptors: What are genetic studies telling us about their role in real life? *Cell, 83,* 859–869.

Keegan, B. R., Feldman, J. L., Begemann, G., Ingham, P. W., & Yelon, D. (2005). Retinoic acid signaling restricts the cardiac progenitor pool. *Science, 307,* 247–249.

Kochhar, D. M. (1973). Limb development in mouse embryos. I. Analysis of teratogenic effects of retinoic acid. *Teratology, 7,* 289–298.

Kochhar, D. M., Jiang, H., Pennera, J. D., Beard, R. L., & Chandraratna, R. A. S. (1996). Differential teratogenic response of mouse embryos to receptor selective analogs of retinoic acid. *Chemico-Biological Interactions, 100,* 1–12.

Kondoh, H., & Takemoto, T. (2024). The origin and regulation of neuromesodermal progenitors (NMPs) in embryos. *Cells, 13,* 21.

Krumlauf, R. (1993). Hox genes and pattern formation in the branchial region of the vertebrate head. *Trend in Genetics, 9,* 106–112.

Kumar, S., Cunningham, T. J., & Duester, G. (2016). Nuclear receptor corepressors *Ncor1* and *Ncor2* (*Smrt*) are required for retinoic acid-dependent repression of *Fgf8* during somitogenesis. *Developmental Biology, 418,* 204–215.

Kumar, S., & Duester, G. (2010). Retinoic acid signaling in perioptic mesenchyme represses Wnt signaling via induction of Pitx2 and Dkk2. *Developmental Biology, 340,* 67–74.

Kumar, S., & Duester, G. (2014). Retinoic acid controls body axis extension by directly repressing *Fgf8* transcription. *Development, 141,* 2972–2977.

Lammer, G. J., Chen, D. T., Hoar, R. M., Agnish, N. D., Benke, P. J., Braun, J. T., et al. (1985). Retinoic acid embryopathy. *The New England Journal of Medicine, 313,* 837–841.

Lanctôt, C., Moreau, A., Chamberland, M., Tremblay, M. L., & Drouin, J. (1999). Hindlimb patterning and mandible development require the *Ptx1* gene. *Development, 126*, 1805–1810.

Lewandoski, M., Sun, X., & Martin, G. R. (2000). Fgf8 signalling from the AER is essential for normal limb development. *Nature Genetics, 26*, 460–463.

Linville, A., Radtke, K., Waxman, J. S., Yelon, D., & Schilling, T. F. (2009). Combinatorial roles for zebrafish retinoic acid receptors in the hindbrain, limbs and pharyngeal arches. *Developmental Biology, 325*, 60–70.

Logan, M., & Tabin, C. J. (1999). Role of Pitx1 upstream of Tbx4 in specification of hindlimb identity. *Science, 283*, 1736–1739.

Lohnes, D., Mark, M., Mendelsohn, C., Dollé, P., Dierich, A., Gorry, P., et al. (1994). Function of the retinoic acid receptors (RARs) during development. (I) Craniofacial and skeletal abnormalities in RAR double mutants. *Development, 120*, 2723–2748.

Maden, M. (1982). Vitamin A and pattern formation in the regenerating limb. *Nature, 295*, 672–675.

Maden, M., Gale, E., Kostetskii, I., & Zile, M. H. (1996). Vitamin A-deficient quail embryos have half a hindbrain and other neural defects. *Current Biology, 6*, 417–426.

Mariani, F. V., Ahn, C. P., & Martin, G. R. (2008). Genetic evidence that FGFs have an instructive role in limb proximal-distal patterning. *Nature, 453*, 401–405.

Marshall, H., Studer, M., Pöpperl, H., Aparicio, S., Kuroiwa, A., Brenner, S., et al. (1994). A conserved retinoic acid response element required for early expression of the homeobox gene *Hoxb-1*. *Nature, 370*, 567–571.

Matt, N., Dupé, V., Garnier, J.-M., Dennefeld, C., Chambon, P., Mark, M., et al. (2005). Retinoic acid-dependent eye morphogenesis is orchestrated by neural crest cells. *Development, 132*, 4789–4800.

Mendelsohn, C., Lohnes, D., Décimo, D., Lufkin, T., LeMeur, M., Chambon, P., et al. (1994). Function of the retinoic acid receptors (RARs) during development. (II) Multiple abnormalities at various stages of organogenesis in RAR double mutants. *Development, 120*, 2749–2771.

Mercader, N., Leonardo, E., Piedra, M. E., Martínez-A, C., Ros, M. A., & Torres, M. (2000). Opposing RA and FGF signals control proximodistal vertebrate limb development through regulation of Meis genes. *Development, 127*, 3961–3970.

Meyers, E. N., & Martin, G. R. (1999). Differences in left-right axis pathways in mouse and chick: functions of FGF8 and SHH. *Science, 285*, 403–406.

Mic, F. A., Haselbeck, R. J., Cuenca, A. E., & Duester, G. (2002). Novel retinoic acid generating activities in the neural tube and heart identified by conditional rescue of *Raldh2* null mutant mice. *Development, 129*, 2271–2282.

Mic, F. A., Molotkov, A., Benbrook, D. M., & Duester, G. (2003). Retinoid activation of retinoic acid receptor but not retinoid X receptor is sufficient to rescue lethal defect in retinoic acid synthesis. *Proceedings of the National Academy of Sciences of the United States of America, 100*, 7135–7140.

Mic, F. A., Molotkov, A., Fan, X., Cuenca, A. E., & Duester, G. (2000). RALDH3, a retinaldehyde dehydrogenase that generates retinoic acid, is expressed in the ventral retina, otic vesicle and olfactory pit during mouse development. *Mechanisms of Development, 97*, 227–230.

Minguillon, C., Nishimoto, S., Wood, S., Vendrell, E., Gibson-Brown, J. J., & Logan, M. P. (2012). Hox genes regulate the onset of Tbx5 expression in the forelimb. *Development, 139*, 3180–3188.

Molotkov, A., Molotkova, N., & Duester, G. (2006). Retinoic acid guides eye morphogenetic movements via paracrine signaling but is unnecessary for retinal dorsoventral patterning. *Development, 133*, 1901–1910.

Molotkova, N., Molotkov, A., & Duester, G. (2007). Role of retinoic acid during forebrain development begins late when Raldh3 generates retinoic acid in the ventral subventricular zone. *Developmental Biology, 303*, 601–610.

Molotkova, N., Molotkov, A., Sirbu, I. O., & Duester, G. (2005). Requirement of mesodermal retinoic acid generated by Raldh2 for posterior neural transformation. *Mechanisms of Development, 122*, 145–155.

Mory, A., Ruiz, F. X., Dagan, E., Yakovtseva, E. A., Kurolap, A., Pares, X., et al. (2014). A missense mutation in ALDH1A3 causes isolated microphthalmia/anophthalmia in nine individuals from an inbred Muslim kindred. *European Journal of Human Genetics, 22*, 419–422.

Moutier, E., Ye, T., Choukrallah, M. A., Urban, S., Osz, J., Chatagnon, A., et al. (2012). Retinoic acid receptors recognize the mouse genome through binding elements with diverse spacing and topology. *Journal of Biological Chemistry, 287*, 26328–26341.

Naiche, L. A., Holder, N., & Lewandoski, M. (2011). FGF4 and FGF8 comprise the wavefront activity that controls somitogenesis. *Proceedings of the National Academy of Sciences of the United States of America, 108*, 4018–4023.

Naiche, L. A., & Papaioannou, V. E. (2003). Loss of Tbx4 blocks hindlimb development and affects vascularization and fusion of the allantois. *Development, 130*, 2681–2693.

Nau, H. (1995). Chemical structure—Teratogenicity relationships, toxicokinetics and metabolism in risk assessment of retinoids. *Toxicology Letters, 82-83*, 975–979.

Nedelec, B., Rozet, J. M., & Fares Taie, L. (2019). Genetic architecture of retinoic-acid signaling-associated ocular developmental defects. *Human Genetics, 138*, 937–955.

Niederreither, K., Subbarayan, V., Dollé, P., & Chambon, P. (1999). Embryonic retinoic acid synthesis is essential for early mouse post-implantation development. *Nature Genetics, 21*, 444–448.

Niederreither, K., Vermot, J., Messaddeq, N., Schuhbaur, B., Chambon, P., & Dollé, P. (2001). Embryonic retinoic acid synthesis is essential for heart morphogenesis in the mouse. *Development, 128*, 1019–1031.

Niederreither, K., Vermot, J., Schuhbaur, B., Chambon, P., & Dollé, P. (2000). Retinoic acid synthesis and hindbrain patterning in the mouse embryo. *Development, 127*, 75–85.

Nishimoto, S., Minguillon, C., Wood, S., & Logan, M. P. (2014). A combination of activation and repression by a colinear Hox code controls forelimb-restricted expression of Tbx5 and reveals Hox protein specificity. *PLoS Genetics, 10*, e1004245.

Nishimoto, S., Wilde, S. M., Wood, S., & Logan, M. P. (2015). RA acts in a coherent feed-forward mechanism with Tbx5 to control limb bud induction and initiation. *Cell Reports, 12*, 879–891.

Noji, S., Nohno, T., Koyama, E., Muto, K., Ohyama, K., Aoki, Y., et al. (1991). Retinoic acid induces polarizing activity but is unlikely to be a morphogen in the chick limb bud. *Nature, 350*, 83–86.

Novitch, B. G., Wichterle, H., Jessell, T. M., & Sockanathan, S. (2003). A requirement for retinoic acid-mediated transcriptional activation in ventral neural patterning and motor neuron specification. *Neuron, 40*, 81–95.

Oliveira, C. R., Knapp, D., Elewa, A., Gerber, T., Gonzalez Malagon, S. G., Gates, P. B., et al. (2022). Tig1 regulates proximo-distal identity during salamander limb regeneration. *Nature Communications, 13*, 1141.

Park, E., Tjia, M., Zuo, Y., & Chen, L. (2018). Postnatal ablation of synaptic retinoic acid signaling impairs cortical information processing and sensory discrimination in mice. *Journal of Neuroscience, 38*, 5277–5288.

Patel, N. S., Rhinn, M., Semprich, C. I., Halley, P. A., Dolle, P., Bickmore, W. A., et al. (2013). FGF signalling regulates chromatin organisation during neural differentiation via mechanisms that can be uncoupled from transcription. *PLoS Genetics, 9*, e1003614.

Penkov, D., San Martin, D. M., Fernandez-Diaz, L. C., Rossello, C. A., Torroja, C., Sanchez-Cabo, F., et al. (2013). Analysis of the DNA-binding profile and function of TALE homeoproteins reveals their specialization and specific interactions with hox genes/proteins. *Cell Reports, 3*, 1321–1333.

Pennimpede, T., Cameron, D. A., MacLean, G. A., Li, H., Abu-Abed, S., & Petkovich, M. (2010). The role of CYP26 enzymes in defining appropriate retinoic acid exposure during embryogenesis. *Birth Defects Research, 88*, 883–894.

Perlmann, T., Rangarajan, P. N., Umesono, K., & Evans, R. M. (1993). Determinants for selective RAR and TR recognition of direct repeat HREs. *Genes & Development, 7*, 1411–1422.

Plaisancie, J., Bremond-Gignac, D., Demeer, B., Gaston, V., Verloes, A., Fares-Taie, L., et al. (2016). Incomplete penetrance of biallelic ALDH1A3 mutations. *European Journal of Medical Genetics, 59*, 215–218.

Pradhan, A., Zeng, X. I., Sidhwani, P., Marques, S. R., George, V., Targoff, K. L., et al. (2017). FGF signaling enforces cardiac chamber identity in the developing ventricle. *Development, 144*, 1328–1338.

Probst, S., Kraemer, C., Demougin, P., Sheth, R., Martin, G. R., Shiratori, H., et al. (2011). SHH propagates distal limb bud development by enhancing CYP26B1-mediated retinoic acid clearance via AER-FGF signalling. *Development, 138*, 1913–1923.

Rada-Iglesias, A., Bajpai, R., Swigut, T., Brugmann, S. A., Flynn, R. A., & Wysocka, J. (2011). A unique chromatin signature uncovers early developmental enhancers in humans. *Nature, 470*, 279–283.

Rallis, C., Bruneau, B. G., Del Buono, J., Seidman, C. E., Seidman, J. G., Nissim, S., et al. (2003). *Tbx5* is required for forelimb bud formation and continued outgrowth. *Development, 130*, 2741–2751.

Rhinn, M., & Dolle, P. (2012). Retinoic acid signalling during development. *Development, 139*, 843–858.

Rhinn, M., Schuhbaur, B., Niederreither, K., & Dolle, P. (2011). Involvement of retinol dehydrogenase 10 in embryonic patterning and rescue of its loss of function by maternal retinaldehyde treatment. *Proceedings of the National Academy of Sciences of the United States of America, 108*, 16687–16692.

Rossant, J., Zirngibl, R., Cado, D., Shago, M., & Giguère, V. (1991). Expression of a retinoic acid response element-*hsplacZ* transgene defines specific domains of transcriptional activity during mouse embryogenesis. *Genes & Development, 5*, 1333–1344.

Ryckebusch, L., Wang, Z., Bertrand, N., Lin, S.-C., Chi, X., Schwartz, R., et al. (2008). Retinoic acid deficiency alters second heart field formation. *Proceedings of the National Academy of Sciences of the United States of America, 105*, 2913–2918.

Sandell, L. L., Sanderson, B. W., Moiseyev, G., Johnson, T., Mushegian, A., Young, K., et al. (2007). RDH10 is essential for synthesis of embryonic retinoic acid and is required for limb, craniofacial, and organ development. *Genes & Development, 21*, 1113–1124.

Sirbu, I. O., & Duester, G. (2006). Retinoic acid signaling in node ectoderm and posterior neural plate directs left-right patterning of somitic mesoderm. *Nature Cell Biology, 8*, 271–277.

Sirbu, I. O., Gresh, L., Barra, J., & Duester, G. (2005). Shifting boundaries of retinoic acid activity control hindbrain segmental gene expression. *Development, 132*, 2611–2622.

Sirbu, I. O., Zhao, X., & Duester, G. (2008). Retinoic acid controls heart anteroposterior patterning by down-regulating *Isl1* through the *Fgf8* pathway. *Developmental Dynamics, 237*, 1627–1635.

Slavotinek, A. (2019). Genetics of anophthalmia and microphthalmia. Part 2: Syndromes associated with anophthalmia-microphthalmia. *Human Genetics, 138*, 831–846.

Sorrell, M. R., & Waxman, J. S. (2011). Restraint of Fgf8 signaling by retinoic acid signaling is required for proper heart and forelimb formation. *Developmental Biology, 358*, 44–55.

Stratford, T., Horton, C., & Maden, M. (1996). Retinoic acid is required for the initiation of outgrowth in the chick limb bud. *Current Biology, 6*, 1124–1133.
Studer, M., Pöpperl, H., Marshall, H., Kuroiwa, A., & Krumlauf, R. (1994). Role of a conserved retinoic acid response element in rhombomere restriction of *Hoxb-1*. *Science, 265*, 1728–1732.
Takebayashi, H., Nabeshima, Y., Yoshida, S., Chisaka, O., & Ikenaka, K. (2002). The basic helix-loop-helix factor Olig2 is essential for the development of motoneuron and oligodendrocyte lineages. *Current Biology, 12*, 1157–1163.
Tickle, C., Alberts, B. M., Wolpert, L., & Lee, J. (1982). Local application of retinoic acid to the limb bud mimics the action of the polarizing region. *Nature, 296*, 564–565.
Vermot, J., Llamas, J. G., Fraulob, V., Niederreither, K., Chambon, P., & Dollé, P. (2005). Retinoic acid controls the bilateral symmetry of somite formation in the mouse embryo. *Science, 308*, 563–566.
Vivanco Ruiz, M. D. M., Bugge, T. H., Hirschmann, P., & Stunnenberg, H. G. (1991). Functional characterization of a natural retinoic acid responsive element. *EMBO Journal, 10*, 3829–3838.
Wanek, N., Gardiner, D. M., Muneoka, K., & Bryant, S. V. (1991). Conversion by retinoic acid of anterior cells into ZPA cells in the chick wing bud. *Nature, 350*, 81–83.
Warkany, J., & Schraffenberger, S. (1946). Congenital malformations induced in rats by maternal vitamin A deficiency. I. Defects of the eye. *Arch Ophthalmol, 35*, 150–169.
Weisschuh, N., Dressler, P., Schuettauf, F., Wolf, C., Wissinger, B., & Gramer, E. (2006). Novel mutations of FOXC1 and PITX2 in patients with Axenfeld-Rieger malformations. *Investigative Ophthalmology & Visual Science, 47*, 3846–3852.
White, J. C., Shankar, V. N., Highland, M., Epstein, M. L., DeLuca, P. F., & Clagett-Dame, M. (1998). Defects in embryonic hindbrain development and fetal resorption resulting from vitamin A deficiency in the rat are prevented by feeding pharmacological levels of all-*trans*-retinoic acid. *Proceedings of the National Academy of Sciences of the United States of America, 95*, 13459–13464.
Williams, A. L., & Bohnsack, B. L. (2019). What's retinoic acid got to do with it? Retinoic acid regulation of the neural crest in craniofacial and ocular development. *Genesis, 57*, e23308.
Wilson, V., Olivera-Martinez, I., & Storey, K. G. (2009). Stem cells, signals and vertebrate body axis extension. *Development, 136*, 1591–1604.
Yashiro, K., Zhao, X., Uehara, M., Yamashita, K., Nishijima, M., Nishino, J., et al. (2004). Regulation of retinoic acid distribution is required for proximodistal patterning and outgrowth of the developing limb. *Developmental Cell, 6*, 411–422.
Zhao, X., Sirbu, I. O., Mic, F. A., Molotkova, N., Molotkov, A., Kumar, S., et al. (2009). Retinoic acid promotes limb induction through effects on body axis extension but is unnecessary for limb patterning. *Current Biology, 19*, 1050–1057.

CHAPTER TWO

Multiple roles for retinoid signaling in craniofacial development

Masahiro Nakamura and Lisa L. Sandell[*]

Department of Oral Immunology and Infectious Diseases, University of Louisville School of Dentistry, Louisville, KY, United States
*Corresponding author. e-mail address: lisa.sandell@louisville.edu

Contents

1. Spatio-temporal regulation of RA signaling	35
2. Retinoid transport	35
3. RA generating enzymes	37
4. RA signal transduction through nuclear receptors	38
5. RA degrading enzymes	39
6. RA role in axial patterning of hindbrain and NCC	39
6.1 Midface	42
7. RA role in early eye development	44
8. RA role in nasal airway and upper lip morphogenesis	45
9. Salivary gland morphogenesis requires RA	47
10. Secondary palate	48
11. Bone and sutures	50
12. Stage-specific RA roles	51
References	52

Abstract

Retinoic acid (RA) signaling plays multiple essential roles in development of the head and face. Animal models with mutations in genes involved in RA signaling have enabled understanding of craniofacial morphogenic processes that are regulated by the retinoid pathway. During craniofacial morphogenesis RA signaling is active in spatially restricted domains defined by the expression of genes involved in RA production and RA breakdown. The spatial distribution of RA signaling changes with progressive development, corresponding to a multiplicity of craniofacial developmental processes that are regulated by RA. One important role of RA signaling occurs in the hindbrain. There RA contributes to specification of the anterior-posterior (AP) axis of the developing CNS and to the neural crest cells (NCC) which form the bones and nerves of the face and pharyngeal region. In the optic vesicles and frontonasal process RA orchestrates development of the midface, eyes, and nasal airway. Additional roles for RA in craniofacial development include regulation of submandibular salivary gland development and maintaining patency in the sutures of the cranial vault.

The vertebrate craniofacial complex—the head and face—is an intricate structure composed of cellular derivatives from all three germ layers; endoderm, mesoderm and ectoderm. Of particular importance are ectoderm-derived neural crest cells (NCC), which give rise to most of the bone, connective tissue, and some nerves of the face (Dash & Trainor, 2020; Szabó & Mayor, 2018). Growth and morphogenesis of developing craniofacial structures are orchestrated by cell interactions mediated by a small number of key signaling pathways, one of which involves retinoic acid (RA). "RA signaling" is a pathway based on a family of small lipophilic molecules, collectively known as retinoids, that are metabolically related to retinol, commonly known as "Vitamin A" (Redfern, 2020). In the developing head and face, as in the whole embryo, RA signaling occurs in spatially restricted domains that progressively change over time. The tightly regulated dynamic distribution of RA signaling corresponds to a multiplicity of craniofacial structures that are regulated by RA over the course of development including the midface, eyes, nasal airway, secondary palate, external and middle ears, salivary glands, and sutures of the cranial vault.

The important role of the retinoid pathway in craniofacial morphogenesis was first identified experimentally by nutritional experiments performed in the 1930s and 1940s in which animals were raised on diets devoid of Vitamin A (Hale, 1935; Warkany, 1945; Warkany et al., 1943). Malformations resulting from such diets included craniofacial abnormalities such as cleft palate and anophthalmia, as well as defects in other body systems.

Craniofacial malformations occur not only under conditions of Vitamin A deficiency. Exposure to excess Vitamin A or its bioactive metabolite, RA, is similarly deleterious. The effects of excess RA signaling have been assessed in numerous studies typically involving teratogenic doses of retinoids in the range of 1–2 μM, which is approximately 1000-fold excess of endogenous levels in serum or tissue (2–10 pmol/gram in mouse) (Kane et al., 2005). Such extreme perturbation of RA levels can yield confounding or even paradoxical effects (Lee et al., 2012). In contrast, loss-of-function mutations that disrupt enzymes or transcription factors involved in RA production, clearance, or signal transduction alter endogenous levels of RA activity. Here we review the roles of RA in craniofacial development, with a primary focus on insights gained from genetic loss-of-function mutations that perturb RA production, breakdown, or signaling activity in vivo.

1. Spatio-temporal regulation of RA signaling

During craniofacial morphogenesis RA signaling is active in a dynamic spatio-temporal pattern (Fig. 1A–D). In mouse embryos the tissue distribution of RA signaling activity can be detected using a reporter transgene that is responsive to RA transcriptional regulation (Rossant et al., 1991). RA signaling is first detected in the headfold stage embryo at E7.5, with an anterior boundary in the hindbrain that sharpens at E7.6 to the rhombomere2/3 (r2/3) border (Sirbu et al., 2005). A new domain of RA signaling arises at E8.3 in the optic pits that begin to form in the forebrain (Rossant et al., 1991; Wagner et al., 2000). From E8.5 to E9.5, as the neural tube closes and NCC migration commences, RA distribution remains dynamic. Hindbrain RA activity shifts caudally while forebrain/eye RA signaling expands, initially being confined to the neuroepithelium of the optic vesicles and subsequently encompassing the surrounding perioptic mesenchyme and the telencephalon (Bok et al., 2011; Ribes et al., 2006). By E10.5, RA is active within the lambdoid junction—the interface zone between the lateral and medial nasal processes and the maxillary portion of pharyngeal arch (PA) 1—which contributes to the upper lip, anterior airway and primary palate. At this stage RA is also active in the primordia of the submandibular salivary glands in the interior mandible (Wright et al., 2015).

The dynamic distribution of RA signaling is controlled by site-specific expression of enzymes involved synthesis and clearance of RA. Genes important for regulating RA activity include those whose products are involved in entry of the precursor retinol into the cells, the two-step enzymatic conversion of the precursor retinol into bioactive RA, transduction of the signal via nuclear transcription factors, and finally, breakdown of RA to inactive products (Fig. 2). Mutations in these genes can alter the level and distribution of RA and disrupt craniofacial morphogenesis. Because RA signaling regulates multiple processes that occur at progressive stages of craniofacial development, disruption of RA signaling at different embryonic stages can yield distinctive phenotypes corresponding to the role of RA at the time of disturbance.

2. Retinoid transport

Cells receive retinol from the bloodstream (Fig. 2). Retinol is carried in the serum by retinol binding protein 4 (RBP4) (Quadro et al., 2005).

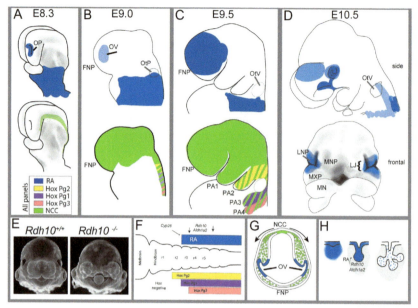

Fig. 1 Multiple roles for RA signaling throughout craniofacial morphogenesis. (A–D) Dynamic RA signaling distribution in craniofacial tissues throughout gastrulation and early organogenesis phases of development in mice. (A) At E8.3 RA signaling is active in the optic pits and in the hindbrain with an anterior boundary at the preotic sulcus. At this stage neural crest cells are induced at the lateral margins of the neural plate. (B) At E9.0 the anterior boundary of RA activity is shifted posteriorly and the optic vesicles that possess RA signaling are enclosed within the neural tube. Neural crest cells begin to migrate from the posterior forebrain, midbrain and anterior hindbrain. Hindbrain *Hox* genes are expressed. (C) At E9.5 the domain of RA signaling in the optic region and frontonasal process expands. Neural crest cells migrate to fill the pharyngeal arches. (D) At E10.5 RA signaling is active in the eye and lambdoid junction of the developing midfacial region. (E) Development of the facial midline depends on early RA signaling. Multiple mutant mouse models that disrupt RA production or RA signaling display midline facial cleft. Example shown is $Rdh10^{-/-}$ null mutant partially rescued with retinaldehyde. (F) RA is crucial for proper expression of *Hox* gene family members, which mediate A-P patterning of the hindbrain and hindbrain-derived NCC that generate facial skeleton and nerves. (G) RA signaling in the optic vesicles is needed for formation of the optic cup. At later stages RA signaling expands to include surrounding periocular mesenchyme, which is needed for anterior eye development. (H) Morphogenesis of submandibular salivary glands requires RA at initiation and during epithelial branching morphogenesis. FNP, Frontonasal process; Hox Pg, *Hox* paralog group; LJ, lambdoid junction; LNP, lateral nasal process; MN, mandibular process of PA1; MNP, medial nasal process; MXP, maxillary process of PA1; OP, optic pit; OtP, otic pit; OtV, otic vesicle; OV, optic vesicle; r, rhombomere.

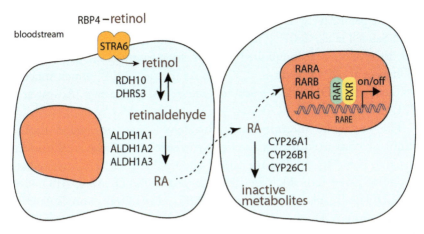

Fig. 2 Production of RA and transduction of RA signal. Gene products essential for RA signaling activity during craniofacial morphogenesis include those involved in retinoid transport, conversion of retinol to bioactive RA, nuclear transcription factors, and RA degrading enzymes. RA signaling can occur in a paracrine manner with production and signaling activity in separate cells (shown), or production and signaling can occur within the same cell.

Being a lipophilic molecule retinol can diffuse through membranes (Noy et al., 1992), however, entry into cells is facilitated by transporters, particularly the channel transmembrane protein encoded by STRA6 (Stimulated by retinoic acid gene 6) (Kelly & von Lintig, 2015). Null mutation of *Rbp4* in mice compromises delivery of retinol to the fetus and exacerbates RA signaling deficiency in conditions of dietary retinoid deprivation (Quadro et al., 2005). In humans and zebrafish loss of *STRA6* function owing to mutation or morpholino knockdown causes phenotypes associated with reduced RA signaling and deficient vitamin A (Gerth-Kahlert et al., 2013; Golzio et al., 2007; Isken et al., 2008; Kelly & von Lintig, 2015; Pasutto et al., 2007; White et al., 2008). In mice, a loss-of-function *Stra6* mutation also produces retinoid deficiency phenotypes, although milder than those observed in humans (Amengual et al., 2014; Berry et al., 2013).

3. RA generating enzymes

Inside cells retinol is converted to bioactive RA via two sequential oxidation steps (Fig. 2). Initially retinol is reversibly oxidized to the intermediate retinaldehyde. In embryos oxidization of retinol to retinaldehyde is mediated primarily by retinol dehydrogenase 10 (RDH10) (Sandell et al., 2012; Sandell et al., 2007). The reverse reaction, reduction of retinaldehyde back to retinol, is accomplished by dehydrogenase/reductase 3 (DHRS3) (Billings et al., 2013; Feng et al., 2010; Kam et al., 2013), which functions together with RDH10 as a heterotetrametric oxidoreductase (Adams et al., 2021, 2014).

The metabolic intermediate retinaldehyde can be cycled back to retinol or converted forward via an irreversible oxidation reaction into bioactive RA (Fig. 2). The oxidation of retinaldehyde to RA is mediated by one of three retinaldehyde dehydrogenases (ALDH1A1, ALDH1A2, or ALDH1A3), (also known as RALDH1, RALDH2, and RALDH3) (Cunningham & Duester, 2015). Of these, ALDH1A2 is responsible for a majority of RA synthesis within a developing embryo and loss of *Aldh1a2* in mice results in near complete elimination of RA production and embryo lethality by mid-gestation (Niederreither et al., 1999). *Aldh1a1* and *Aldh1a3* are expressed later, after the onset of facial morphogenesis, and loss of these genes produces milder RA deficiency phenotypes in specific tissues (Dupe et al., 2003; Fan et al., 2003). Once generated, RA can regulate transcription within the synthesizing cell, or it can diffuse to a neighboring cell to function in a paracrine fashion (Fig. 2).

Because RA signaling is essential for early heart development, mutations that severely disrupt RA production such as null mutants of *Rdh10* or *Aldh1a2* impair heart morphogenesis and are lethal for an embryo prior to full development of the head and face (Niederreither et al., 1999; Sandell et al., 2012). As a result of lethality, mutations that eliminate or severely reduce RA activity provide little information about the role of RA in craniofacial development. Moreover, tissue-specific conditional knockout models of genes involved in RA synthesis are difficult to interpret owing to the ability of retinoids to diffuse between cells and robust mechanisms of feedback regulation of RA homeostasis (O'Connor et al., 2022). Instead, analysis of the requirement for RA generating genes during craniofacial development has been aided by hypomorphic alleles that retain residual RA or by conditional global knockouts that are inactivated in a stage-specific manner.

4. RA signal transduction through nuclear receptors

RA transduces its signal by binding as ligand to retinoic acid receptor (RAR) nuclear transcription factors (Fig. 2). In mammals the RAR family has three members: RARA, RARB, and RARG (Rhinn & Dollé, 2012). Being transcription factors, these are typically localized within the nucleus where they form heterodimers with related retinoid X receptors (RXR), activating or repressing transcription of target genes by binding to DNA regulatory sequences known as Retinoic Acid Response Elements (RARE) (Berenguer et al., 2020; Moutier et al., 2012; Rhinn & Dollé, 2012). Presence of RA ligand modulates activity of RAR-RXR complexes by causing a conformational change that switches the binding of nuclear receptor coactivators or nuclear receptor corepressors (le Maire et al., 2019). Because RA signaling is transduced via RAR, mutation or loss of RARs can attenuate RA signaling activity, however, owing to functional redundancy between family members, mutations of individual single RAR genes often have mild or no phenotype. The requirement for RAR in mediating RA signaling is instead revealed by compound knockout models that eliminate two or more of the RAR genes or by expression of dominant negative alleles that block function RAR function.

5. RA degrading enzymes

The finely controlled spatial distribution and level of RA signaling is delineated by a balance between sites of RA production and sites of RA degradation. The breakdown of bioactive RA to inactive metabolites is mediated by CYP26 family enzymes (Fig. 2), which are members of the cytochrome P450 superfamily of proteins (Dubey et al., 2018; Ross & Zolfaghari, 2011). Two CYP26 enzymes, CYP26A1 and CYP26B1 are essential for spatially restricted clearance of RA in developing vertebrate embryos. Mutation of *Cyp26a1* or *Cyp26b1* in mice results in excessive and ectopic RA signaling (Abu-Abed et al., 2001; Sakai et al., 2001; Uehara et al., 2007). A third paralog, *Cyp26c1*, is not needed on its own, but works cooperatively with *Cyp26a1* in mice (Uehara et al., 2007). Spontaneous mutations of *Cyp26c1* reveal a role for this gene in facial morphogenesis in cows and humans (Lee et al., 2018; Sieck et al., 2020; Slavotinek et al., 2013).

6. RA role in axial patterning of hindbrain and NCC

Early craniofacial development begins at neurulation with elevation of the head folds and subsequent closure of the neural tube, which occurs from embryonic day 7.5 (E7.5) through E9.5 in mouse. As the head folds elevate and optic pits form on the anterior neuroepithelium, NCC are induced at the lateral margins of the neural ectoderm (Fig. 1A). With progressive uplifting, the headfolds ultimately meet and fuse at the dorsal midline forming a neural tube and enclosing the optic vesicles within the lumen (Fig. 1B and G). Concomitantly, the hindbrain becomes segmented into rhombomeres along the anterior-posterior (A-P) axis. As this happens, the hindbrain rhombomeres are specified according to their A-P position by the pattern of expression of homeobox (*Hox*) genes, a family of transcription factors that, in vertebrates, are organized as four distinct paralogy groups (Bedois et al., 2021; Parker et al., 2018). Within each of the paralogy groups, *Hox* genes are expressed in a nested manner with overlapping domains in the posterior embryo and distinct anterior boundaries, a pattern that forms a combinatorial "*Hox* code" (Parker et al., 2018). While the majority of the hindbrain and trunk express at least one *Hox* gene, the forebrain, midbrain, and rhombomere 1 of the hindbrain lack *Hox* gene expression.

The first NCC to exit the dorsal neural tube emigrate from *Hox*-negative caudal forebrain and midbrain (Fig. 1B). Once delaminated, they move in a sheet toward the rostral tip of the embryo, migrating beneath the surface ectoderm, streaming around the expanding optic vesicles, and ultimately populating the presumptive frontonasal process as the neural tube closes (Williams & Bohnsack, 2020). The *Hox*-negative NCC of the frontonasal process contribute to the frontal bone, nasal bone, and philtrum of the upper lip in the facial midline. A more posterior stream of NCC emigrating from the caudal midbrain and rhombomere 1, also *Hox*-negative, migrates into PA1. The PA1 NCC are fated to produce the maxillary bones, zygomatic bones, and mandible, as well as the malleus, incus, tympanic ring, and auditory canal of the ear. In a rostral-caudal progression NCC emigrate from more posterior levels of the hindbrain, which are *Hox*-positive (Toro-Tobon et al., 2023). Those NCC streaming into PA2 will contribute to the stapes of the middle ear, styloid process, lesser horns of the hyoid bone, and outer ear pinnae. NCC in PA3 and 4 produce the greater horn and lower half of the hyoid bone, and the thyroid and cricoid cartilages, respectively.

The importance of segmental *Hox* gene expression and A-P patterning for craniofacial development is demonstrated by phenotypes resulting when *Hox* genes are mutated or mis-expressed. In mouse, *Xenopus,* and zebrafish, loss or knockdown of *Hoxa2* (the anterior-most *Hox* gene) expands the *Hox*-negative domain in a posterior direction, which produces homeotic transformations such that normal PA2 derivatives are replaced with mirror duplications of structures normally derived from PA1 (Baltzinger et al., 2005; Gendron-Maguire et al., 1993; Hunter & Prince, 2002; Minoux et al., 2013; Rijli et al., 1993; Santagati et al., 2005). Conversely, ectopic expression of *Hoxa2* in anterior NCC that normally *Hox*-negative in mice or chick causes duplication of PA2 elements, disrupts development of anterior frontonasal skeletal elements, middle ear bones, and jaw(Grammatopoulos et al., 2000; Kitazawa et al., 2015, 2022; Minoux et al., 2013). The *Hox* code and A-P patterning also control formation NCC-derived cranial nerves (Carpenter et al., 1993; Gavalas et al., 1998; Goddard et al., 1996; Jungbluth et al., 1999; Mark et al., 1993; Studer et al., 1996).

A major role for RA signaling in craniofacial morphogenesis is control of A-P patterning via transcriptional regulation of *Hox* genes within the hindbrain. The pattern of RA in the hindbrain, with an anterior RA-free zone and posterior RA-positive zone is defined by expression of enzymes that mediate RA synthesis and degradation. Expression of *Rdh10* and *Aldh1a2* in presomitic mesoderm of the gastrulating embryo initiates production of RA that diffuses from the mesoderm to the neural tube in a paracrine manner. The diffusing RA activates signaling to an anterior limit at the r2/3 border (Sirbu et al., 2005), with the anterior hindbrain being maintained as an RA-free zone by RA degrading CYP26 enzymes present in the head (Sirbu et al., 2005; White & Schilling, 2008).

The role of RA in regulating hindbrain A-P pattern is demonstrated by malformations that occur from mutations that disrupt RA regulatory control of specific *Hox* genes. *Hoxa1* and *Hoxb1* have associated RA responsive RARE cis-regulatory sequences. Targeted mutation of these RARE elements in mice attenuates expression of associated *Hox* genes, diminishes their responsiveness to RA, and produces A-P patterning defects (Dupé et al., 1997; Gavalas et al., 1998; Studer et al., 1998). The $Hoxa1^{RARE}$ and $Hoxb1^{RARE}$ mutations alone, or in combination with corresponding null mutations, cause or exacerbate defects in development of structures dependent on *Hoxa1* or *Hoxb1*, including the external ear, middle ear, pharyngeal cartilages, and cranial nerves V, VII, VIII, and IX.

Deficient RA production, such as occurs by mutation or knockdown of *Rdh10* or *Aldh1a2*, reduces or causes a posterior shift in hindbrain expression boundaries of *Hox* gene family members (D'Aniello et al., 2015; Friedl et al., 2019; Niederreither et al., 1999, 2000). Conversely, excess RA, as happens when *Cyp26* genes or *Dhrs3* are mutated or knocked down, yields upregulation or anterior expansion of *Hox* gene expression (Billings et al., 2013; Feng et al., 2010; Hernandez et al., 2007; Sirbu et al., 2005; Uehara et al., 2007; White & Schilling, 2008). Disruption of RA signaling by compound mutation of RARα and RARβ alters *Hox* gene expression pattern, disturbs rhombomere A-P identity, and yields disorganized post-otic cranial nerves (Dupé et al., 1999).

Craniofacial elements disrupted by mutation in RA genes are similar to those disrupted by mutation of *Hox* genes. For example, mutations of RA genes or knockout of anterior *Hox* genes in mice both produce malformations of PA2 derivatives including the tympanic ring, squamosal bone cranial nerve IX, hyoid, and styloid, stapes, and pinnae (Dupé & Pellerin, 2009; Gavalas et al., 1998; Gendron-Maguire et al., 1993; Ghyselinck & Duester, 2019; Lohnes et al., 1994; Minoux et al., 2013; Rijli et al., 1993). In humans, microtia and conductive hearing loss resulting from malformations in the external and/or middle ear have been associated with mutation of *Hoxa2* (Alasti et al., 2008; Si et al., 2020), and with mutation of *CYP26A1* or *CYP26B1* (Grand et al., 2021; Guo et al., 2020; Morton et al., 2016; Silveira et al., 2023). In mice, *Cyp26b1* mutation also causes reduction of the external ear, although *Hox* gene expression defects were not detected in that model (Maclean et al., 2009). Ectopic expression of *Hox* genes in *Hox*-negative domains disrupts formation of the PA1-derived jaw (Alexandre et al., 1996; Pasqualetti et al., 2000). Similarly, knockdown of *cyp26b1* in zebrafish results in defective or absent jaw cartilages (Reijntjes et al., 2007) and mutation of *CYP26C1* causes malformation of the mandible in Herford cattle (Sieck et al., 2020).

6.1 Midface

Midfacial structures such as the nose, upper lip, frontal bone, premaxilla, and nasal cartilages develop from derivatives of the frontonasal process (FNP), while the lateral face and secondary palate derive from PA1 (Farlie et al., 2016; Som & Naidich, 2013). As the neural tube closes, the bulbous FNP is formed by growth and expansion of the anterior neuroepithelium and underlying frontonasal mesenchyme, including NCC, which push the rostralmost facial tissue forward and outward (Fig. 1B and C). On the surface of the FNP, nasal placodes form on either side of the midline. During this phase of facial development patterning is orchestrated by reciprocal signaling interactions between NCC and various surrounding epithelia including surface ectoderm and pharyngeal endoderm (Helms et al., 2005). Midface development depends upon proper regulation of Hedgehog signaling as dysregulation of sonic hedgehog (SHH) results in phenotypes such as hyper- or hypotelorism, holoprosencephaly, and midline clefts (Abramyan, 2019; Brugmann et al., 2010; Hu & Marcucio, 2009; Muenke & Beachy, 2000). Fibroblast growth factor (FGF) signaling is likewise crucial for midface development. Altered gene dosage of fibroblast growth factor 8 (*Fgf8*) produces severed midfacial phenotypes including midfacial clefting (Griffin et al., 2013). *Alx* family genes are also important for midfacial development as mutations produce midfacial malformations in mice (Beverdam et al., 2001; Iyyanar et al., 2022), zebrafish (Dee et al., 2013), and humans (Farlie et al., 2016).

In the frontonasal region, RA signaling is first active in the ventral rim of the optic pit as the neural folds elevate and fuse (mouse E8.3–E9.0) (Fig. 1A). Progressively RA activity expands to include the surface ectoderm and presumptive lens placode, rostral forebrain, optic vesicles and nasal placodes (Dollé et al., 2010; Mic et al., 2004; Ribes et al., 2006; Wagner et al., 2000). Enzymes responsible for production of RA in the frontonasal region include RDH10, ALDH1A1, ALDH1A2, and ALDH1A3 (Dupe et al., 2003; Mic et al., 2004; Molotkov et al., 2006; Sandell et al., 2007).

Morphogenesis of the frontonasal midline depends on properly regulated RA activity. Midfacial clefts occur in multiple mouse models with deficient RA production or signaling: stage-specific knockout or hypomorphic mutation of *Rdh10* (Sandell et al., 2007; Wu et al., 2022) compound mutants lacking *Aldh1A2;Aldh1A3* (Halilagic et al., 2007), *Rara/g* double knockout (Lohnes et al., 1994), E9.5 stage-specific triple *Rara/b/g* knockout (Teletin et al., 2023), or loss of RAR activity by expression of a dominant-negative mutant *RAR* (Damm et al., 1993; Gao et al., 2021). Widening of the developing face, similar to hypertelorism, is also noted in RA deficient embryos (Wu et al., 2022). RA signaling is needed cell-autonomously within NCC for development of midfacial structures but is not required for their migration. NCC-specific *Rara/g* double knockout or *Rara/b/g* triple knockout causes midline clefts without impacting NCC migration (Dupé & Pellerin, 2009). Excessive RA signaling also disrupts midfacial development. Mouse *Cyp26b1* knockout produces craniofacial phenotypes including truncated nasal process (Maclean et al., 2009). In humans, phenotypes of *CYP26B1*-related disorders include midface hypoplasia and midface retrusion (Grand et al., 2021; Morton et al., 2016; Silveira et al., 2023).

Molecularly, the RA role in frontonasal and midfacial development likely involves regulation of SHH signaling. Midfacial clefts in Tamoxifen

E7.0 conditional *Rdh10* knockout are associated with elevated *Shh* and phenotypes can be partially rescued by pharmacologic inhibition of SHH (Wu et al., 2022). Downstream targets of RA signaling relevant for midfacial development may include *Alx* family members as downregulation of *Alx1* and *Alx3* are observed in the Tamoxifen E7.0 conditional *Rdh10* knockout with reduced RA.

7. RA role in early eye development

Whereas eyes are rightly considered an extension of the CNS, they also act as organizers within the developing facial region and defects such as anophthalmia or microphthalmia are considered severe craniofacial malformations (Kish et al., 2011). Early eye development begins with optic vesicles pouching outward from the forebrain, which starts at E8.5 in mouse. Once a vesicle nears the surface ectoderm, the outer portion begins invaginating back towards the forebrain, transforming the optic vesicle into an optic cup at E10.5. The optic vesicles and cups are surrounded by perioptic mesenchyme, which is derived from NCC and mesodermal derivatives. NCC in the perioptic mesenchyme interact with the neuroepithelial optic tissue, influencing morphogenesis of the optic cup and contributing to anterior eye structures including cornea and iris (Bryan et al., 2020; Williams & Bohnsack, 2020).

RA signaling, acting as a paracrine signal involved in interactions between ocular and surrounding tissues, is essential for eye development (Cvekl & Wang, 2009; Duester, 2022; Nedelec et al., 2019; Williams & Bohnsack, 2019). In the eye field RA reporter activity is first detected in mouse embryos at E8.3, later expanding to include the periocular mesenchyme and the frontonasal process (Dollé et al., 2010; Mic et al., 2004; Molotkov et al., 2006; Ribes et al., 2006; Wagner et al., 2000).

Reduction or elimination of RA in the eye field can disrupt eye formation at early stages leading to severe defects such absence of externally visible eye. Mutation of *Rdh10, Aldh1a2* or triple knockout of *Aldh1a1, Aldh1a2*, and *Aldh1a3*, compromises optic cup formation in mice (Molotkov et al., 2006; Mic et al., 2004; Sandell et al., 2007). Various *Rar* double knockouts produce phenotypes ranging from reduced palpebral aperture to anophthalmia (Ghyselinck et al., 1997; Lohnes et al., 1994). Some RA essential function in eye development occurs within NCC as NCC-specific triple *Rara/b/g* produces anophthalmia (Dupé & Pellerin, 2009). Downstream targets of RA signaling important for eye development include *Foxc1* and *Pitx2*; two eye development genes whose expression is reduced by loss of RA signaling (Matt et al., 2005; Molotkov et al., 2006). Presence of RARE sequences upstream of *Pitx2* suggest that RA may directly regulate transcription of this gene (Kumar & Duester, 2010).

After formation of the optic cup, RA continues to be essential for formation of anterior eye structures such as eyelid and cornea. Double knockout of *Aldh1a1* and *Aldh1a3* in mice, which eliminates RA signaling in the eye field after 9.5, produces eyelid and cornea malformations (Matt et al., 2005; Molotkov, Molotkova, & Duester, 2006). Similar defects are observed in a late embryonic nutritional vitamin A deficiency model in rat (See et al., 2008). Double NCC-specific knockout of *Rarb/g* disrupts development of anterior eye elements such as eyelid and cornea (Matt et al., 2008). RA signaling is also crucial for correct patterning of extraocular muscles as demonstrated by the phenotype of *Rdh10* null mutant mice (Comai et al., 2020).

In addition to experimental models, mutations in human RA-related genes such as *RBP4, STRA6, ALDH1A3,* and *RARB* can lead to anophthalmia or microphthalmia (Duester, 2022; Harding & Moosajee, 2019; Nedelec et al., 2019). A heterogeneous array of *RARB* pathogenic variants, including gain-of-function alleles, dominant negative alleles, and possibly haploinsufficient loss-of-function alleles, underlies a syndromic form of microphthalmia known as MCOPS12 (Caron et al., 2023). Human *STRA6* mutations produce anophthalmia/microphthalmia along with with other craniofacial dysmorphologies such as micrognathia and broad nasal bridge (Chassaing et al., 2009; Pasutto et al., 2007; Sadowski et al., 2017; Seller et al., 1996).

8. RA role in nasal airway and upper lip morphogenesis

Development of the nasal airway and upper lip begins with formation of a bilateral pair of nasal placodes on frontonasal process (Som & Naidich, 2013). As the frontonasal tissue expands, the placodes sink below the surface forming nasal pits and then nasal sacs. Invagination of the nasal pits subdivides each side of the frontonasal process into two distinct regions—a lateral nasal process (LNP) and a medial nasal process (MNP)—which are evident at E10.5 in mouse and week 5 in human. On each side of the face the maxillary portion of PA1 (MXP) grows forward to meet the LNP and

MNP forming a three-way intersection known as the lambdoid junction (Fig. 1D). Fusion of the lambdoid junction forms the tissue of the primary palate, philtrum, upper lip, and anterior roof of the mouth. The fused zone between the MXP and MNP, creates a "nasal fin" that closes the floor of the nasal sacs, separating the nasal cavity from the oral cavity. Subsequent thinning and rupture of the dorsal aspect of the nasal fins, which occurs at E11.5 in mouse, forms the primitive posterior choanae and re-establishes an open connection between the oral and nasal cavity (Tamarin, 1982).

Molecularly, FGF signaling has been shown to influence development of the frontonasal region and airway in experimental animals. Mutations disrupting FGF signaling are linked with human syndromes involving the nasal region such as Apert syndrome (OMIM #101200), Pfeiffer syndrome (OMIM #101600) and Muenke syndrome (OMIM #602849) (Griffin et al., 2013; Gupta et al., 2020; Shao et al., 2015; Hehr & Muenke, 1999). The transcription factor GATA3 is also likely important for upper lip nasal airway development as bilateral cleft lip and choanal atresia have been reported in human patients with the syndrome known as hypoparathyroidism, deafness, and renal dysplasia (HDR) that is associated with mutation or duplication of *GATA3* (Bernardini et al., 2009; Kita et al., 2019; Van Esch et al., 2000).

RA signaling is active in, and required for, morphogenesis of upper lip and airway. In mouse, RA signaling reporter activity is present in the lambdoid junction region at E10.5 and much of the nasal epithelium and mesenchyme are positively labeled by a lineage reporter for RA signaling activity at E11.5 (Dollé et al., 2010; Kurosaka et al., 2017). RA in the developing nasal region is produced by RDH10 and ALDH1A3. Choanal atresia and cleft lip are observed in stage-specific knockout of *Rdh10* by administration of tamoxifen at E7.5 (Kurosaka et al., 2017), knockout of *Aldh1a3* (Dupe et al., 2003), double knockout of *Aldh1a2/Aldh1a3* with partial rescue to overcome early lethality (Halilagic et al., 2007), or stage-specific triple knockout of *Rara/b/g* following administration of tamoxifen at E8.5 or E9.5 (Teletin et al., 2023). Underdevelopment of the nasal region is also noted in a rat nutritional model of late embryonic stage Vitamin A deficiency (See et al., 2008). In frog, knockdown of *aldh1a2* or *rarg* causes cleft upper lip and primary palate (Kennedy & Dickinson, 2012).

Molecularly, the nasal phenotypes in stage-specific mouse *Rdh10* or *Aldh1a3* null embryos are associated with ectopic expression of *Fgf8* in the residual nasal fin (Dupe et al., 2003; Kurosaka et al., 2017). The excess expression of *Fgf8* in these models demonstrates that RA is needed to repress *Fgf8* expression, a relationship that parallels the known mutually inhibitory interaction between RA and FGF8 in body axis extension (Diez Del Corral et al., 2003), forelimb bud initiation (Cunningham et al., 2013; Zhao et al., 2009) and interdigital cell death required for separation of the digits (Hernandez-Martinez et al., 2009). Importantly, RA is known to repress *Fgf8* by direct binding interaction of RA-liganded RAR to a RARE upstream of *Fgf8* (Berenguer et al., 2020; Kumar & Duester, 2014). However, in Vitamin A deficient quail embryos abnormal frontonasal development is associated with absence of FGF8 in the nasal pit and nasal fin (Halilagic et al., 2007). The contradictory *Fgf8* responses in these different experimental systems of frontonasal development remains to be reconciled. An additional factor downstream of RA signaling in frontonasal development is *Gata3*. *Gata3* is reduced in stage-specific *Rdh10* mutant mice and knockdown of *Gata3* expression results in a partial choanal atresia and stenosis, a phenotype that is exacerbated by dietary Vitamin A deficiency (Kurosaka et al., 2021).

Formation of the nasal airway is sensitive not only to RA deficiency but also to RA excess. Choanal stenosis and atresia have been reported in situations of excess RA signaling in the context of human patients with mutation in the RA degrading enzyme *Cyp26b1* (Grand et al., 2021).

Whereas the midface develops by fusion of the maxillary and nasal processes at the lambdoid junction, the lateral face requires fusion of maxillary and mandibular portions of PA1. Incomplete fusion of these processes during embryogenesis results in a condition known as focal facial dermal dysplasia, characterized by ectodermal lesions of the temporal or preauricular region. The human condition focal facial dermal dysplasia, type IV, (MIM614974), is associated with mutation of *CYP26C1*, demonstrating a role for RA clearance in PA1 process fusion (Lee et al., 2018; Slavotinek et al., 2013). The condition is characterized by scar-like lesions on the facial skin extending from external ear toward the lateral commissure of the mouth.

9. Salivary gland morphogenesis requires RA

Salivary glands are important exocrine organs in orofacial anatomy. Submandibular salivary glands (SMG) initiate development at the presumptive base of the tongue at E11.0 in mouse. On the oral surface of PA1 the glands initiate as epithelial placodes that thicken and then invaginate into the underlying NCC-derived mesenchyme to form an initial bud, expressing *Sox9* as it does so (Chatzeli et al., 2017; Tucker, 2007). The simple initial bud grows, undergoes branching morphogenesis to produce acini and ducts, and cavitates to form lumens and ducts. Development of the gland depends upon interactions between the oral epithelium and underlying mesenchyme (Jaskoll et al., 2005).

Mouse experimental studies have revealed RA signaling regulates submandibular salivary gland development at the initiation stages and during branching morphogenesis (Fig. 1H). At E10.5, prior to initiation of salivary gland development, RA signaling is active on each side of the midline of PA1 in domains that presage the sites of salivary placode formation (Wright et al., 2015). *Rdh10* and *Aldha2* are expressed in the presumptive salivary gland mesenchyme as early as E10.5 and E11.5, respectively (Metzler et al., 2018; Wright et al., 2015). At initiation, RA signaling is active in the underlying mesenchyme and overlying thickening epithelium. Analysis of salivary gland initiation in cultured mandible explants demonstrates *Rdh10* and RA are required for formation of the initial bud and expression of the salivary epithelial marker *Sox9* (Metzler et al., 2018). To date RA is the only signaling pathway known to regulate the transition from naïve oral ectoderm to salivary gland epithelium at the initiation of gland development.

After initiation, subsequent growth and branching of SMG epithelium continue to require RA signaling. $Rdh10^{trex}$ mutants with sub-optimal retinal supplementation form a small initial bud but have impaired growth and branching of salivary epithelium (Wright et al., 2015). Salivary gland agenesis is noted in *Rara/g* double null mutant mouse fetuses (Lohnes et al., 1994) and glands are hypoplastic or absent in a rat nutritional model of late embryonic stage Vitamin A deficiency (See et al., 2008). At the initial bud stage RA acts in a paracrine manner with RA production occurring in the NCC-derived mesenchyme and RA signaling activity being present and active predominantly in the epithelium (Abashev et al., 2017; Wright et al., 2015).

10. Secondary palate

The secondary palate is formed by growth of tissue from the MXP of PA1, which form the palate shelves that meet at the midline inside the mouth to separate the oral cavity and the nasal cavity (Bush & Jiang, 2012; Dixon et al., 2011). The shelves grow initially downward toward the mandible on either side of the elevating tongue. With progressive expansion of the facial complex the mandible and tongue are displaced downward and forward, and, at around E14.5 in mouse, the fetus begins activating jaw and tongue movement, which opens space the palate shelves to re-orient for horizontal growth toward the midline (Asada et al., 1997; Condie et al., 1997; Culiat et al., 1995; Homanics et al., 1997; Oh et al., 2010; Tsunekawa et al., 2005). Once they meet at the midline the palate shelves adhere and fuse, creating the roof of the mouth and separating the oral and nasal cavities. Secondary palate morphogenesis is influenced by both intrinsic processes, such as growth and fusion of the palate shelves, and extrinsic processes, such as outgrowth of the mandible or movement of the fetal jaw and tongue. Factors required intrinsically for palate morphogenesis include *FGF10* (Rice et al., 2004), a secreted signaling molecule, and *TBX1*, a transcription factor, loss of which is a major factor in human *22q11.2* deletion syndrome in humans (Funato, 2022; Funato et al., 2012; Gao et al., 2013; Haddad et al., 2019). Factors required extrinsically for palate development include *Hoxa2*, which is needed for A-P patterning of the pharyngeal musculoskeletal anatomy (Barrow & Capecchi, 1999).

Although there is little or no RA signaling activity present within developing secondary palate tissues (Dollé et al., 2010; Okano et al., 2012), properly regulated levels of RA are essential for palate formation; RA deficiency or RA excess can each cause clefts. The importance of sufficient retinoid for secondary palate formation was first noted in observations of cleft palate in early studies of vitamin A nutrition in experimental animals (Hale, 1935; Warkany, 1945; Warkany et al., 1943; Wilson et al., 1953) and supported by genetic RA deficiency in *Rara/g* double knockout mice (Lohnes et al., 1994). While the association between retinoid deficiency and cleft secondary palate was well-established, the underlying mechanism of the cleft phenotype remained elusive for many years. Analysis of the Tamoxifen 8.5 stage-specific *Rdh10* knockout mouse embryos revealed the major etiology of the cleft secondary palate in RA deficiency is the disruption of fetal jaw and tongue movement (Friedl et al., 2019). Mutant fetuses have no intrinsic palate defects but display malformations in oropharyngeal nerves and cartilages and lacked the fetal tongue movement. Molecularly, genes dysregulated by this model of deficient retinoid signaling include *Hoxa1 and Hoxb1*, which are important for anterior-posterior patterning and the *Tbx1* gene, which plays an important in *22q11.2* deletion syndrome.

Elevated RA signaling activity also produce cleft palate phenotypes. Mutation of *Dhrs3*, which acts to prevent overproduction of RA, results in elevated RA signaling and cleft secondary palate (Billings et al., 2013). Similarly, cleft palate is observed in *Cyp26b1* mutant mice that have impaired clearance of RA (Maclean et al., 2009; Okano et al., 2012). Cleft palate in $Cyp26b1^{-/-}$ embryos is associated with reduced expression of *Fgf10* and *Tbx1* in the bend region palate shelf and also tongue height abnormalities, the latter suggesting a possible defect in tongue withdrawal (Okano et al., 2012).

11. Bone and sutures

Development and maintenance of bones, including those of the craniofacial skeleton involves a balance between osteogenesis and bone resorption, mediated by osteoblasts and osteoclasts, respectively. The facial skeleton or viscerocranium, derived from NCC, develops by chondrogenesis through a cartilage intermediate. The bones of the skull or braincase, known as the neurocranium or calvarium, are formed by intramembranous ossification without a cartilage intermediate. Calvarial bones are initially connected by flexible cranial sutures—the metopic, sagittal, coronal, and lambdoid—fibrous joints that allow the skull to flex during birth and to expand for growth of the brain. Fusion of the sutures occurs in an age-dependent sequence starting within the first year of postnatal life and continuing into adulthood (Stanton et al., 2022). Premature fusion of cranial vault sutures results in craniosynostosis, a major craniofacial disorder that restricts growth of the brain.

RA signaling is needed for bone formation in the craniofacial skeleton. Inhibition of RAR activity specifically in osteoblasts by tissue-specific expression of a dominant-negative RAR disrupts morphogenesis and ossification of facial and skull bones (Dai et al., 2023). While RA activity is needed in osteoblasts for bone formation, properly regulated clearance of RA from suture tissues is needed to ensure that these regions remain unossified. Human patients with pathogenic variants of *CYP26B1* have *CYP26B1*-related disorder (OMIM 605207). The condition is characterized by craniosynostosis, unossified areas of the skull (cranium bifidum), encephalocele, and facial dysmorphology (Grand et al., 2021; Laue et al., 2011; Morton et al., 2016; Silveira et al., 2023). Phenotypes of this condition vary with severity of mutation. Hypomorphic *CYP26B1* mutations are associated with calvarial bone defects and craniosynostosis but are compatible with survival to adulthood. Null alleles have severe calvarial bone hypoplasia (precluding suture formation), encephalocele, and are prenatal lethal. *Cyp26b1* knockout mice have hypomineralization of calvarial bones (Maclean et al., 2009), reminiscent of null mutations in human. Analysis of zebrafish *cyp26b1* reveals expression in the osteogenic cells at the edges of the suture plates, and a hypomorphic *cyp26b1* mutation is associated with cranial suture synostosis (Laue et al., 2008). Collectively, these observations demonstrate that properly regulated RA signaling activity and RA clearance are essential both for cranial vault ossification and to maintain patency of sutures to prevent premature ossification.

12. Stage-specific RA roles

Throughout progressive craniofacial development RA plays multifarious roles, some of which have been identified using Tamoxifen to induce Cre-mediated ablation of genes for RA production or signal transduction in mice. To interpret the timing and developmental processes impacted by such conditional inactivation methods, it is important to consider that the reduction of RA signaling would not be immediate with administration of Tamoxifen and may vary with gene target locus and function. Tamoxifen-induced ablation of *Rara* and *Rarg* on the *Rarb*$^{-/-}$ background was shown be near complete within a 24-h time frame (Vernet et al., 2020). In contrast, inactivation of the *Rdh10* gene was shown to be complete within 48 h (E10.5), with a small amount of residual RA reporter RNA remaining detectable at 72 h (Metzler et al., 2018). A less complete elimination of RA activity following *Rdh10* ablation is expected because *Rdh10* null mutants still produce a small amount of RA (Sandell et al., 2012) and because blocking RA production could potentially leave residual RA to signal through the RAR until this RA is degraded.

Reduction or inactivation of RA signaling at different stages of craniofacial morphogenesis allows distinct functional windows of RA activity to be identified. When the timing of RA disruption occurs at or near E9.5, midfacial cleft and choanal atresia are observed (Kurosaka et al., 2017; Teletin et al., 2023; Wu et al., 2022), indicating an essential role for RA in frontonasal process and nasal airway morphogenesis occurs between E8.5–E9.5. RA attenuation/elimination near E10.5 disrupts formation of submandibular salivary glands and causes mild shortening of the snout (Metzler et al., 2018; Teletin et al., 2023), indicating that RA role in morphogenesis of these tissues occurs in the E9.5–E10.5 interval.

Disruption of RA at this stage also reveals a cleft secondary palate phenotype secondary to defects in pharyngeal motor nerves and cartilages needed for fetal mouth movement (Friedl et al., 2019), suggesting RA signaling within the E9.5–E10.5 window is needed for development of motor nerves and cartilages of the pharyngeal region.

References

Abashev, T. M., Metzler, M. A., Wright, D. M., & Sandell, L. L. (2017). *Developmental Dynamics, 246*, 135–147.
Abramyan, J. (2019). *Journal of Developmental Biology, 7*, 9.
Abu-Abed, S., Dolle, P., Metzger, D., Beckett, B., Chambon, P., & Petkovich, M. (2001). *Genes & Development, 15*, 226–240.
Adams, M. K., Belyaeva, O. V., Wu, L., & Kedishvili, N. Y. (2014). *Journal of Biological Chemistry, 289*, 14868–14880.
Adams, M. K., Belyaeva, O. V., Wu, L., Chaple, I. F., Dunigan-Russell, K., Popov, K. M., & Kedishvili, N. Y. (2021). *The Biochemical Journal, 478*, 3597–3611.
Alasti, F., Sadeghi, A., Sanati, M. H., Farhadi, M., Stollar, E., Somers, T., & Van Camp, G. (2008). *American Journal of Human Genetics, 82*, 982–991.
Alexandre, D., Clarke, J. D. W., Oxtoby, E., Yan, Y.-L., Jowett, T., & Holder, N. (1996). *Development (Cambridge, England), 122*, 735–746.
Amengual, J., Zhang, N., Kemerer, M., Maeda, T., Palczewski, K., & Von Lintig, J. (2014). *Human Molecular Genetics, 23*, 5402–5417.
Asada, H., Kawamura, Y., Maruyama, K., Kume, H., Ding, R. G., Kanbara, N., ... Obata, K. (1997). *Proceedings of the National Academy of Sciences of the United States of America, 94*, 6496–6499.
Baltzinger, M., Ori, M., Pasqualetti, M., Nardi, I., & Rijli, F. M. (2005). *Developmental Dynamics: An Official Publication of the American Association of Anatomists, 234*, 858–867.
Barrow, J., & Capecchi, M. (1999). *Development (Cambridge, England), 126*, 5011–5026.
Bedois, A. M. H., Parker, H. J., & Krumlauf, R. (2021). *Diversity, 13*, 398.
Berenguer, M., Meyer, K. F., Yin, J., & Duester, G. (2020). *PLoS Biology, 18*, e3000719.
Bernardini, L., Sinibaldi, L., Capalbo, A., Bottillo, I., Mancuso, B., Torres, B., ... Dallapiccola, B. (2009). *Clinical Genetics, 76*, 117–119.
Berry, D. C., Jacobs, H., Marwarha, G., Gely-Pernot, A., O'Byrne, S. M., DeSantis, D., ... Ghyselinck, N. B. (2013). *The Journal of Biological Chemistry, 288*, 24528–24539.
Beverdam, A., Brouwer, A., Reijnen, M., Korving, J., & Meijlink, F. (2001). *Development (Cambridge, England), 128*, 3975–3986.
Billings, S. E., Pierzchalski, K., Butler Tjaden, N. E., Pang, X. Y., Trainor, P. A., Kane, M. A., & Moise, A. R. (2013). *FASEB Journal, 27*, 4877–4889.
Bok, J., Raft, S., Kong, K. A., Koo, S. K., Dräger, U. C., & Wu, D. K. (2011). *Proceedings of the National Academy of Sciences of the United States of America, 108*, 161–166.
Brugmann, S. A., Allen, N. C., James, A. W., Mekonnen, Z., Madan, E., & Helms, J. A. (2010). *Human Molecular Genetics, 19*, 1577–1592.
Bryan, C. D., Casey, M. A., Pfeiffer, R. L., Jones, B. W., & Kwan, K. M. (2020). *Development (Cambridge, England), 147*.
Bush, J. O., & Jiang, R. (2012). *Development (Cambridge, England), 139*, 231–243.
Caron, V., Chassaing, N., Ragge, N., Boschann, F., Ngu, A. M., Meloche, E., ... Michaud, J. L. (2023). *Genetics in Medicine: Official Journal of the American College of Medical Genetics, 25*, 100856.
Carpenter, E. M., Goddard, J. M., Chisaka, O., Manley, N. R., & Capecchi, M. R. (1993). *Development (Cambridge, England), 118*, 1063–1075.

Chassaing, N., Golzio, C., Odent, S., Lequeux, L., Vigouroux, A., Martinovic-Bouriel, J., ... Calvas, P. (2009). *Human Mutation, 30*, E673–E681.
Chatzeli, L., Gaete, M., & Tucker, A. S. (2017). *Development (Cambridge, England), 144*, 2294–2305.
Comai, G. E., Tesařová, M., Dupé, V., Rhinn, M., Vallecillo-García, P., da Silva, F., ... Tajbakhsh, S. (2020). *PLoS Biology, 18*, e3000902.
Condie, B. G., Bain, G., Gottlieb, D. I., & Capecchi, M. R. (1997). *Proceedings of the National Academy of Sciences of the United States of America, 94*, 11451–11455.
Culiat, C. T., Stubbs, L. J., Woychik, R. P., Russell, L. B., Johnson, D. K., & Rinchik, E. M. (1995). *Nature Genetics, 11*, 344–346.
Cunningham, T. J., & Duester, G. (2015). *Nature Reviews. Molecular Cell Biology, 16*, 110–123.
Cunningham, T. J., Zhao, X., Sandell, L. L., Evans, S. M., Trainor, P. A., & Duester, G. (2013). *Cell Reports, 3*, 1503–1511.
Cvekl, A., & Wang, W. L. (2009). *Experimental Eye Research, 89*, 280–291.
D'Aniello, E., Ravisankar, P., & Waxman, J. S. (2015). *PLoS One, 10*, e0138588.
Dai, Q., Sun, S., Jin, A., Gong, X., Xu, H., Yang, Y., ... Jiang, L. (2023). *Journal of Dental Research, 102*, 667–677.
Damm, K., Heyman, R. A., Umesono, K., & Evans, R. M. (1993). *Proceedings of the National Academy of Sciences of the United States of America, 90*, 2989–2993.
Dash, S., & Trainor, P. A. (2020). *Bone, 137*, 115409.
Dee, C. T., Szymoniuk, C. R., Mills, P. E., & Takahashi, T. (2013). *Human Molecular Genetics, 22*, 239–251.
Diez Del Corral, R., Olivera-Martinez, I., Goriely, A., Gale, E., Maden, M., & Storey, K. (2003). *Neuron, 40*, 65–79.
Dixon, M. J., Marazita, M. L., Beaty, T. H., & Murray, J. C. (2011). *Nature Reviews. Genetics, 12*, 167–178.
Dollé, P., Fraulob, V., Gallego-Llamas, J., Vermot, J., & Niederreither, K. (2010). *Developmental Dynamics, 239*, 3260–3274.
Dubey, A., Rose, R. E., Jones, D. R., & Saint-Jeannet, J.-P. (2018). *Genesis (New York, N. Y.: 2000), 56*, e23091.
Duester, G. (2022). *Cells, 11*, 322.
Dupe, V., Matt, N., Garnier, J. M., Chambon, P., Mark, M., & Ghyselinck, N. B. (2003). *Proceedings of the National Academy of Sciences of the United States of America, 100*, 14036–14041.
Dupé, V., Davenne, M., Brocard, J., Dollé, P., Mark, M., Dierich, A., ... Rijli, F. M. (1997). *Development (Cambridge, England), 124*, 399–410.
Dupé, V., Ghyselinck, N. B., Wendling, O., Chambon, P., & Mark, M. (1999). *Development (Cambridge, England), 126*, 5051–5059.
Dupé, V., & Pellerin, I. (2009). *Developmental Dynamics, 238*, 2701–2711.
Fan, X., Molotkov, A., Manabe, S., Donmoyer, C. M., Deltour, L., Foglio, M. H., ... Duester, G. (2003). *Molecular and Cellular Biology, 23*, 4637–4648.
Farlie, P. G., Baker, N. L., Yap, P., & Tan, T. Y. (2016). *Molecular Syndromology, 7*, 312–321.
Feng, L., Hernandez, R. E., Waxman, J. S., Yelon, D., & Moens, C. B. (2010). *Developmental Biology, 338*, 1–14.
Friedl, R. M., Raja, S., Metzler, M. A., Patel, N. D., Brittian, K. R., Jones, S. P., & Sandell, L. L. (2019). *Disease Models and Mechanisms, 12*, dmm039073.
Funato, N. (2022). *Journal of Developmental Biology, 10*.
Funato, N., Nakamura, M., Richardson, J. A., Srivastava, D., & Yanagisawa, H. (2012). *Human Molecular Genetics, 21*, 2524–2537.
Gao, S., Li, X., & Amendt, B. A. (2013). *Current Allergy and Asthma Reports, 13*, 613–621.

Gao, T., Wright-Jin, E. C., Sengupta, R., Anderson, J. B., & Heuckeroth, R. O. (2021). *JCI Insight, 6*.
Gavalas, A., Studer, M., Lumsden, A., Rijli, F. M., Krumlauf, R., & Chambon, P. (1998). *Development (Cambridge, England), 125*, 1123–1136.
Gendron-Maguire, M., Mallo, M., Zhang, M., & Gridley, T. (1993). *Cell, 75*, 1317–1331.
Gerth-Kahlert, C., Williamson, K., Ansari, M., Rainger, J. K., Hingst, V., Zimmermann, T., ... Fitzpatrick, D. R. (2013). *Molecular Genetics & Genomic Medicine, 1*, 15–31.
Ghyselinck, N. B., & Duester, G. (2019). *Development (Cambridge, England), 146*, dev167502.
Ghyselinck, N. B., Dupe, V., Dierich, A., Messaddeq, N., Garnier, J. M., Rochette-Egly, C., ... Mark, M. (1997). *The International Journal of Developmental Biology, 41*.
Goddard, J. M., Rossel, M., Manley, N. R., & Capecchi, M. R. (1996). *Development (Cambridge, England), 122*, 3217–3228.
Golzio, C., Martinovic-Bouriel, J., Thomas, S., Mougou-Zrelli, S., Grattagliano-Bessières, B., Bonnière, M., ... Etchevers, H. C. (2007). *American Journal of Human Genetics, 80*, 1179–1187.
Grammatopoulos, G. A., Bell, E., Toole, L., Lumsden, A., & Tucker, A. S. (2000). *Development (Cambridge, England), 127*, 5355–5365.
Grand, K., Skraban, C. M., Cohen, J. L., Dowsett, L., Mazzola, S., Tarpinian, J., ... Deardorff, M. A. (2021). *American Journal of Medical Genetics. Part A, 185*, 2766–2775.
Griffin, J. N., Compagnucci, C., Hu, D., Fish, J., Klein, O., Marcucio, R., & Depew, M. J. (2013). *Developmental Biology, 374*, 185–197.
Guo, P., Ji, Z., Jiang, H., Huang, X., Wang, C., & Pan, B. (2020). *International Journal of Pediatric Otorhinolaryngology, 139*, 110488.
Gupta, P., Tripathi, T., Singh, N., Bhutiani, N., Rai, P., & Gopal, R. (2020). *Journal of Family Medicine and Primary Care, 9*, 1825–1833.
Haddad, R. A., Clines, G. A., & Wyckoff, J. A. (2019). *Clinical Diabetes and Endocrinology, 5*, 13.
Hale, F. (1935). *American Journal of Ophthamology, 18*, 1087–1092.
Halilagic, A., Ribes, V., Ghyselinck, N. B., Zile, M. H., Dollé, P., & Studer, M. (2007). *Developmental Biology, 303*, 362–375.
Harding, P., & Moosajee, M. (2019). *Journal of Developmental Biology, 7*.
Hehr, U., & Muenke, M. (1999). *Molecular Genetics and Metabolism, 68*, 139–151.
Helms, J. A., Cordero, D., & Tapadia, M. D. (2005). *Development (Cambridge, England), 132*, 851–861.
Hernandez, R. E., Putzke, A. P., Myers, J. P., Margaretha, L., & Moens, C. B. (2007). *Development (Cambridge, England), 134*, 177–187.
Hernandez-Martinez, R., Castro-Obregon, S., & Covarrubias, L. (2009). *Development (Cambridge, England), 136*, 3669–3678.
Homanics, G. E., DeLorey, T. M., Firestone, L. L., Quinlan, J. J., Handforth, A., Harrison, N. L., ... Olsen, R. W. (1997). *Proceedings of the National Academy of Sciences of the United States of America, 94*, 4143–4148.
Hu, D., & Marcucio, R. S. (2009). *Development (Cambridge, England), 136*, 107–116.
Hunter, M. P., & Prince, V. E. (2002). *Developmental Biology, 247*, 367–389.
Isken, A., Golczak, M., Oberhauser, V., Hunzelmann, S., Driever, W., Imanishi, Y., ... von Lintig, J. (2008). *Cell Metabolism, 7*, 258–268.
Iyyanar, P. P. R., Wu, Z., Lan, Y., Hu, Y.-C., & Jiang, R. (2022). *Frontiers in Cell and Developmental Biology, 10*.
Jaskoll, T., Abichaker, G., Witcher, D., Sala, F., Bellusci, S., Hajihosseini, M., & Melnick, M. (2005). *BMC Developmental Biology, 5*, 11.
Jungbluth, S., Bell, E., & Lumsden, A. (1999). *Development (Cambridge, England), 126*, 2751–2758.

Kam, R. K., Shi, W., Chan, S. O., Chen, Y., Xu, G., Lau, C. B., ... Zhao, H. (2013). *The Journal of Biological Chemistry, 288,* 31477–31487.

Kane, M. A., Chen, N., Sparks, S., & Napoli, J. L. (2005). *The Biochemical Journal, 388,* 363–369.

Kelly, M., & von Lintig, J. (2015). *Hepatobiliary Surgery and Nutrition, 4,* 229–242.

Kennedy, A. E., & Dickinson, A. J. (2012). *Developmental Biology, 365,* 229–240.

Kish, P. E., Bohnsack, B. L., Gallina, D., Kasprick, D. S., & Kahana, A. (2011). *Genesis (New York, N. Y.: 2000), 49,* 222–230.

Kita, M., Kuwata, Y., & Usui, T. (2019). *Auris, Nasus, Larynx, 46,* 808–812.

Kitazawa, T., Fujisawa, K., Narboux-Nême, N., Arima, Y., Kawamura, Y., Inoue, T., ... Kurihara, H. (2015). *Developmental Biology, 402,* 162–174.

Kitazawa, T., Minoux, M., Ducret, S., & Rijli, F. M. (2022). *Journal of Developmental Biology, 10.*

Kumar, S., & Duester, G. (2010). *Developmental Biology, 340,* 67–74.

Kumar, S., & Duester, G. (2014). *Development (Cambridge, England), 141,* 2972–2977.

Kurosaka, H., Mushiake, J., Mithun, S., Wu, Y., Wang, Q., Kikuchi, M., ... Yamashiro, T. (2021). *Human Molecular Genetics, 30,* 2383–2392.

Kurosaka, H., Wang, Q., Sandell, L., Yamashiro, T., & Trainor, P. A. (2017). *Human Molecular Genetics, 26,* 1268–1279.

Laue, K., Jänicke, M., Plaster, N., Sonntag, C., & Hammerschmidt, M. (2008). *Development (Cambridge, England), 135,* 3775–3787.

Laue, K., Pogoda, H. M., Daniel, P. B., van Haeringen, A., Alanay, Y., von Ameln, S., ... Breuning, M. H. (2011). *The American Journal of Human Genetics.*

le Maire, A., Teyssier, C., Balaguer, P., Bourguet, W., & Germain, P. (2019). *Cells, 8.*

Lee, L. M. Y., Leung, C.-Y., Tang, W. W. C., Choi, H.-L., Leung, Y.-C., McCaffery, P. J., ... Shum, A. S. W. (2012). *Proceedings of the National Academy of Sciences, 109,* 13668–13673.

Lee, B. H., Morice-Picard, F., Boralevi, F., Chen, B., & Desnick, R. J. (2018). *Journal of Human Genetics, 63,* 257–261.

Lohnes, D., Mark, M., Mendelsohn, C., Dolle, P., Dierich, A., Gorry, P., ... Chambon, P. (1994). *Development (Cambridge, England), 120,* 2723–2748.

Maclean, G., Dollé, P., & Petkovich, M. (2009). *Developmental Dynamics, 238,* 732–745.

Mark, M., Lufkin, T., Vonesch, J. L., Ruberte, E., Olivo, J. C., Dollé, P., ... Chambon, P. (1993). *Development (Cambridge, England), 119,* 319–338.

Matt, N., Dupe, V., Garnier, J. M., Dennefeld, C., Chambon, P., Mark, M., & Ghyselinck, N. B. (2005). *Development (Cambridge, England), 132,* 4789–4800.

Matt, N., Ghyselinck, N. B., Pellerin, I., & Dupe, V. (2008). *Developmental Biology, 320,* 140–148.

Metzler, M. A., Raja, S., Elliott, K. H., Friedl, R. M., Tran, N. Q. H., Brugmann, S. A., ... Sandell, L. L. (2018). *Development (Cambridge, England), 145.*

Mic, F. A., Molotkov, A., Molotkova, N., & Duester, G. (2004). *Developmental Dynamics: An Official Publication of the American Association of Anatomists, 231,* 270–277.

Minoux, M., Kratochwil, C. F., Ducret, S., Amin, S., Kitazawa, T., Kurihara, H., ... Rijli, F. M. (2013). *Development (Cambridge, England), 140,* 4386–4397.

Molotkov, A., Molotkova, N., & Duester, G. (2006). *Development (Cambridge, England), 133,* 1901–1910.

Morton, J. E., Frentz, S., Morgan, T., Sutherland-Smith, A. J., & Robertson, S. P. (2016). *American Journal of Medical Genetics. Part A, 170*(10), 2706.

Moutier, E., Ye, T., Choukrallah, M.-A., Urban, S., Osz, J., Chatagnon, A., ... Davidson, I. (2012). *Journal of Biological Chemistry, 287,* 26328–26341.

Muenke, M., & Beachy, P. A. (2000). *Current Opinion in Genetics & Development, 10,* 262–269.

Nedelec, B., Rozet, J. M., & Fares Taie, L. (2019). *Human Genetics, 138*, 937–955.
Niederreither, K., Subbarayan, V., Dollé, P., & Chambon, P. (1999). *Nature Genetics, 21*, 444–448.
Niederreither, K., Vermot, J., Schuhbaur, B., Chambon, P., & Dollé, P. (2000). *Development (Cambridge, England), 127*, 75–85.
Noy, N., Slosberg, E., & Scarlata, S. (1992). *Biochemistry, 31*, 11118–11124.
O'Connor, C., Varshosaz, P., & Moise, A. R. (2022). *Nutrients, 14*.
Oh, W. J., Westmoreland, J. J., Summers, R., & Condie, B. G. (2010). *PLoS One, 5*, e9758.
Okano, J., Kimura, W., Papaionnou, V. E., Miura, N., Yamada, G., Shiota, K., & Sakai, Y. (2012). *Developmental Dynamics, 241*, 1744–1756.
Parker, H. J., Pushel, I., & Krumlauf, R. (2018). *Developmental Biology, 444*, S67–S78.
Pasqualetti, M., Ori, M., Nardi, I., & Rijli, F. M. (2000). *Development (Cambridge, England), 127*, 5367–5378.
Pasutto, F., Sticht, H., Hammersen, G., Gillessen-Kaesbach, G., Fitzpatrick, D. R., Nurnberg, G., ... Rauch, A. (2007). *American Journal of Human Genetics, 80*, 550–560.
Quadro, L., Hamberger, L., Gottesman, M. E., Wang, F., Colantuoni, V., Blaner, W. S., & Mendelsohn, C. L. (2005). *Endocrinology, 146*, 4479–4490.
Redfern, C. P. F. (2020). Chapter One - Vitamin A and its natural derivatives. In E. Pohl (Ed.). *Methods in Enzymology* (pp. 1–25). Academic Press.
Reijntjes, S., Rodaway, A., & Maden, M. (2007). *The International Journal of Developmental Biology, 51*, 351–360.
Rhinn, M., & Dollé, P. (2012). *Development (Cambridge, England), 139*, 843–858.
Ribes, V., Wang, Z., Dollé, P., & Niederreither, K. (2006). *Development (Cambridge, England), 133*, 351–361.
Rice, R., Spencer-Dene, B., Connor, E. C., Gritli-Linde, A., McMahon, A. P., Dickson, C., ... Rice, D. P. C. (2004). *The Journal of Clinical Investigation, 113*, 1692–1700.
Rijli, F. M., Mark, M., Lakkaraju, S., Dierich, A., Dolle, P., & Chambon, P. (1993). *Cell, 75*, 1333–1349.
Ross, A. C., & Zolfaghari, R. (2011). *Annual Review of Nutrition, 31*, 65–87.
Rossant, J., Zirngibl, R., Cado, D., Shago, M., & Goguere, V. (1991). *Genes & Development, 5*, 1333–1344.
Sadowski, S., Chassaing, N., Gaj, Z., Czichos, E., Wilczynski, J., & Nowakowska, D. (2017). *Birth Defects Research, 109*, 251–253.
Sakai, Y., Meno, C., Fujii, H., Nishino, J., Shiratori, H., Saijoh, Y., ... Hamada, H. (2001). *Genes & Development, 15*, 213–225.
Sandell, L. L., Lynn, M. L., Inman, K. E., McDowell, W., & Trainor, P. A. (2012). *PLoS One, 7*, e30698.
Sandell, L. L., Sanderson, B. W., Moiseyev, G., Johnson, T., Mushegian, A., Young, K., ... Trainor, P. A. (2007). *Genes And Development, 21*, 1113–1124.
Santagati, F., Minoux, M., Ren, S. Y., & Rijli, F. M. (2005). *Development (Cambridge, England), 132*, 4927–4936.
See, A. W., Kaiser, M. E., White, J. C., & Clagett-Dame, M. (2008). *Developmental Biology, 316*, 171–190.
Seller, M. J., Davis, T. B., Fear, C. N., Flinter, F. A., Ellis, I., & Gibson, A. G. (1996). *American Journal of Medical Genetics, 62*, 227–229.
Shao, M., Liu, C., Song, Y., Ye, W., He, W., Yuan, G., ... Gu, S. (2015). *Journal of Molecular Cell Biology, 7*, 441–454.
Si, N., Meng, X., Lu, X., Zhao, X., Li, C., Yang, M., ... Jiang, H. (2020). *Gene, 757*, 144945.
Sieck, R. L., Fuller, A. M., Bedwell, P. S., Ward, J. A., Sanders, S. K., Xiang, S. H., ... Steffen, D. J. (2020). *Genes (Basel), 11*.

Silveira, K. C., Fonseca, I. C., Oborn, C., Wengryn, P., Ghafoor, S., Beke, A., ... Kannu, P. (2023). *Human Genetics, 142*, 1571–1586.

Sirbu, I. O., Gresh, L., Barra, J., & Duester, G. (2005). *Development (Cambridge, England), 132*, 2611–2622.

Slavotinek, A. M., Mehrotra, P., Nazarenko, I., Tang, P. L., Lao, R., Cameron, D., ... Desnick, R. J. (2013). *Human Molecular Genetics, 22*, 696–703.

Som, P. M., & Naidich, T. P. (2013). *American Journal of Neuroradiology, 34*, 2233–2240.

Stanton, E., Urata, M., Chen, J.-F., & Chai, Y. (2022). *Disease Models & Mechanisms, 15*.

Studer, M., Gavalas, A., Marshall, H., Ariza-McNaughton, L., Rijli, F. M., Chambon, P., & Krumlauf, R. (1998). *Development (Cambridge, England), 125*, 1025–1036.

Studer, M., Lumsden, A., Ariza-McNaughton, L., Bradley, A., & Krumlauf, R. (1996). *Nature, 384*, 630–634.

Szabó, A., & Mayor, R. (2018). *Annual Review of Genetics, 52*, 43–63.

Tamarin, A. (1982). *The American Journal of Anatomy, 165*, 319–337.

Teletin, M., Mark, M., Wendling, O., Vernet, N., Féret, B., Klopfenstein, M., ... Ghyselinck, N. B. (2023). *Biomedicines, 11*, 198.

Toro-Tobon, S., Manrique, M., Paredes-Gutierrez, J., Mantilla-Rivas, E., Oh, H., Ahmad, L., ... Rogers, G. F. (2023). *The Journal of Craniofacial Surgery, 34*, 2237–2241.

Tsunekawa, N., Arata, A., & Obata, K. (2005). *The European Journal of Neuroscience, 21*, 173–178.

Tucker, A. S. (2007). *Seminars in Cell & Developmental Biology, 18*, 237–244.

Uehara, M., Yashiro, K., Mamiya, S., Nishino, J., Chambon, P., Dolle, P., & Sakai, Y. (2007). *Developmental Biology, 302*, 399–411.

Van Esch, H., Groenen, P., Nesbit, M. A., Schuffenhauer, S., Lichtner, P., Vanderlinden, G., ... Devriendt, K. (2000). *Nature, 406*, 419–422.

Vernet, N., Condrea, D., Mayere, C., Féret, B., Klopfenstein, M., Magnant, W., ... Ghyselinck, N. B. (2020). *Science Advances, 6*, eaaz1139.

Wagner, E., McCaffery, P., & Dräger, U. C. (2000). *Developmental Biology, 222*, 460–470.

Warkany, J. (1945). *The Milbank Memorial Fund Quarterly, 23*, 66–77.

Warkany, J., Nelson, R. C., & Schraffenberger, E. (1943). *American Journal of Diseases of Children, 65*, 882–894.

White, T., Lu, T., Metlapally, R., Katowitz, J., Kherani, F., Wang, T. Y., ... Young, T. L. (2008). *Molecular Vision, 14*, 2458–2465.

White, R. J., & Schilling, T. F. (2008). *Developmental Dynamics: An Official Publication of the American Association of Anatomists, 237*, 2775–2790.

Williams, A. L., & Bohnsack, B. L. (2019). *Genesis (New York, N. Y.: 2000), 57*, e23308.

Williams, A. L., & Bohnsack, B. L. (2020). *Frontiers in Cell and Developmental Biology, 8*, 595896.

Wilson, J. G., Roth, C. B., & Warkany, J. (1953). *American Journal of Anatomy, 92*, 189–217.

Wright, D. M., Buenger, D. E., Abashev, T. M., Lindeman, R. P., Ding, J., & Sandell, L. L. (2015). *Developmental Biology, 407*, 57–67.

Wu, Y., Kurosaka, H., Wang, Q., Inubushi, T., Nakatsugawa, K., Kikuchi, M., ... Yamashiro, T. (2022). *Journal of Dental Research, 101*, 686–694.

Zhao, X., Sirbu, I. O., Mic, F. A., Molotkova, N., Molotkov, A., Kumar, S., & Duester, G. (2009). *Current Biology: CB, 19*, 1050–1057.

CHAPTER THREE

Meiosis and retinoic acid in the mouse fetal gonads: An unforeseen twist

Giulia Perrotta[a,b,c,d], Diana Condrea[a,b,c,d], and Norbert B. Ghyselinck[a,b,c,d,*]

[a]Université de Strasbourg, IGBMC UMR 7104, Illkirch, France
[b]CNRS, UMR 7104, Illkirch, France
[c]Inserm, UMR-S 1258, Illkirch, France
[d]IGBMC, Institut de Génétique et de Biologie Moléculaire et Cellulaire, Illkirch, France
*Corresponding author. e-mail address: norbert@igbmc.fr

Contents

1. Introduction	60
2. ATRA and CYP26B1 as meiotic inducing and preventing substances	62
3. Mutant mouse models disqualify ATRA as the MIS	63
4. How to reconcile seemingly irreconcilable observations	67
5. What else, if not ATRA?	69
6. Mutant mouse models support the idea that CYP26B1 acts as an MPS	70
7. The evidence that *Cyp26b1* ablation increases testis ATRA is all refutable	71
8. Meiosis initiates in CYP26B1-deficient testes lacking ALDH1A1	73
9. If it does not degrade ATRA, what does CYP26B1 do?	76
10. What about ATRA in the human fetal gonad?	79
11. Concluding remark	80
Acknowledgments	81
Competing interests statement	81
Funding	81
Data availability statement	81
References	81

Abstract

In mammals, differentiation of germ cells is crucial for sexual reproduction, involving complex signaling pathways and environmental cues defined by the somatic cells of the gonads. This review examines the long-standing model positing that all-*trans* retinoic acid (ATRA) acts as a meiosis-inducing substance (MIS) in the fetal ovary by inducing expression of STRA8 in female germ cells, while CYP26B1 serves as a meiosis-preventing substance (MPS) in the fetal testis by degrading ATRA and preventing STRA8 expression in the male germ cells until postnatal development. Recent genetic studies in the mouse challenge this paradigm, revealing that meiosis initiation in female germ cells can occur independently of ATRA signaling, with key roles played by other intrinsic factors like DAZL and DMRT1, and extrinsic signals such as BMPs and vitamin C. Thus, ATRA can no longer be considered as 'the' long-searched MIS. Furthermore, evidence indicates that CYP26B1 does not prevent meiosis by degrading ATRA in the fetal testis, but acts by degrading an unidentified MIS or synthesizing an equally unknown MPS. By emphasizing the necessity of genetic loss-of-function approaches to accurately delineate the roles of signaling molecules such ATRA in vivo, this chapter calls for a reevaluation of the mechanisms instructing and preventing meiosis initiation in the fetal ovary and testis, respectively. It highlights the need for further research into the molecular identities of the signals involved in these processes.

Abbreviations

AGN193109	inverse agonistic ligand for all RAR isotypes, from Allergan, Inc.
ATRA	all-*trans* retinoic acid.
BMP	bone morphogenetic protein.
BMS-194753	agonistic ligand selective for RARA, from Bristol-Myers Squibb pharmaceuticals.
BMS-213309	agonistic ligand selective for RARB, from Bristol-Myers Squibb pharmaceuticals.
BMS-189394	agonistic ligand selective for RARG, from Bristol-Myers Squibb pharmaceuticals.
BMS-204493	inverse agonistic ligand for all RAR isotypes, from Bristol-Myers Squibb pharmaceuticals.
CYP	cytochrome P450 enzymes.
CYP26	cytochrome P450 family 26 enzymes.
E	embryonic day.
MIS	meiosis-inducing substance.
MPS	meiosis-preventing substance.
PGC	primordial germ cells.
RAR	retinoic acid receptor.
RARE	retinoic acid-responsive element.
VAD	vitamin A deficient.
WIN-18	446, bis(dichloroacetyl)diamine, inhibitor of retinaldehyde dehydrogenase.

1. Introduction

In mammals, the hallmark of sex reproduction is the joining of a female and a male haploid germ cells that reconstitute a diploid embryo at the basis of the next generation. The germ cells of both sexes arise from a common cell-type, called the primordial germ cells (PGC), that either follow the female or the male pathway of differentiation named oogenesis or spermatogenesis, respectively. The PGC are formed as a small cohort of proximal epiblast cells that move into the extraembryonic tissues at the base of the allantois (Ginsburg, Snow & McLaren, 1990; Lawson & Hage, 1994). They proliferate and migrate along the hindgut of the embryo to reach the genital ridges. These structures arise from the intermediate mesoderm and are initially bipotential. Under the influence of the WNT4/RSPO1/CTNNB1 signaling pathways, the female genital ridges develop as ovaries (Chassot, Gillot & Chaboissier, 2014), while under the influence of the SRY/SOX9/FGF9 signaling pathways, the male genital ridges develop as testes (Kobayashi, Chang, Chaboissier, Schedl & Behringer, 2005). In the mouse, this occurs at around embryonic day (E) 10.5 to E11.5 (McLaren, 1988). PGC continue to proliferate for a few days, so that about 25,000 germ cells reside in the fetal mouse gonads at E13.5 (Tam & Snow, 1981). It is only when sex determination of the gonadal soma has occurred that germ cells commit to the female, or the male differentiation pathway (McLaren & Southee, 1997). Indeed, the first morphological sign of sex-specific germ cell differentiation is seen at around E13.5, when female germ cells stop proliferating and initiate meiosis in the fetal ovary (McLaren & Southee, 1997; McLaren, 2003). This is done thanks to the expression of STRA8 and MEIOSIN proteins (reviewed in Ishiguro, 2023). The oocytes then proceed through the leptotene, zygotene and pachytene stages of the first meiotic prophase division. They arrest at diplotene stage and remain in this state until just before ovulation postnatally, at which time they complete the first meiotic division, begin the second one and arrest again. Meiosis is completed only after fertilization (Colas & Guerrier, 1995). In contrast, male germ cells do not initiate meiosis in the fetal testis. They arrest in the G0 or G1 phase of the mitotic cycle between E13.5 and E14.5, and enter a period of quiescence (Hilscher et al., 1974). After birth, the male germ cells resume mitosis and some differentiate as spermatogonia stem cells. It is only at puberty that some daughter spermatogonia initiate meiosis (Clermont, 1972).

For a long time, it was unknown whether germ cells were primed to enter meiosis by default, or if they were instructed by signals coming from the environment generated by the somatic cells surrounding them (Donovan, Stott, Cairns, Heasman & Wylie, 1986). Studies with chimeric mice have shown that germ cells, whether female or male, enter meiosis when they are embedded into a fetal ovary, but do not if in a fetal testis. Thus, the initial sex-specific differentiation of a germ cell is determined by its gonadic environment, but not by its chromosomal content (Evans, Ford & Lyon, 1977; McLaren, 1983; McLaren, 1995). Three possible scenarios could account for this observation: (i) somatic feminizing factors in the ovary promote meiotic initiation, while quiescence is established by default in the testis; (ii) somatic masculinizing factors in the testis promote quiescence, while meiotic entry occurs spontaneously and cell-autonomously in the ovary; (iii) both feminizing and masculinizing factors instruct meiosis in ovaries and quiescence in testes, respectively (Bowles & Koopman, 2007; Kocer, Reichmann, Best & Adams, 2009). Earlier studies involving co-culture of gonads have suggested that a diffusible "meiosis-inducing substance" (MIS) originating from the cranial part of the mesonephros (i.e., *rete* ovary/*rete* testis) triggers male germ cells to enter meiosis (Byskov & Saxén, 1976; Byskov, 1974). They also showed that a "meiosis-preventing substance" (MPS) diffusing from the fetal testis itself blocks meiosis in the female germ cells (Byskov, 1978). Accordingly, the prevailing model was that a MIS is present both in the ovary and the fetal testis, while an MPS is active only in the fetal testis (Bowles & Koopman, 2007; Kocer, Reichmann, Best & Adams, 2009). However, the nature of both the MIS and the MPS remained unknown for years.

2. ATRA and CYP26B1 as meiotic inducing and preventing substances

The discovery that Stimulated by retinoic acid, clone number 8 (*Stra8*) is specifically expressed in female germ cells at E12.5, just prior to entering meiosis (Menke, Koubova & Page, 2003), while *Cyp26b1* is restricted to testis somatic cells at E13.5, just prior to quiescence of male germ cells (Menke & Page, 2002) has given a new impetus. Actually, because (i) *Stra8* expression is induced by pharmacological levels of ATRA in cultured cells (Oulad-Abdelghani et al., 1996), (ii) CYP26B1 is identified as an ATRA degrading enzyme (White et al., 2000), (iii) the mRNA coding the ATRA-producing enzyme ALDH1A2 is expressed in the mesonephros (Niederreither et al., 1997), and (iv) STRA8 is essential to trigger meiotic entry (Baltus et al., 2006; Mark et al., 2008), an elegant model was devised, whereby ATRA plays the role of MIS and CYP26B1 plays the role of MPS (Bowles et al., 2006; Koubova et al., 2006). According to this new paradigm, ATRA is synthesized by the mesonephros in both sexes and it diffuses into the ovary and the testis through the rete ovary/rete testis. Within the ovary, ATRA drives the expression of *Stra8*, which triggers meiotic initiation. This fits perfectly well to the characteristics expected for a MIS. Within the testis, Sertoli cells encasing germ cells express CYP26B1, and form thereby a catabolic barrier that degrades ATRA coming from mesonephros. As a result, germ cells in the fetal testis are not exposed to the MIS, and they become quiescent. In this scenario the MPS is not a secreted molecule but rather a cytosolic enzyme (Bowles & Koopman, 2007).

The model according which ATRA and CYP26B1 are the long-searched MIS and MPS, respectively, was set up in the mouse, deduced essentially from expression patterns, and functionally tested by using exogenous retinoids added to organotypic culture experiments at levels much higher than endogenous ATRA, with the exception of CYP26B1, the role of which was studied in part by analyzing *Cyp26b1*-null fetal testes. Subsequently, it was extensively popularized during almost two decades, every two to three years (Bowles & Koopman, 2007; Bowles & Koopman, 2010; Feng, Bowles & Koopman, 2014; Spiller & Bowles, 2015; Spiller & Bowles, 2019; Spiller & Bowles, 2022; Spiller, Bowles & Koopman, 2012; Spiller, Koopman & Bowles, 2017). The same causes inevitably producing the same effects, organotypic culture experiments and/or expression pattern analyses also exist to support the model in multiple vertebrate species (Childs, Cowan, Kinnell, Anderson & Saunders, 2011; Feng et al., 2015; Lau, Lee & Chang, 2013; Li et al., 2016; Piprek et al., 2013; Rodríguez-Marí et al., 2013; Smith, Roeszler, Bowles, Koopman & Sinclair, 2008; Wallacides, Chesnel, Chardard, Flament & Dumond, 2009). However, conclusions based on treatment of cell lines, organotypic cultures or animals with ATRA or agonists/antagonists have often not been supported by animal studies in which genes encoding for ATRA metabolism enzymes or for ATRA receptors are deleted (Cunningham & Duester, 2015). Accordingly, the model was challenged in 2011 by a first, elegant, genetic study (Kumar et al., 2011), that was the matter of a fiery debate (Griswold, Hogarth, Bowles & Koopman, 2012; Kumar, Cunningham & Duester, 2013). To solve it, some authors called for the analysis of fetal ovaries when all three ALDH1A enzymes or all three ATRA-receptors (RAR) were genetically rendered nonfunctional simultaneously (Agrimson & Hogarth, 2016). In the following section, we discuss the evidence drawn from such mutant mouse models.

3. Mutant mouse models disqualify ATRA as the MIS

The first blow to the established model came from the analysis of mutant mice lacking ALDH1A2 or ALDH1A2 and ALDH1A3 (Kumar et al., 2011).

The *Aldh1a2*-null mutants die during early embryogenesis, preventing examination of ATRA-dependent processes later in fetal development. To circumvent this lethality, low doses of ATRA were given to pregnant females from E6.25 to E9.5, allowing thereby ALDH1A2-deficient embryos to develop through E14.5. As ATRA is cleared within 12–24 h (Mic, Haselbeck, Cuenca & Duester, 2002), such a protocol allows the mutant fetuses to reach a developmental stage at which the gonads can be investigated in the absence of ALDH1A2 activity and the absence of ATRA administered several days earlier. Besides, by using the ATRA-sensitive reporter transgene *Tg(RARE-Hspa1b/lacZ)12Jrt* (Rossant, Zirngibl, Cado, Shago & Giguère, 1991), it was demonstrated that the rescued *Aldh1a2*-null fetuses totally lacked ATRA activity in the mesonephros and the ovaries at the time meiosis initiates. But against the odds, normal expression of *Stra8* was detected, indicating meiotic initiation was normal. The same holds true when ALDH1A3 was simultaneously lacking along with ALDH1A2 in double mutants. Based on these findings it was proposed that *Stra8* expression does not depend on ATRA-signaling, meiotic initiation can take place in the absence of ATRA and even though ATRA is synthesized in the mesonephros, it is totally dispensable to both of these events (Kumar et al., 2011). The proponents of the established model argued that the ALDH1A1 enzyme was present in the ovaries of these mutants, ensuring that some ATRA produced locally acted redundantly to induce meiosis (Griswold, Hogarth, Bowles & Koopman, 2012). In agreement with this proposal, they showed that the expression of *Stra8* and the onset of meiosis were delayed of about 1 day in *Aldh1a1*-null ovaries. They concluded that ALDH1A1 is the main ATRA-synthesizing enzyme in the ovary, and considered as resolved the conundrum posed by the conflicting data sets (Bowles et al., 2016). However, the absence of ATRA-dependent activity in the mesonephros and ovary of mutants lacking both ALDH1A2 and ALDH1A3 (Kumar et al., 2011), casts doubts on the possibility that ALDH1A1 actually generates ATRA for ovary.

In 2020, two other genetic studies put a cat among the pigeons. In the first one, three distinct gene deletion models were generated, in which ablation of *Aldh1a1*, *Aldh1a2* and *Aldh1a3* was simultaneously induced from E9.5, E10.5 or E11.5 (Chassot et al., 2020). Importantly, the efficient loss of ALDH1A activity was assessed by using two sensitive GFP-based reporter systems, and by measuring the residual amount of ATRA by mass spectrometry. These controls showed that the mutant ovaries contained less ATRA than the control testes (Chassot et al., 2020), in which ATRA is supposedly absent, because it is cleared by CYP26B1 (Bowles & Koopman, 2007). The lack of ATRA was also attested by the morphological defects displayed by the fetuses, which were characteristic of a functional state of ATRA-deficiency (Ghyselinck & Duester, 2019; Mark, Ghyselinck & Chambon, 2006; Niederreither & Dollé, 2008). At E13.5, *Stra8* was expressed at a lower level than in controls, but at E16.5 most, if not all, germ cells were meiotic in the ovaries of such mutants. This genetic study shows therefore that deletion of all three *Aldh1a* genes slightly delays but does not impair meiosis entry, indicating that ATRA signaling is dispensable for instructing meiosis in female germ cells (Chassot et al., 2020).

In the second study, another gene deletion model was generated, in which *Rara* and *Rarg* ablation was induced from either E9.5 or E10.5 in the context of mice carrying a *Rarb*-null mutation that can survive and reproduce (Vernet et al., 2020). Efficient loss of RARs was assessed at the gene level by using PCR, at the mRNA level by using in situ hybridization, and at the protein level by using immunohistochemistry and western blots with anti-RAR antibodies. The loss of RAR functioning was also attested by the multiple morphological defects displayed by the mutant fetuses, which were, here again, characteristic of a functional state of ATRA-deficiency (Ghyselinck & Duester, 2019; Mark, Ghyselinck & Chambon, 2006; Niederreither & Dollé, 2008). At E14.5, many germ cells expressed STRA8, the cohesin REC8 and the synaptonemal complex protein SYCP3 in the ovaries of these mutants. Moreover, the expression of mRNAs encoding several other meiotic genes was evidenced in individual germ cells devoid of *Rara*, *Rarb* and *Rarg* mRNAs. Among them, *Rec8* mRNA was normally expressed, although it is considered as an ATRA-regulated gene (Koubova et al., 2014). A notable exception was *Stra8* mRNA, whose expression was delayed by 24 h. Nonetheless, most, if not all, the RAR-null oocytes reached the zygotene stage of meiosis at E15.5 (Vernet et al., 2020). These findings indicate that RARs are dispensable for both ovary differentiation and meiosis initiation during the fetal life. At best, they are instrumental to the timely expression of *Stra8*, which can however recover within 1 day of development. By using grafting experiments, this study further demonstrated that the ovaries of mutants lacking RARs produce functional oocytes which are able to complete meiosis and fertilization postnatally, indicating thereby that the entire meiotic process can take place in the total absence of RARs (Vernet et al., 2020). The fact that ablation of all 3 *Rar*-coding genes in the somatic cells of the female urogenital ridge from E10.5 does not impair granulosa cell specification, differentiation, and reproductive function further demonstrates that ATRA-signaling is dispensable for ovary physiology (Minkina et al., 2017).

Recently, a third genetic study was published aimed at identifying the ATRA-responsive elements (RAREs) controlling expression of *Stra8* (Feng et al., 2021). It showed that subtle mutation of two functional RAREs located upstream of *Stra8* promoter was sufficient to impair the ATRA-dependent expression of a reporter gene in embryonic carcinoma cells in vitro. Mutations of the same two sites in vivo resulted in the loss of about 50 % of *Stra8* expression, but normal expression of many other meiotic genes at E14.5. Not surprisingly, these RARE mutant females were fertile. The remaining expression of *Stra8* was optimistically ascribed by the authors to other, yet unidentified, RAREs possibly located upstream of the regulatory region they studied (Feng et al., 2021). However, another explanation could be that it rather represented the expression level induced independently of ATRA-activated RARs, as evidenced in RAR- and ALDH1A-deficient ovaries (Chassot et al., 2020; Vernet et al., 2020). In keeping with alternative possibility, deletion of the proximal regulatory region that encompassed a DMRT1-binding site totally abrogated *Stra8* expression both in vitro and in vivo (Feng et al., 2021).

In the end, the genetic studies altogether point to the following conclusions. First, in rescued *Aldh1a2*- or *Aldh1a2;Aldh1a3*-null mutants (Kumar et al., 2011) meiosis initiation is normal. Second, in mutant females null for *Aldh1a1* alone (Bowles et al., 2016), for all 3 *Aldh1a* simultaneously (Chassot et al., 2020), for all 3 *Rars* simultaneously (Vernet et al., 2020), or in mutants lacking the RAREs of *Stra8* gene (Feng et al., 2021), female germ cells entered meiosis in the fetal ovaries, despite a slightly reduced and delayed expression of *Stra8*. Therefore, we propose a simplest explanation to account for these findings: (i) meiosis initiation per se does not depend on ATRA signaling; (ii) ATRA synthesized by ALDH1A1 in the ovary itself activates RAR signaling to amplify the timely expression of STRA8 at E13.5, not to initiate it; (iii) ATRA synthesized in the mesonephros by ALDH1A2 is useless to this process. Accordingly, ATRA should no longer be considered as the long-searched MIS, because meiosis can clearly occur in ovaries devoid of either ATRA or RARs. Quite interestingly, such a conclusion made for oocytes echoes the recent observation that both expression of *Stra8* and initiation of meiosis can also occur in spermatocytes devoid of ATRA (Kirsanov et al., 2023; Teletin et al., 2019). In the end, the reasoning that pharmacological ATRA induces *Stra8* expression in P19 embryonal carcinoma cells (Oulad-Abdelghani et al., 1996), and *Stra8* controls meiotic entry in the mouse (Baltus et al., 2006), implies that ATRA control of meiotic entry looks like a syllogism, which no longer stands up to the genetic facts.

4. How to reconcile seemingly irreconcilable observations

The genetic studies apparently contradict the admitted model (Bowles & Koopman, 2007; Griswold, Hogarth, Bowles & Koopman, 2012), but several clarifications can be proposed to explain the discrepancies. First, the conclusion made from mutant mice that *Stra8* expression and meiotic initiation are not instructed by, but simply facilitated by ATRA is not incompatible with the finding that germ cells failed to enter meiosis in the ovaries of vitamin A-deficient (VAD) female rat fetuses (Li & Clagett-Dame, 2009). In this model, *Stra8* expression is reduced but not abolished at E15.5, and the number of SYCP3-positive oocytes is decreased but not absent at E18.5, and the germ cells that do enter meiosis progress through the meiotic prophase until E21.5 (Li & Clagett-Dame, 2009). In rat ovaries, the onset of meiosis is much less synchronous than in mice. Only 10 % of germ cells enter the preleptotene stage between E18 and E18.5 (Hilscher & Hilscher, 1972; Prépin, Gibello-Kervran, Charpentier & Jost, 1985). The proportion of meiotic cells raises to 75 % E19.5 % and 100 % at E21.5, with still 10 % of oocytes being in leptonema (Prépin, Gibello-Kervran, Charpentier & Jost, 1985). Thus, one can easily imagine that meiosis was delayed by VAD, instead of being arrested, as are most of the other morphogenetic vitamin A-dependent processes in VAD rats (reviewed in Mark, Ghyselinck & Chambon, 2006). In keeping with this interpretation, some germ cells still expressed the PGC marker POU5F1 in the ovaries of VAD rats at E18.5, indicating that their differentiation is delayed (Li & Clagett-Dame, 2009). Analyzing VAD ovaries between E19.5 and E20.5 may have revealed that germ cells indeed entered meiosis, but later than in vitamin A-sufficient conditions. Along the same vein, the fact that treatment of E11.5 mouse urogenital ridges for 48 h with the ALDH1A inhibitor WIN18,446 (Paik et al., 2014) reduced, but did not abolish, *Stra8* expression means by no way that meiosis does not initiate in ATRA-deficient ovaries (Hogarth et al., 2011). Here again, one can propose that meiosis was delayed by WIN18,446, instead of being arrested.

Second, the experiments done to prevent ATRA or RAR functioning in cultured fetal ovaries were all performed by using BMS-204493 (Koubova et al., 2006; Koubova et al., 2014) or AGN193109 (Bowles et al., 2006; Bowles et al., 2016), which are not RAR neutral antagonists, but actually RAR inverse agonists. Such ligands are capable of repressing RAR basal activity by favoring and stabilizing recruitment of corepressors at RARE-containing loci, even in the absence of endogenous ATRA (Germain et al., 2009; Klein et al., 1996). As the *Stra8* promoter contains RAR-binding sites on which RARs are actually bound in vivo (Chatagnon et al., 2015; Feng et al., 2021; Kumar et al., 2011), adding a RAR inverse agonist necessarily tethers nuclear receptor corepressors at the locus, promotes chromatin compaction and thereby *Stra8* extinction, irrespective of the presence of endogenous ATRA in the ovary. The same holds true for the *Rec8* gene, which also contains a RARE (Chatagnon et al., 2015). The fact that expression of these two genes is shut down upon exposure to the RAR inverse agonists therefore means by no way that it is normally controlled by endogenous ATRA. The use of a RAR neutral antagonist would have been much more informative (Germain et al., 2009). Then, assuming that *Rec8* and *Stra8* are actually not regulated by endogenous ATRA, it is not surprising to find them expressed in ovaries lacking ATRA (Chassot et al., 2020) or RARs (Vernet et al., 2020).

Third, it has been emphasized that alterations generated by exogenously administered ATRA at high levels do not necessarily reflect physiological processes operating in real life with endogenous ATRA which exists at a much lower level (Horton & Maden, 1995; Mark, Ghyselinck & Chambon, 2006). However, the paradigm according to which the physiological effects of a substance can be identified by studying the consequences of an excess of that substance administered to living beings is still alive and kicking in biology. By comparison, it's hard to imagine researchers in economics or in social sciences determining the purchasing power of a population by studying the behavior of lottery winners. Yet this is exactly the kind of approach biologists are using, literally flooding organs or cells with huge, non-physiological doses of ATRA, in order to determine the effects of endogenous ATRA, that normally acts at very low doses and only in given cell-types. As a matter of fact, conclusions based on treatment of cell lines or animals with ATRA, agonists or inverse agonists have often not been supported by animal studies in which genes encoding ALDH1As or RARs are deleted (Cunningham & Duster, 2015). In the case of the gonads, the simple fact that RAREs are present in *Stra8* gene can explain its artefactually forced expression by ATRA or agonists added to gonads cultured in vitro (Bowles et al., 2006; Koubova et al., 2006). Then, one could ask why the *Stra8* gene contains RAREs if its expression does not need to be induced by ATRA for meiotic initiation? The sole situation in which *Stra8* is actually induced by ATRA in vivo appears to be prior to meiosis during spermatogonia differentiation, at the transition between the A aligned (undifferentiated) and A1 (differentiated) states (Busada & Geyer, 2016; Griswold, 2022; Zhou et al., 2008). Ironically, the role of STRA8 in this process is largely unknown.

5. What else, if not ATRA?

Since ATRA cannot be the MIS, entering meiosis is either an autonomous, intrinsic, property of female germ cells (discussed in Kocer, Reichmann, Best & Adams, 2009), or it requires one or several other extrinsic cues acting as a MIS. As to the intrinsic properties of germ cells, expression of the RNA-binding protein DAZL is key for their acquisition of competence (also called licensing) to undergo either the female- or the male-specific path of differentiation (Gill, Hu, Lin & Page, 2011; Lin, Gill, Koubova & Page, 2008). In addition, the DMRT1 transcription factor contributes to the switch from mitosis to meiosis by directly regulating *Stra8* expression in female germ cells (Krentz et al., 2011). Then, some epigenetic factors are also instrumental to the expression of *Stra8*, and consequently, the initiation of meiosis (Wang & Tilly, 2010). Examples of such factors are (i) the DNMT1 and TET1 enzymes, that catalyze DNA methylation and demethylation of meiotic genes at specific stages, respectively (Hargan-Calvopina et al., 2016; Yamaguchi et al., 2012), and (i) the polycomb repressive complex PRC1 or the SWI/SNF complex, that promote structural modifications of chromatin (Ito et al., 2021; Yokobayashi et al., 2013).

As to the extrinsic cues, recent data showed that vitamin C is required during gestation to induce DNA demethylation of key genes in germ cells, and therefore the timed initiation of meiosis (DiTroia et al., 2019). The WNT/β-catenin (CTNNB1) signaling pathway also plays a critical role in meiosis because a proportion of PGCs neither expressed *Stra8* nor entered meiosis in the absence of RSPO1, an activator of this pathway in the fetal ovary (Chassot et al., 2011). Signaling by bone morphogenetic proteins (BMPs) is another pathway instrumental to meiotic initiation. In female PGC-like cells derived from embryonic stem cells, ATRA and STRA8 are not sufficient on their own to induce premeiotic DNA replication, unless BMPs are added, indicating that ATRA signaling serves to enhance a preexisting situation (Miyauchi et al., 2017). This finding is tied with earlier studies, showing that ATRA accelerated entry in meiosis in cultured rat fetal ovaries (Livera, Rouiller-Fabre, Valla & Habert, 2000). In response to BMP signaling, (i) the MSX1 and MSX2 homeodomain transcription factors activate *Stra8* expression and promote thereby meiosis initiation in female germ cells (Le Bouffant et al., 2011); and (ii) the transcriptional regulator with GATA-like zinc fingers ZGLP1 activates key genes involved in meiosis initiation, independently of ATRA (Nagaoka et al., 2020). In response to activin A (INHBA homodimers), expression of STRA8 was increased and progression of meiosis was accelerated (Liang et al., 2015). Together, these studies indicate that several convergent pathways are able to initiate *Stra8* expression and the transition from mitosis to meiosis cycle, independently of ATRA signaling.

6. Mutant mouse models support the idea that CYP26B1 acts as an MPS

While the action of ATRA in the fetal ovary is contested, there is unquestionable genetic evidence that CYP26B1 is a testis-specific factor acting either by promoting male germ cells to enter a transient mitotic arrest in G0/G1 phase of cell cycle (quiescence) between E12.5 and E14.5, or by preventing them from entering meiosis during fetal development. The analysis of fetal testes from a first *Cyp26b1*-null mouse line (Yashiro et al., 2004) actually showed expression of *Stra8* et *Sycp3* at E13.5, indicating that germ cells entered meiotic prophase I at an early stage of development (Bowles et al., 2006). This finding was extended by the analysis of two other *Cyp26b1*-null mouse lines, showing that germ cells entered meiosis between E13.5 and E15.5: they expressed *Rec8*, *Sycp3*, *Dmc1* and *Spo11* mRNAs as well as STRA8 protein, and proceeded through meiosis until the zygotene or pachytene stages of prophase I (Bowles et al., 2010; Koubova et al., 2014; MacLean, Li, Metzger, Chambon & Petkovich, 2007). This phenotype is not due to a sex-reversal of the testicular somatic environment, because both Sertoli and Leydig remained normally differentiated, as attested by their expression of anti-Müllerian hormone (AMH) and 3β-hydroxysteroid dehydrogenase (HSD3B1), respectively (MacLean, Li, Metzger, Chambon & Petkovich, 2007). In addition, starting from E13.5 a number of germ cells were lost by apoptosis in *Cyp26b1*-null testes, as attested by terminal deoxynucleotidyl transferase dUTP nick-end labeling and cleaved caspase-3 staining (MacLean, Li, Metzger, Chambon & Petkovich, 2007). As ablation of *Cyp26b1* in Sertoli cells induces a similar outcome, it can be concluded that CYP26B1 acts in Sertoli cells, but not in interstitial cells (Li, MacLean, Cameron, Clagett-Dame & Petkovich, 2009).

By deleting *Cyp26b1* specifically in Sertoli cells by E15.5, it was additionally shown that abolishing CYP26B1 activity caused male germ cells to exit from G0, to re-enter the mitotic cycle, and then to initiate the meiotic prophase (Li, MacLean, Cameron, Clagett-Dame & Petkovich, 2009). In keeping with this, it was shown that pharmacological inhibition of CYP26B1 activity in cultured testes prevented male germ cell mitotic arrest, and allowed commitment to the male program of development (Trautmann et al., 2008). Later, it was further shown that CYP26B1 suppresses two distinct programs in male germ cells: a STRA8-dependent meiotic pathway and a STRA8-independent mitotic pathway (Saba, Wu & Saga, 2014). As CYP26B1 currently has no other known substrates than ATRA (Isoherranen & Zhong, 2019; Thatcher & Isoherranen, 2009; Topletz et al., 2012), the admitted idea is that ATRA is the trigger for the defects displayed by *Cyp26b1*-null testes. It is hypothesized that CYP26B1 acts by forming a catabolic barrier which prevents the exposure of male germ cells to ATRA. However, is the current evidence that ablation of CYP26B1 increases ATRA in the fetal testis solid?

7. The evidence that *Cyp26b1* ablation increases testis ATRA is all refutable

First, experiments showed that *Stra8* expression was not induced when fetal testes were cultured in the presence of a potent, but non-specific, CYP inhibitor (ketoconazole), combined with the RAR inhibitor BMS-204493. The same effect was observed when a more selective inhibitor of CYP26s (Ro115866) was tested in combination with BMS-204493 (Koubova et al., 2006). Assuming that *Stra8* expression is necessarily controlled by ATRA (see above), these findings supposedly constitute the crux demonstrating that ATRA was present in fetal testes when CYP26B1 was inhibited. However, as discussed above, BMS-204493 is not a RAR neutral antagonist, but a RAR inverse agonist, which does not compete with endogenous ATRA but represses RAR activity, even the absence of ATRA (Germain et al., 2009). This casts serious doubts on the conclusion that ATRA levels were actually increased upon inhibition of CYP26B1. In keeping with this, the finding that administration of ketoconazole was unable to increase activity of the *Tg(RARE-Hspa1b/lacZ)12Jrt* transgene in control fetal testes led to the conclusion that ATRA degradation is not the mechanism by which CYP26B1 acts (Kumar et al., 2011).

Second, because the germ cells of wild-type fetal testes treated with Am580, an RAR agonist that is resistant to the action of CYP26B1, promptly expressed *Stra8* and entered meiosis, it was concluded that ATRA, rather than another metabolite synthesized by CYP26B1, was responsible for phenotypes observed in *Cyp26b1*-null testes (MacLean, Li, Metzger, Chambon & Petkovich, 2007). This was, however, not unexpected, given that treating wild-type fetal testes with either ATRA or RAR agonists selective for each of the three RAR isotypes (i.e., BMS-194753 for RARA, BMS-213309 for RARB and BMS-270394 for RARG) similarly activated *Stra8* expression and meiotic initiation (Koubova et al., 2006). If the correct interpretation of such experiments is that CYP26B1 was not clearing the testes from synthetic ligands because they are resistant to CYP26B1-mediated metabolism (Bowles & Koopman, 2007; Griswold, Hogarth, Bowles & Koopman, 2012; MacLean, Li, Metzger, Chambon & Petkovich, 2007), why would ATRA treatment produce exactly the same results although it is sensitive to CYP26B1-mediated metabolism (Bowles et al., 2006; Koubova et al., 2006)?

Third, on the basis of ATRA-dependent reporter cells cultured with testicular extracts, *Cyp26b1*-null fetal testes were shown to contain three-times more ATRA than control testes. However, the absolute quantities detected in each extract were not specified (MacLean, Li, Metzger, Chambon & Petkovich, 2007). This should not represent large quantities, since control testes are devoid of ATRA, based on the activity of the *Tg (RARE-Hspa1b/lacZ)12Jrt* transgene (Bowles et al., 2006; Kumar et al., 2011). Consistent with this, the amount of ATRA measured in control testes was almost equal to the amount of ATRA remaining in the ovaries lacking *Aldh1a1*, *Aldh1a2* and *Aldh1a3* from E10.5 (Chassot et al., 2020). In the end, three times little is still little, or nothing at all.

Fourth, the expression of *Aldh1a1* is three-fold lower in *Cyp26b1*-null testes than in controls (Bowles et al., 2009; see also below). Knowing that ATRA cannot originate from the mesonephros (Kumar et al., 2011), this decrease of *Aldh1a1* expression rather evokes a decreased production of ATRA in the seminiferous cords of *Cyp26b1*-null mutants.

8. Meiosis initiates in CYP26B1-deficient testes lacking ALDH1A1

These considerations collectively suggest that ATRA degradation may not be the mechanism underlying CYP26B1 action in the fetal testis. To expand on this provocative notion, we thought it would be interesting to study the fate of germ cells in testes lacking both CYP26B1 and ATRA-synthesizing enzymes. We first determined by immunohistochemistry the expression pattern of ALDH1A1, ALDH1A2 and ALDH1A3 at E15.5, the developmental stage when STRA8 protein expression is at its climax in *Cyp26b1*-null testes (Koubova et al., 2014). The antibody against ALDH1A1 gave a robust signal in the cytoplasm of Sertoli cells surrounding the germ cells (Fig. 1A and D), and in some NR2F2-negative cells of the interstitial tissue (asterisks, Fig. 1A and D). The expression of ALDH1A2 was limited to the surface epithelium lining the testis (large arrowheads, Fig. 1B and E), while that of ALDH1A3 was restricted to the Sertoli cells of the *rete* testis region (arrows, Fig. 1C and F). This pattern is essentially similar to the one observed at E14.5 (Teletin, Vernet, Ghyselinck & Mark, 2017). As it is unlikely that ATRA acts as a long-distance diffusing molecule to reach the germ cells, which are encapsulated in the seminiferous cords, we reasoned that a null mutation of *Aldh1a1* was sufficient to make the environment of germ cells devoid of ATRA. Thus, we generated $Cyp26b1^{-/-};Aldh1a1^{-/-}$ compound mutant fetuses and analyzed their testes.

In the control testis at E15.5, ALDH1A1 was detected in Sertoli and interstitial cells (arrowhead, Fig. 2A). The signal in Sertoli cells was no longer detected in the $Aldh1a1^{-/-}$ background (Fig. 2B and D), while it persisted in the interstitial cells (asterisks). The signal in these cells may correspond to either a non-specific staining, or to ALDH1A7, whose peptide sequence displays 90% identities with ALDH1A1 (Hsu, Chang, Hoffmann & Duester, 1999) and whose mRNA is expressed in fetal testes (Bowles et al., 2009). This enzyme cannot however contribute to the amount of ATRA build up in the fetal testis because it does not synthesize ATRA (Hsu, Chang, Hoffmann & Duester, 1999). The ALDH1A1 protein was detected at a lower level in Sertoli cells of the $Cyp26b1^{-/-}$ testis (arrowhead, Fig. 2C), in agreement with the reported decreased level of *Aldh1a1* mRNA in *Cyp26b1*-null testes (Bowles et al., 2009). Importantly, Sertoli cells remained normally differentiated in the absence of CYP26B1, singly or in combination with loss of ALDH1A1. This was assessed by detecting SOX9, which appeared unchanged between control and mutant

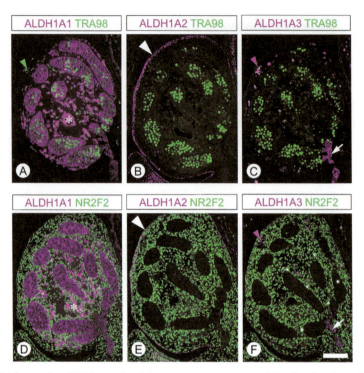

Fig. 1 **The major ATRA-synthesizing enzyme expressed in the fetal testis and contacting germ cells is ALDH1A1.** Transverse, consecutive, histological sections of E15.5 fetuses immunolabeled with antibodies against ALDH1A1 (A, D), ALDH1A2 (B, E) or ALDH1A3 (C, E; magenta signals) and TRA98 (A–C), a nuclear protein specifically expressed in germ cells, or NR2F2 (D–F), a nuclear receptor formerly called COUPTF-II (green signals). ALDH1A1 is restricted to the Sertoli cells, which are encapsulating germ cells, and to some somatic cells, which were NR2F2-negative and located in the interstitial tissue of the testis (asterisks). ALDH1A2 is restricted to the surface epithelium of the testis (large arrowheads). ALDH1A3 is restricted to the *rete* testis (arrows). Red blood cells are sometimes artefactually stained in green (small green arrowhead in A) or in magenta (small magenta arrowhead in C and F). Scale bar in (F): 100 μm (A–F).

testes (Fig. 2E–H). This indicated that the somatic environment of the germ cells did not experience ATRA-induced sex-reversal, contrasting thereby with the observation that the $Cyp26b1^{-/-}$ testis displayed a mild ovotestis phenotype at E13.5 (Bowles et al., 2018). Then, as previously shown (Koubova et al., 2014; MacLean, Li, Metzger, Chambon & Petkovich, 2007), numerous germ cells were STRA8-positive and/or SYCP3-positive in the testis of $Cyp26b1^{-/-}$ mutants at E15.5 (arrows, Fig. 2K and O), indicating they were meiotic. In contrast, none of them

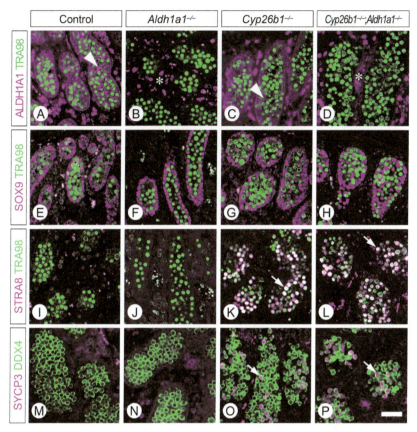

Fig. 2 Germ cells are meiotic in the testis of mouse fetuses lacking either CYP26B1 alone or CYP26B1 and ALDH1A1 together. Transverse histological sections of E15.5 fetuses immunolabeled with antibodies against ALDH1A1 (A–D), SOX9 (E–F), STRA8 (I–L) or SYCP3 (M–P; magenta signals) and TRA98 (A–L), a nuclear protein specifically expressed in germ cells, or DDX4 (M–P), a cytoplasmic ATP-dependent RNA helicase specific to germ cells (green signals). ALDH1A1 is restricted to the Sertoli cells (arrowheads), which are encapsulating germ cells (A). Note that intensity of the Sertoli cell signal is reduced in the CYP26B1-deficient testis (C). The signal observed in some somatic cells located in the interstitial tissue of the testis (asterisks) is unspecific as it is detected also in the ALDH1A1-deficient testis (B, D). SOX9 is detected in the nuclei of Sertoli cells of both control and mutant testes (E–H). STRA8 is not detected in control and in ALDH1A1-deficient testes (I, J). In contrast, it is detected in a large proportion of germ cell nuclei (arrows) in mutants lacking CYP26B1 alone, or lacking both CYP26B1 and ALDH1A1 (K, L). The superimposition of magenta and green signals makes the germ cells appearing white (K, L). SYCP3 is not detected in control and in ALDH1A1-deficient testes (M, N). In contrast, it is detected in several germ cell nuclei (arrows) in mutants lacking CYP26B1 alone, or lacking both CYP26B1 and ALDH1A1 (O, P). Scale bar in (P): 50 μm (A–P).

initiated meiosis in control and $Aldh1a1^{-/-}$ testes (Fig. 2I, J, M, and N). Most importantly, numerous germ cells were STRA8-positive and/or SYCP3-positive also in the testis of $Cyp26b1^{-/-};Aldh1a1^{-/-}$ compound mutants (Fig. 2L and P), in proportions similar to that observed in $Cyp26b1^{-/-}$ mutants (compare Fig. 2K with L, and O with P). Considering that ALDH1A1 is the major source of ATRA in the fetal testis (see above), this indicates that the defects displayed by the CYP26B1-deficient testes cannot be ascribed to an excess of uncleared ATRA because they are not rescued by ablation of ALDH1A1. We may anticipate that the proponents of the established model will cry foul because ALDH1A2 and ALDH1A3 were still present in $Cyp26b1^{-/-};Aldh1a1^{-/-}$ mutants. However, we could argue that ALDH1A1 is the sole, functional, source of ATRA in the fetal ovary (see above), despite the expression of ALDH1A2 in the coelomic epithelium (Teletin, Vernet, Ghyselinck & Mark, 2017). Therefore, there is no obvious reason that a similar expression pattern of ALDH1A2 in the surface epithelium of the fetal testis (Fig. 1) may provoke a different effect. In the end, the present genetic study and data from two other studies (Bellutti et al., 2019; Kumar et al., 2011) support the notion that CYP26B1 does not prevent germ cells to enter meiosis by degrading ATRA in the fetal testis.

Consistent with this proposal, it was recently shown that the ectopic expression of CYP26B1 in fetal ovaries significantly decreased the number of STRA8-positive germ cells. In contrast, the ectopic expression of CYP26A1 was unable to abolish *Stra8* expression and to impair meiotic initiation (Bellutti et al., 2019). Considering that CYP26A1 is much more (at least 10-fold more) potent than CYP26B1 to degrade ATRA (Isoherranen & Zhong, 2019; Topletz et al., 2012), it is evident that clearing ATRA efficiently had no effect on meiotic initiation (Bellutti et al., 2019). This finding further confirms that ATRA is not the signal initiating meiosis in the fetal ovary (see above), and indicates that CYP26B1 displays an activity other than degrading ATRA in the fetal testis (Bellutti et al., 2019).

9. If it does not degrade ATRA, what does CYP26B1 do?

Instead of degrading ATRA, CYP26B1 likely functions either by degrading an as yet unknown MIS derived from the mesonephros, or by synthesizing an MPS in the fetal testis. The former hypothesis is supported by explant studies suggesting that a factor other than ATRA travels from the mesonephros to the testis to induce meiosis (Kumar et al., 2011). The latter hypothesis finds support from older studies suggesting the existence of a meiosis- inhibiting factor produced in the fetal testis (McLaren & Southee, 1997). This, of course, does not fit to CYP26B1 itself, since it is a cytoplasmic enzyme located in Sertoli cells. It may only apply to one of its substrates or products, respectively, that can diffuse to the germ cells.

During evolution, the phylogenetic separation caused diversification of CYP26A1 from CYP26B1 and CYP26C1, resulting in two independent clades within the CYP26 family, and an additional gene duplication led to separation of CYP26B1 and CYP26C1 genes (Carvalho et al., 2017). Since CYP120 is also able to catalyze the hydroxylation of ATRA in cyanobacteria (Alder, Bigler, Werck-Reichhart & Al-Babili, 2009; Ke, Baudry, Makris, Schuler & Sligar, 2005), we introduced CYP120 of three different species in the phylogenetic tree. Surprisingly, they all clustered with CYP26A1 (Fig. 3), suggesting that the ancestral enzyme for ATRA metabolism is CYP26A1 rather than CYP26B1 or CYP26C1. It is worth noting that within a given species, sequence conservation among the three CYP26 family members is lower than conservation of a given CYP26 member between two species. For instance, mouse CYP26B1 and CYP26A1 only share 42 % amino acid sequence identity, while mouse and human CYP26B1 share 97 % amino acid sequence identity (MacLean et al., 2001). Nonetheless, all three enzymes are characterized as ATRA hydroxylases that carry out essentially the same metabolic function. Such a functional redundancy is quite unusual with CYP enzymes. For example, CYP19, CYP17 and CYP24 that conduct endogenous metabolism of a single substrate lack multiple family members. On the opposite, small differences in protein sequence, such as those observed between CYP3A4 and CYP3A5 or between CYP2C9, CYP2C19 and CYP2C8, result in distinct substrate specificities and catalytic activities (Isoherranen & Zhong, 2019). It can be speculated that multiple CYP26 enzymes are required to metabolize different retinoids. Alternatively, it is well possible that CYP26B1 also uses substrates other than retinoids. In keeping with this alternative proposal, CYP26B1 is much less efficient than CYP26A1 to clear ATRA (Isoherranen & Zhong, 2019; Topletz et al., 2012). Moreover, the developed homology models based on other CYP crystal structures suggested some differences in the active site architecture between CYP26A1 and CYP26B1, opening possibilities for other substrates (Karlsson, Strid, Sirsjö & Eriksson, 2008). Last, it is assumed that CYP26B1

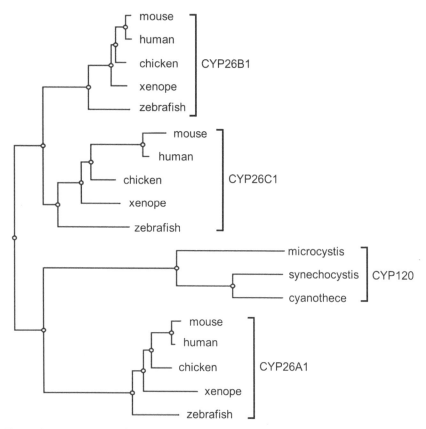

Fig. 3 CYP26 enzyme phylogenetic tree constructed with protein sequences. All CYP26 protein sequences were obtained from NCBI protein database. Evolutionary analyses were conducted in MAFFT alignment. The evolutionary history was inferred using the UPGMA method (Sneath & Sokal, 1973) and the evolutionary distances were computed using the Poisson correction method (Zuckerkandl & Pauling, 1965).

needs to interact with ATRA-bound cellular retinoic acid-binding protein 1 (holo-CRABP1) to accept ATRA as a substrate via substrate channeling (Nelson et al., 2016), but CRABP1 is not expressed in fetal Sertoli cells (Souali-Crespo et al., 2023). Further studies are needed to provide insight on the nature and molecular identity of the actual substrates and products of CYP26B1 in the fetal testis.

Aside from a CYP26B1-degraded substrate or a CYP26B1-synthesized product, other signals can evidently prevent meiosis initiation. Treatment of mouse fetal testes with an inhibitor of secretion and post-Golgi trafficking (brefeldin A) was sufficient to induce germ cells enclosed within the testis cords from entering meiosis, indicating that secreted molecules are important to prevent meiosis initiation (Best, Sahlender, Walther, Peden & Adams, 2008). In agreement with this proposal, conditioned media from cultured fetal mouse testes were shown to contain secreted protein(s) capable of inhibiting meiosis, having a molecular weight greater than 10 kDa (Byskov, 1978; Guerquin et al., 2010). Accordingly, several secreted proteins are known to prevent meiotic initiation, amongst which is FGF9 (Barrios et al., 2010; Bowles et al., 2010), NODAL (Souquet et al., 2012; Wu et al., 2013), activin and transforming growth factor beta (Gustin, Stringer, Hogg, Sinclair & Western, 2016; Miles et al., 2013). It is however difficult to dissociate whether some of these proteins actually inhibit meiosis or induce mitotic arrest (Moreno et al., 2010; Shimada & Ishiguro, 2023; Spiller & Bowles, 2019).

10. What about ATRA in the human fetal gonad?

As to the human fetal gonads, genetic loss-of-function studies are not feasible. Most of the knowledge on the role of ATRA is therefore acquired by using in vitro cultures, with the possible pitfalls described above. Nonetheless, it appears that ATRA has no major role in the human gonad, neither in the female nor in the male. On the one hand, no meiotic cells are detected in human fetal testes treated with ATRA (Lambrot et al., 2006). In keeping with this finding, exogenous ATRA induces STRA8 expression in cultured human fetal testis, but is not sufficient to cause expression of meiosis-associated genes (Childs, Cowan, Kinnell, Anderson & Saunders, 2011). Thus, as in the mouse, mechanisms other than CYP26B1-mediated degradation of ATRA may exist to inhibit meiotic initiation in the human fetal testis. On the other hand, many germ cells never initiate meiosis in cultured human fetal ovaries stimulated by ATRA, suggesting that a meiotic inhibitor, unrelated to ATRA, exists in the human ovary (Le Bouffant et al., 2010). Besides, CYP26B1, whose expression is greater in the human fetal ovary than in testis (Childs, Cowan, Kinnell, Anderson & Saunders, 2011), and FGF9 are both present in the human ovary, where they act simultaneously to repress STRA8 and meiosis in human fetal female germ cells (Frydman et al., 2017). Thus, contrary to the mouse, the human fetal ovary and testis share meiotic preventing pathways.

11. Concluding remark

The use of genetic loss-of-function approaches is essential to identify the processes controlled by a given gene product (Gridley & Murray, 2022; Lloyd, 2024). Global knockouts or recombinase-mediated cell-type restricted ablation of genes coding for the ATRA-synthesizing enzymes (ALDH1As), the ATRA-receptors (RARs) or the ATRA-degrading enzymes (CYP26s) have been particularly useful for discovering the events regulated by ATRA during embryogenesis and fetal development. One may complain that several pitfalls, such as inconsistent, incomplete or mosaic patterns of deletion, can yield confounding interpretations of the results (Spiller & Bowles, 2022). In the case of RARs, however, effective gene invalidation is indisputable, since (i) RARs are no longer detectable at the protein level, (ii) transcriptome analysis shows that single germ cells no longer contain mRNA coding for RARs, and (iii) the resulting overall fetal phenotype corresponds in every respect to a state of functional ATRA deficiency (Vernet et al., 2020). In the case of ALDH1A, loss of ATRA-synthesizing activity is also unquestionable since GFP-based reporter systems and quantitative mass spectrometry showed that mutant ovaries contained less ATRA than control testes, in which ATRA is supposedly absent (Chassot et al., 2020). In opposition to the genetic loss-of-function studies dealing with endogenous ATRA, it has been emphasized that studies using administration of exogenous ATRA do not necessarily reflect physiological processes operating in real life (Horton & Maden, 1995; Mark, Ghyselinck & Chambon, 2006). A recent example is treatment with pharmacological levels of ATRA can alter proximo-distal (PD) patterning during limb regeneration, while endogenous ATRA is not normally required for PD patterning as shown by genetic studies that remove ATRA-synthesizing enzymes (Berenguer & Duester, 2023; Cunningham & Duester, 2015). Evidently, most RAREs located in the genome will respond to pharmacological levels of ATRA and induce ectopic activation or repression of nearby genes (Chatagnon et al., 2015), but such a response does not mean that the genes are regulated the same way by endogenous ATRA which is much lower in concentration. In order to determine whether ATRA normally controls a given gene, one must remove endogenous ATRA, preferably by a genetic loss-of-function approach (Duester, 2017). According to such studies described above (Chassot et al., 2020; Feng et al., 2021; Kumar et al., 2011; Vernet et al., 2020), it is legitimate to propose that ATRA-activated RARs participate to the timed expression of STRA8 in female germ cells, but are dispensable for instructing meiosis initiation. Thus, it cannot be '*the*' meiosis-inducing substance, and the controversy should end.

As to CYP26B1, genetic evidence undoubtedly shows that it is a testis-specific factor acting by preventing male germ cells from entering meiosis during fetal development (Bowles et al., 2006; Bowles et al., 2010; Koubova et al., 2014; Li, MacLean, Cameron, Clagett-Dame & Petkovich, 2009; MacLean, Li, Metzger, Chambon & Petkovich, 2007). However, instead of doing so by degrading ATRA, as widely admitted, CYP26B1 functions either by degrading an as yet unknown 'meiosis-inducing substance' derived from the mesonephros, or by synthesizing an equally unknown 'meiosis-preventing substance' in the fetal testis.

Acknowledgments

We warmly thank Nadège Vernet, Marius Teletin, Manuel Mark, Marie-Christine Chaboissier, Eric Pailhoux and Gabriel Livera for the numerous and passionate scientific discussions we had and for all the advices they provided. We also thank all present and past members of the team at IGBMC, with special thanks to Betty Féret and Muriel Klopfenstein.

Competing interests statement

The author declares no competing interests.

Funding

This work was supported by grants from CNRS, INSERM, UNISTRA, and Agence Nationale pour la Recherche (ANR-10-LABX-0030-INRT, ANR-13-BSV2-0017, ANR-20-CE14-0022 and ANR-22-CE14-0017). This work of the Interdisciplinary Thematic Institute IMCBio+ , as part of the ITI 2021–2028 program of the University of Strasbourg, CNRS and Inserm, was also supported by IdEx Unistra (ANR-10-IDEX-0002), and by SFRI-STRAT'US project (ANR-20-SFRI-0012) and EUR IMCBio (ANR-17-EURE-0023) under the framework of the France 2030 Program.

Data availability statement

No data is provided in this report.

References

Agrimson, K. S., & Hogarth, C. A. (2016). Germ cell commitment to oogenic versus spermatogenic pathway: The role of retinoic acid. *Results and Problems in Cell Differentiation, 58*, 135–166.

Alder, A., Bigler, P., Werck-Reichhart, D., & Al-Babili, S. (2009). In vitro characterization of Synechocystis CYP120A1 revealed the first nonanimal retinoic acid hydroxylase. *FEBS Journal, 276*, 5416–5431.

Baltus, A. E., Menke, D. B., Hu, Y. C., Goodheart, M. L., Carpenter, A. E., de Rooij, D. G., & Page, D. C. (2006). In germ cells of mouse embryonic ovaries, the decision to enter meiosis precedes premeiotic DNA replication. *Nature Genetics, 38*, 1430–1434.

Barrios, F., Filipponi, D., Pellegrini, M., Paronetto, M. P., Di Siena, S., Geremia, R., ... Dolci, S. (2010). Opposing effects of retinoic acid and FGF9 on Nanos2 expression and meiotic entry of mouse germ cells. *Journal of Cell Science, 123,* 871–880.

Bellutti, L., Abby, E., Tourpin, S., Messiaen, S., Moison, D., Trautmann, E., ... Livera, G. (2019). Divergent roles of CYP26B1 and endogenous retinoic acid in mouse fetal gonads. *Biomolecules, 9,* 536.

Berenguer, M., & Duester, G. (2023). Genetic loss-of-function does not support gain-of-function studies suggesting retinoic acid controls limb bud timing and scaling. *Frontiers in Cell and Developmental Biology, 11,* 1149009.

Best, D., Sahlender, D. A., Walther, N., Peden, A. A., & Adams, I. R. (2008). Sdmg1 is a conserved transmembrane protein associated with germ cell sex determination and germline-soma interactions in mice. *Development (Cambridge, England), 135,* 1415–1425.

Bowles, J., Feng, C. W., Ineson, J., Miles, K., Spiller, C. M., Harley, V. R., ... Koopman, P. (2018). Retinoic acid antagonizes testis development in mice. *Cell Reports, 24,* 1330–1341.

Bowles, J., Feng, C. W., Knight, D., Smith, C. A., Roeszler, K. N., Bagheri-Fam, S., ... Koopman, P. (2009). Male-specific expression of Aldh1a1 in mouse and chicken fetal testes: Implications for retinoid balance in gonad development. *Developmental Dynamics, 238,* 2073–2080.

Bowles, J., Feng, C. W., Miles, K., Ineson, J., Spiller, C., & Koopman, P. (2016). ALDH1A1 provides a source of meiosis-inducing retinoic acid in mouse fetal ovaries. *Nature Communications, 7,* 10845.

Bowles, J., Feng, C. W., Spiller, C., Davidson, T. L., Jackson, A., & Koopman, P. (2010). FGF9 suppresses meiosis and promotes male germ cell fate in mice. *Developmental Cell, 19,* 440–449.

Bowles, J., Knight, D., Smith, C., Wilhelm, D., Richman, J., Mamiya, S., ... Koopman, P. (2006). Retinoid signaling determines germ cell fate in mice. *Science (New York, N. Y.), 312,* 596–600.

Bowles, J., & Koopman, P. (2007). Retinoic acid, meiosis and germ cell fate in mammals. *Development (Cambridge, England), 134,* 3401–3411.

Bowles, J., & Koopman, P. (2010). Sex determination in mammalian germ cells: Extrinsic versus intrinsic factors. *Reproduction (Cambridge, England), 139,* 943–958.

Busada, J. T., & Geyer, C. B. (2016). The role of retinoic acid (RA) in spermatogonial differentiation. *Biology of Reproduction, 94,* 10.

Byskov, A. G. (1974). Does the rete ovarii act as a trigger for the onset of meiosis? *Nature, 252,* 396–397.

Byskov, A. G. (1978). Regulation of initiation of meiosis in fetal gonads. *International Journal of Andrology, 1*(s2a), 29–38.

Byskov, A. G., & Saxén, L. (1976). Induction of meiosis in fetal mouse testis in vitro. *Developmental Biology, 52,* 193–200.

Carvalho, J. E., Theodosiou, M., Chen, J., Chevret, P., Alvarez, S., De Lera, A. R., ... Schubert, M. (2017). Lineage-specific duplication of amphioxus retinoic acid degrading enzymes (CYP26) resulted in sub-functionalization of patterning and homeostatic roles. *BMC Evolutionary Biology, 17,* 24.

Clermont, Y. (1972). Kinetics of spermatogenesis in mammals: Seminiferous epithelium cycle and spermatogonial renewal. *Physiological Reviews, 52,* 198–236. PMID:4621362.

Chassot, A. A., Gillot, I., & Chaboissier, M. C. (2014). R-spondin1, WNT4, and the CTNNB1 signaling pathway: Strict control over ovarian differentiation. *Reproduction (Cambridge, England), 148,* R97–R110.

Chassot, A. A., Gregoire, E. P., Lavery, R., Taketo, M. M., de Rooij, D. G., Adams, I. R., & Chaboissier, M. C. (2011). RSPO1/β-catenin signaling pathway regulates oogonia differentiation and entry into meiosis in the mouse fetal ovary. *PLoS One, 6,* e25641.

Chassot, A. A., Le Rolle, M., Jolivet, G., Stevant, I., Guigonis, J. M., Da Silva, F., ... Chaboissier, M. C. (2020). Retinoic acid synthesis by ALDH1A proteins is dispensable for meiosis initiation in the mouse fetal ovary. *Science Advances, 6*, eaaz1261.

Chatagnon, A., Veber, P., Morin, V., Bedo, J., Triqueneaux, G., Sémon, M., ... Benoit, G. (2015). RAR/RXR binding dynamics distinguish pluripotency from differentiation associated cis-regulatory elements. *Nucleic Acids Research, 43*, 4833–4854.

Childs, A. J., Cowan, G., Kinnell, H. L., Anderson, R. A., & Saunders, P. T. (2011). Retinoic acid signalling and the control of meiotic entry in the human fetal gonad. *PLoS One, 6*, e20249.

Colas, P., & Guerrier, P. (1995). The oocyte metaphase arrest. *Progress in Cell Cycle Research, 1*, 299–308.

Cunningham, T. J., & Duester, G. (2015). Mechanisms of retinoic acid signalling and its roles in organ and limb development. *Nature Reviews. Molecular Cell Biology, 16*, 110–123.

DiTroia, S. P., Percharde, M., Guerquin, M. J., Wall, E., Collignon, E., Ebata, K. T., ... Ramalho-Santos, M. (2019). Maternal vitamin C regulates reprogramming of DNA methylation and germline development. *Nature, 573*, 271–275.

Donovan, P. J., Stott, D., Cairns, L. A., Heasman, J., & Wylie, C. C. (1986). Migratory and postmigratory mouse primordial germ cells behave differently in culture. *Cell, 44*, 831–838.

Duester, G. (2017). Retinoic acid's reproducible future. *Science (New York, N. Y.), 358*, 1395. PMID:29242338.

Evans, E. P., Ford, C. E., & Lyon, M. F. (1977). Direct evidence of the capacity of the XY germ cell in the mouse to become an oocyte. *Nature, 267*, 430–431.

Feng, C. W., Burnet, G., Spiller, C. M., Cheung, F. K. M., Chawengsaksophak, K., Koopman, P., & Bowles, J. (2021). Identification of regulatory elements required for *Stra8* expression in fetal ovarian germ cells of the mouse. *Development (Cambridge, England), 148*, dev194977.

Feng, C. W., Bowles, J., & Koopman, P. (2014). Control of mammalian germ cell entry into meiosis. *Molecular and Cellular Endocrinology, 382*, 488–497.

Feng, R., Fang, L., Cheng, Y., He, X., Jiang, W., Dong, R., ... Wang, D. (2015). Retinoic acid homeostasis through aldh1a2 and cyp26a1 mediates meiotic entry in Nile tilapia (Oreochromis niloticus). *Scientific Reports, 5*, 10131.

Frydman, N., Poulain, M., Arkoun, B., Duquenne, C., Tourpin, S., Messiaen, S., ... Livera, G. (2017). Human foetal ovary shares meiotic preventing factors with the developing testis. *Human Reproduction, 32*, 631–642.

Germain, P., Gaudon, C., Pogenberg, V., Sanglier, S., Van Dorsselaer, A., Royer, C. A., ... Gronemeyer, H. (2009). Differential action on coregulator interaction defines inverse retinoid agonists and neutral antagonists. *Chemistry & Biology, 16*, 479–489.

Ghyselinck, N. B., & Duester, G. (2019). Retinoic acid signaling pathways. *Development (Cambridge, England), 146*, dev167502.

Gill, M. E., Hu, Y. C., Lin, Y., & Page, D. C. (2011). Licensing of gametogenesis, dependent on RNA binding protein DAZL, as a gateway to sexual differentiation of fetal germ cells. *Proceedings of the National Academy of Sciences of the USA, 108*, 7443–7448.

Ginsburg, M., Snow, M. H., & McLaren, A. (1990). Primordial germ cells in the mouse embryo during gastrulation. *Development (Cambridge, England), 110*, 521–528.

Gridley, T., & Murray, S. A. (2022). Mouse mutagenesis and phenotyping to generate models of development and disease. *Current Topics in Developmental Biology, 148*, 1–12.

Griswold, M. D. (2022). Cellular and molecular basis for the action of retinoic acid in spermatogenesis. *Journal of Molecular Endocrinology, 69*, T51–T57. PMID:35670629.

Griswold, M. D., Hogarth, C. A., Bowles, J., & Koopman, P. (2012). Initiating meiosis: The case for retinoic acid. *Biology of Reproduction, 86*, 35.

Guerquin, M. J., Duquenne, C., Lahaye, J. B., Tourpin, S., Habert, R., & Livera, G. (2010). New testicular mechanisms involved in the prevention of fetal meiotic initiation in mice. *Developmental Biology, 346*, 320–330.

Gustin, S. E., Stringer, J. M., Hogg, K., Sinclair, A. H., & Western, P. S. (2016). FGF9, activin and TGFβ promote testicular characteristics in an XX gonad organ culture model. *Reproduction (Cambridge, England), 152*, 529–543.

Hargan-Calvopina, J., Taylor, S., Cook, H., Hu, Z., Lee, S. A., Yen, M. R., ... Clark, A. T. (2016). Stage-specific demethylation in primordial germ cells safeguards against precocious differentiation. *Developmental Cell, 39*, 75–86.

Hilscher, W., & Hilscher, B. (1972). Comparative study on oogénesis and prespermatogenesis in the Wistar rat under normal and pathological conditions. *Annales de Biologie Animale, Biochimie, Biophysique, 13*, 128–136.

Hilscher, B., Hilscher, W., Bülthoff-Ohnolz, B., Krämer, U., Birke, A., Pelzer, H., & Gauss, G. (1974). Kinetics of gametogenesis. I. Comparative histological and autoradiographic studies of oocytes and transitional prospermatogonia during oogenesis and prespermatogenesis. *Cell and Tissue Research, 154*, 443–470.

Hogarth, C. A., Evanoff, R., Snyder, E., Kent, T., Mitchell, D., Small, C., ... Griswold, M. D. (2011). Suppression of Stra8 expression in the mouse gonad by WIN 18,446. *Biology of Reproduction, 84*, 957–965.

Horton, C., & Maden, M. (1995). Endogenous distribution of retinoids during normal development and teratogenesis in the mouse embryo. *Developmental Dynamics, 202*, 312–323.

Hsu, L. C., Chang, W. C., Hoffmann, I., & Duester, G. (1999). Molecular analysis of two closely related mouse aldehyde dehydrogenase genes: Identification of a role for Aldh1, but not Aldh-pb, in the biosynthesis of retinoic acid. *The Biochemical Journal, 339*, 387–395.

Ishiguro, K. I. (2023). Mechanism of initiation of meiosis in mouse germ cells. *Current Topics in Developmental Biology, 151*, 1–26 PMID:36681467.

Isoherranen, N., & Zhong, G. (2019). Biochemical and physiological importance of the CYP26 retinoic acid hydroxylases. *Pharmacology & Therapeutics, 204*, 107400.

Ito, T., Osada, A., Ohta, M., Yokota, K., Nishiyama, A., Niikura, Y., ... Kimura, T. (2021). SWI/SNF chromatin remodeling complex is required for initiation of sex-dependent differentiation in mouse germline. *Scientific Reports, 11*, 24074.

Karlsson, M., Strid, Å., Sirsjö, A., & Eriksson, L. A. (2008). Homology models and molecular modeling of human retinoic acid metabolizing enzymes cytochrome P450 26A1 (CYP26A1) and P450 26B1 (CYP26B1). *Journal of Chemical Theory and Computation, 4*, 1021–1027.

Ke, N., Baudry, J., Makris, T. M., Schuler, M. A., & Sligar, S. G. (2005). A retinoic acid binding cytochrome P450: CYP120A1 from Synechocystis sp. PCC 6803. *Archives of Biochemistry and Biophysics, 436*, 110–120.

Kirsanov, O., Johnson, T. A., Niedenberger, B. A., Malachowski, T. N., Hale, B. J., Chen, Q., ... Geyer, C. B. (2023). Retinoic acid is dispensable for meiotic initiation but required for spermiogenesis in the mammalian testis. *Development (Cambridge, England), 150*, dev201638.

Klein, E. S., Pino, M. E., Johnson, A. T., Davies, P. J., Nagpal, S., Thacher, S. M., ... Chandraratna, R. A. (1996). Identification and functional separation of retinoic acid receptor neutral antagonists and inverse agonists. *The Journal of Biological Chemistry, 271*, 22692–22696.

Kobayashi, A., Chang, H., Chaboissier, M. C., Schedl, A., & Behringer, R. R. (2005). Sox9 in testis determination. *Annals of the New York Academy of Sciences, 1061*, 9–17.

Kocer, A., Reichmann, J., Best, D., & Adams, I. R. (2009). Germ cell sex determination in mammals. *Molecular Human Reproduction, 15*, 205–213.

Koubova, J., Hu, Y. C., Bhattacharyya, T., Soh, Y. Q., Gill, M. E., Goodheart, M. L., ... Page, D. C. (2014). Retinoic acid activates two pathways required for meiosis in mice. *PLoS Genetics, 10*, e1004541.

Koubova, J., Menke, D. B., Zhou, Q., Capel, B., Griswold, M. D., & Page, D. C. (2006). Retinoic acid regulates sex-specific timing of meiotic initiation in mice. *Proceedings of the National Academy of Sciences of the USA, 103*, 2474–2479.

Krentz, A. D., Murphy, M. W., Sarver, A. L., Griswold, M. D., Bardwell, V. J., & Zarkower, D. (2011). DMRT1 promotes oogenesis by transcriptional activation of Stra8 in the mammalian fetal ovary. *Developmental Biology, 356*, 63–70.

Kumar, S., Chatzi, C., Brade, T., Cunningham, T. J., Zhao, X., & Duester, G. (2011). Sex-specific timing of meiotic initiation is regulated by Cyp26b1 independent of retinoic acid signalling. *Nature Communications, 2*, 151.

Kumar, S., Cunningham, T. J., & Duester, G. (2013). Resolving molecular events in the regulation of meiosis in male and female germ cells. *Science Signaling, 6*, pe25.

Lambrot, R., Coffigny, H., Pairault, C., Donnadieu, A. C., Frydman, R., Habert, R., & Rouiller-Fabre, V. (2006). Use of organ culture to study the human fetal testis development: effect of retinoic acid. *The Journal of Clinical Endocrinology and Metabolism, 91*, 2696–2703.

Lau, E. L., Lee, M. F., & Chang, C. F. (2013). Conserved sex-specific timing of meiotic initiation during sex differentiation in the protandrous black porgy Acanthopagrus schlegelii. *Biology of Reproduction, 88*, 150.

Lawson, K. A., & Hage, W. J. (1994). Clonal analysis of the origin of primordial germ cells in the mouse. *Ciba Foundation Symposium, 182*, 68–84. PMID:7835158.

Le Bouffant, R., Guerquin, M. J., Duquenne, C., Frydman, N., Coffigny, H., Rouiller-Fabre, V., ... Livera, G. (2010). Meiosis initiation in the human ovary requires intrinsic retinoic acid synthesis. *Human Reproduction, 25*, 2579–2590.

Le Bouffant, R., Souquet, B., Duval, N., Duquenne, C., Hervé, R., Frydman, N., ... Livera, G. (2011). Msx1 and Msx2 promote meiosis initiation. *Development (Cambridge, England), 138*, 5393–5402.

Li, H., & Clagett-Dame, M. (2009). Vitamin A deficiency blocks the initiation of meiosis of germ cells in the developing rat ovary in vivo. *Biology of Reproduction, 81*, 996–1001.

Li, M., Feng, R., Ma, H., Dong, R., Liu, Z., Jiang, W., ... Wang, D. (2016). Retinoic acid triggers meiosis initiation via stra8-dependent pathway in Southern catfish, Silurus meridionalis. *General and Comparative Endocrinology, 232*, 191–198.

Li, H., MacLean, G., Cameron, D., Clagett-Dame, M., & Petkovich, M. (2009). Cyp26b1 expression in murine Sertoli cells is required to maintain male germ cells in an undifferentiated state during embryogenesis. *PLoS One, 4*, e7501.

Liang, G. J., Zhang, X. F., Wang, J. J., Sun, Y. C., Sun, X. F., Cheng, S. F., ... Shen, W. (2015). Activin A accelerates the progression of fetal oocytes throughout meiosis and early oogenesis in the mouse. *Stem Cells and Development, 24*, 2455–2465.

Lin, Y., Gill, M. E., Koubova, J., & Page, D. C. (2008). Germ cell-intrinsic and -extrinsic factors govern meiotic initiation in mouse embryos. *Science (New York, N. Y.), 322*, 1685–1687.

Livera, G., Rouiller-Fabre, V., Valla, J., & Habert, R. (2000). Effects of retinoids on the meiosis in the fetal rat ovary in culture. *Molecular and Cellular Endocrinology, 165*, 225–231.

Lloyd, K. C. K. (2024). Commentary: The International Mouse Phenotyping Consortium: High-throughput in vivo functional annotation of the mammalian genome. *Mammalian Genome* published online Sep 10. PMID:39254744.

MacLean, G., Abu-Abed, S., Dollé, P., Tahayato, A., Chambon, P., & Petkovich, M. (2001). Cloning of a novel retinoic-acid metabolizing cytochrome P450, Cyp26B1, and comparative expression analysis with Cyp26A1 during early murine development. *Mechanisms of Development, 107*, 195–201.

MacLean, G., Li, H., Metzger, D., Chambon, P., & Petkovich, M. (2007). Apoptotic extinction of germ cells in testes of Cyp26b1 knockout mice. *Endocrinology, 148,* 4560–4567.
Mark, M., Ghyselinck, N. B., & Chambon, P. (2006). Function of retinoid nuclear receptors: Lessons from genetic and pharmacological dissections of the retinoic acid signaling pathway during mouse embryogenesis. *Annual Review of Pharmacology and Toxicology, 46,* 451–480. PMID:16402912.
Mark, M., Jacobs, H., Oulad-Abdelghani, M., Dennefeld, C., Féret, B., Vernet, N., ... Ghyselinck, N. B. (2008). STRA8-deficient spermatocytes initiate, but fail to complete, meiosis and undergo premature chromosome condensation. *Journal of Cell Science, 121,* 3233–3242. PMID:18799790.
McLaren, A. (1983). Studies on mouse germ cells inside and outside the gonad. *The Journal of Experimental Zoology, 28,* 167–171. PMID:6663255.
McLaren, A. (1988). Sex determination in mammals. *Trends in Genetics, 4,* 153–157. PMID:3076297.
McLaren, A. (1995). Germ cells and germ cell sex. *Philosophical Transactions of the Royal Society of London. Series B, Biological Sciences, 350,* 229–233.
McLaren, A. (2003). Primordial germ cells in the mouse. *Developmental Biology, 262,* 1–15. PMID:14512014.
McLaren, A., & Southee, D. (1997). Entry of mouse embryonic germ cells into meiosis. *Developmental Biology, 187,* 107–113.
Menke, D. B., Koubova, J., & Page, D. C. (2003). Sexual differentiation of germ cells in XX mouse gonads occurs in an anterior-to-posterior wave. *Developmental Biology, 262,* 303–312.
Menke, D. B., & Page, D. C. (2002). Sexually dimorphic gene expression in the developing mouse gonad. *Gene Expression Patterns, 2,* 359–367.
Mic, F. A., Haselbeck, R. J., Cuenca, A. E., & Duester, G. (2002). Novel retinoic acid generating activities in the neural tube and heart identified by conditional rescue of Raldh2 null mutant mice. *Development (Cambridge, England), 129,* 2271–2282.
Miles, D. C., Wakeling, S. I., Stringer, J. M., van den Bergen, J. A., Wilhelm, D., Sinclair, A. H., & Western, P. S. (2013). Signaling through the TGF beta-activin receptors ALK4/5/7 regulates testis formation and male germ cell development. *PLoS One, 8,* e54606. https://doi.org/10.1371/journal.pone.0054606 Epub 2013 Jan 16. PMID:23342175.
Minkina, A., Lindeman, R. E., Gearhart, M. D., Chassot, A. A., Chaboissier, M. C., Ghyselinck, N. B., ... Zarkower, D. (2017). Retinoic acid signaling is dispensable for somatic development and function in the mammalian ovary. *Developmental Biology, 424,* 208–220.
Miyauchi, H., Ohta, H., Nagaoka, S., Nakaki, F., Sasaki, K., Hayashi, K., ... Saitou, M. (2017). Bone morphogenetic protein and retinoic acid synergistically specify female germ-cell fate in mice. *The EMBO Journal, 36,* 3100–3119.
Moreno, S. G., Attali, M., Allemand, I., Messiaen, S., Fouchet, P., Coffigny, H., ... Habert, R. (2010). TGFbeta signaling in male germ cells regulates gonocyte quiescence and fertility in mice. *Developmental Biology, 342,* 74–84.
Nagaoka, S. I., Nakaki, F., Miyauchi, H., Nosaka, Y., Ohta, H., Yabuta, Y., ... Saitou, M. (2020). ZGLP1 is a determinant for the oogenic fate in mice. *Science (New York, N. Y.), 367,* eaaw4115.
Nelson, C. H., Peng, C. C., Lutz, J. D., Yeung, C. K., Zelter, A., & Isoherranen, N. (2016). Direct protein-protein interactions and substrate channeling between cellular retinoic acid binding proteins and CYP26B1. *FEBS Letter, 590,* 2527–2535.
Niederreither, K., & Dollé, P. (2008). Retinoic acid in development: Towards an integrated view. *Nature Reviews. Genetics, 9,* 541–553.

Niederreither, K., McCaffery, P., Dräger, U. C., Chambon, P., & Dollé, P. (1997). Restricted expression and retinoic acid-induced downregulation of the retinaldehyde dehydrogenase type 2 (RALDH-2) gene during mouse development. *Mechanisms of Development, 62*, 67–78.

Oulad-Abdelghani, M., Bouillet, P., Décimo, D., Gansmuller, A., Heyberger, S., Dollé, P., ... Chambon, P. (1996). Characterization of a premeiotic germ cell-specific cytoplasmic protein encoded by Stra8, a novel retinoic acid-responsive gene. *The Journal of Cell Biology, 135*, 469–477.

Paik, J., Haenisch, M., Muller, C. H., Goldstein, A. S., Arnold, S., Isoherranen, N., ... Amory, J. K. (2014). Inhibition of retinoic acid biosynthesis by the bisdichloroacetyldiamine WIN 18,446 markedly suppresses spermatogenesis and alters retinoid metabolism in mice. *The Journal of Biological Chemistry, 289*, 15104–15117. PMID:24711451.

Piprek, R. P., Pecio, A., Laskowska-Kaszub, K., Kloc, M., Kubiak, J. Z., & Szymura, J. M. (2013). Retinoic acid homeostasis regulates meiotic entry in developing anuran gonads and in Bidder's organ through Raldh2 and Cyp26b1 proteins. *Mechanisms of Development, 130*, 613–627.

Prépin, J., Gibello-Kervran, C., Charpentier, G., & Jost, A. (1985). Number of germ cells and meiotic prophase stages in fetal rat ovaries cultured in vitro. *Journal of Reproduction and Fertility, 73*, 579–583.

Rodríguez-Marí, A., Cañestro, C., BreMiller, R. A., Catchen, J. M., Yan, Y. L., & Postlethwait, J. H. (2013). Retinoic acid metabolic genes, meiosis, and gonadal sex differentiation in zebrafish. *PLoS One, 8*, e73951.

Rossant, J., Zirngibl, R., Cado, D., Shago, M., & Giguère, V. (1991). Expression of a retinoic acid response element-hsplacZ transgene defines specific domains of transcriptional activity during mouse embryogenesis. *Genes & Development, 5*, 1333–1344.

Saba, R., Wu, Q., & Saga, Y. (2014). CYP26B1 promotes male germ cell differentiation by suppressing STRA8-dependent meiotic and STRA8-independent mitotic pathways. *Developmental Biology, 389*, 173–181.

Shimada, R., & Ishiguro, K. I. (2023). Cell cycle regulation for meiosis in mammalian germ cells. *The Journal of Reproduction and Development, 69*, 139–146.

Souquet, B., Tourpin, S., Messiaen, S., Moison, D., Habert, R., & Livera, G. (2012). Nodal signaling regulates the entry into meiosis in fetal germ cells. *Endocrinology, 153*, 2466–2473.

Smith, C. A., Roeszler, K. N., Bowles, J., Koopman, P., & Sinclair, A. H. (2008). Onset of meiosis in the chicken embryo; evidence of a role for retinoic acid. *BMC Developmental Biology, 8*, 85. PMID:18799012.

Sneath, P. H. A., & Sokal, R. R. (1973). *Numerical taxonomy*. San Francisco: Freeman.

Souali-Crespo, S., Condrea, D., Vernet, N., Féret, B., Klopfenstein, M., Grandgirard, E., ... Ghyselinck, N. B. (2023). Loss of NR5A1 in mouse Sertoli cells after sex determination changes cellular identity and induces cell death by anoikis. *Development (Cambridge, England), 150*, dev201710.

Spiller, C. M., Bowles, J., & Koopman, P. (2012). Regulation of germ cell meiosis in the fetal ovary. *International Journal of Developmental Biology, 56*, 779–787.

Spiller, C. M., & Bowles, J. (2015). Sex determination in mammalian germ cells. *Asian Journal of Andrology, 17*, 427–432.

Spiller, C., Koopman, P., & Bowles, J. (2017). Sex determination in the mammalian germline. *Annual Review of Genetics, 51*, 265–285.

Spiller, C., & Bowles, J. (2019). Sexually dimorphic germ cell identity in mammals. *Current Topics in Developmental Biology, 134*, 253–288.

Spiller, C., & Bowles, J. (2022). Instructing mouse germ cells to adopt a female fate. *Sexual Development, 16*, 342–354.

Tam, P. P., & Snow, M. H. (1981). Proliferation and migration of primordial germ cells during compensatory growth in mouse embryos. *Journal of Embryology and Experimental Morphology, 64,* 133–147.

Teletin, M., Vernet, N., Ghyselinck, N. B., & Mark, M. (2017). Roles of retinoic acid in germ cell differentiation. *Current Topics in Developmental Biology, 125,* 191–225.

Teletin, M., Vernet, N., Yu, J., Klopfenstein, M., Jones, J. W., Féret, B., ... Mark, M. (2019). Two functionally redundant sources of retinoic acid secure spermatogonia differentiation in the seminiferous epithelium. *Development (Cambridge, England), 146,* dev170225.

Thatcher, J. E., & Isoherranen, N. (2009). The role of CYP26 enzymes in retinoic acid clearance. *Expert Opinion on Drug Metabolism & Toxicology, 5,* 875–886.

Topletz, A. R., Thatcher, J. E., Zelter, A., Lutz, J. D., Tay, S., Nelson, W. L., & Isoherranen, N. (2012). Comparison of the function and expression of CYP26A1 and CYP26B1, the two retinoic acid hydroxylases. *Biochemical Pharmacology, 83,* 149–163.

Trautmann, E., Guerquin, M. J., Duquenne, C., Lahaye, J. B., Habert, R., & Livera, G. (2008). Retinoic acid prevents germ cell mitotic arrest in mouse fetal testes. *Cell Cycle (Georgetown, Tex.), 7,* 656–664.

Vernet, N., Condrea, D., Mayere, C., Féret, B., Klopfenstein, M., Magnant, W., ... Ghyselinck, N. B. (2020). Meiosis occurs normally in the fetal ovary of mice lacking all retinoic acid receptors. *Science Advances, 6,* eaaz1139.

White, J. A., Ramshaw, H., Taimi, M., Stangle, W., Zhang, A., Everingham, S., ... Petkovich, M. (2000). Identification of the human cytochrome P450, P450RAI-2, which is predominantly expressed in the adult cerebellum and is responsible for all-trans-retinoic acid metabolism. *Proceedings of the National Academy of Sciences of the USA, 97,* 6403–6408.

Wallacides, A., Chesnel, A., Chardard, D., Flament, S., & Dumond, H. (2009). Evidence for a conserved role of retinoic acid in urodele amphibian meiosis onset. *Developmental Dynamics, 238,* 1389–1398.

Wang, N., & Tilly, J. L. (2010). Epigenetic status determines germ cell meiotic commitment in embryonic and postnatal mammalian gonads. *Cell Cycle (Georgetown, Tex.), 9,* 339–349.

Wu, Q., Kanata, K., Saba, R., Deng, C. X., Hamada, H., & Saga, Y. (2013). Nodal/activin signaling promotes male germ cell fate and suppresses female programming in somatic cells. *Development (Cambridge, England), 140,* 291–300.

Yamaguchi, S., Hong, K., Liu, R., Shen, L., Inoue, A., Diep, D., ... Zhang, Y. (2012). Tet1 controls meiosis by regulating meiotic gene expression. *Nature, 492,* 443–447.

Yashiro, K., Zhao, X., Uehara, M., Yamashita, K., Nishijima, M., Nishino, J., ... Hamada, H. (2004). Regulation of retinoic acid distribution is required for proximodistal patterning and outgrowth of the developing mouse limb. *Developmental Cell, 6,* 411–422.

Yokobayashi, S., Liang, C. Y., Kohler, H., Nestorov, P., Liu, Z., Vidal, M., ... Peters, A. H. (2013). PRC1 coordinates timing of sexual differentiation of female primordial germ cells. *Nature, 495,* 236–240.

Zhou, Q., Li, Y., Nie, R., Friel, P., Mitchell, D., Evanoff, R. M., ... Griswold, M. D. (2008). Expression of stimulated by retinoic acid gene 8 (Stra8) and maturation of murine gonocytes and spermatogonia induced by retinoic acid in vitro. *Biology of Reproduction, 78,* 537–545.

Zuckerkandl, E., & Pauling, L. (1965). Evolutionary divergence and convergence in proteins. In V. Bryson, & H. J. Vogel (Eds.). *Evolving genes and proteins* (pp. 97–166). New York: Academic Press.

CHAPTER FOUR

Retinoids and retinoid-binding proteins: Unexpected roles in metabolic disease

William S. Blaner[a,*], Jisun Paik[b], Pierre-Jacques Brun[a], and Marcin Golczak[c]
[a]Department of Medicine, College of Physicians and Surgeons, Columbia University, New York, NY, United States
[b]Department of Comparative Medicine, University of Washington, Seattle, WA, United States
[c]Department of Pharmacology and Cleveland Center for Membrane and Structural Biology, Case Western Reserve University, Cleveland, OH, United States
*Corresponding author. e-mail address: wsb2@cumc.columbia.edu

Contents

1. Introduction	90
2. Recent advances in understanding the biochemistry of retinoid-binding proteins (RBPs)	91
3. The unexpected role of RBP2 in regulating both retinoid and neutral lipid signaling and its potential consequences of this for metabolic disease development	97
4. Diverse and unanticipated actions of retinol-binding protein 4 (RBP4) in retinoid signaling, metabolism and metabolic disease	104
Acknowledgments	108
References	108

Abstract

Alterations in tissue expression levels of both retinol-binding protein 2 (RBP2) and retinol-binding protein 4 (RBP4) have been associated with metabolic disease, specifically with obesity, glucose intolerance and hepatic steatosis. Our laboratories have shown that this involves novel pathways not previously considered as possible linkages between impaired retinoid metabolism and metabolic disease development. We have established both biochemically and structurally that RBP2 binds with very high affinity to very long-chain unsaturated 2-monoacylglycerols like the canonical endocannabinoid 2-arachidonoyl glycerol (2-AG) and other endocannabinoid-like substances. Binding of retinol or 2-MAGs involves the same binding pocket and 2-MAGs are able to displace retinol binding. Consequently, RBP2 is a physiologically relevant binding protein for endocannabinoids and endocannabinoid-like substances and is a nexus where the very potent retinoid and endocannabinoid signaling pathways converge. When *Rbp2*-null mice are challenged orally with fat, this gives rise to elevated levels in the proximal small intestine of both 2-AG and the incretin hormone glucose-dependent insulinotropic polypeptide (GIP) in the proximal small intestine. We propose that elevation of GIP concentrations upon high fat diet feeding gives rise to obesity and the other elements of metabolic disease seen in *Rbp2*-null mice. Unexpectedly, we observed that RBP4 is present in secretory granules of the GIP-secreting intestinal K-cells and is co-secreted with GIP in response to a stimulus that provokes GIP secretion. Moreover, RBP4 is co-secreted along with glucagon from pancreatic alpha-cells in response to a secretory stimulus. The association during the secretory process of RBP4 with potent hormones that regulate metabolism (GIP and glucagon) accounts for at least some of the metabolic disease seen upon over-expression of *Rbp4*.

1. Introduction

For the last 20 years, there has been much research interest aimed at explaining how retinoids (vitamin A and its metabolites) affect metabolic disease development, specifically of obesity, insulin resistance, hepatic disease, and cardiovascular disease. There is now a relatively large literature on this topic but there is still no consensus across this literature which allows general mechanistic conclusions to be drawn regarding disease causality or prevention. It is understood that all-*trans*-retinoic acid (ATRA)-regulated gene transcription must play a role in disease development and that too little or too much ATRA synthesis/degradation underlies the transcriptional dysregulation. Yet, how this may affect the development of specifically one *versus* another form of metabolic disease remains poorly understood. Readers are referred to several extensive reviews on retinoids and metabolic disease for a broad consideration of these topics (Blaner, 2019; Steinhoff, Lass, & Schupp, 2022).

Our laboratories have been studying the involvement of two retinol-binding proteins (RBPs) in metabolic disease development in the adult. This work will be the sole focus of our review. One of these proteins is an intracellular protein, retinol-binding protein 2 (RBP2; referred to in the older literature as cellular retinol-binding protein, type II (CRBPII)) (Demmer et al., 1986; Li, Demmer, Sweetser, Ong, & Gordon, 1986; Napoli, 2017; Ong, 1984). *Rbp2* has been thought to be expressed solely in the adult small intestine (Li et al., 1986; Napoli, 2017; Ong, 1984). *Rbp2*-null mice develop obesity, become insulin resistant, and develop hepatic steatosis as they reach 6- to 7-months of age when maintained on a conventional chow diet, or at a much younger age when they are challenged with a high fat diet for several months (Blaner, Brun, Calderon, & Golczak, 2020; Calderon et al., 2022; Lee et al., 2020; Silvaroli et al., 2021). The second protein is an extracellular protein found in the blood, retinol-binding protein 4 (RBP4; referred to in the older literature as simply RBP) (Quadro et al., 1999). Overexpression of *Rbp4* in mice or elevated circulating levels of RBP4 in humans is associated with obesity, insulin resistance, and hepatic steatosis (Blaner, 2019; Steinhoff et al., 2022).

This article will summarize new findings obtained in the last 5 years regarding the structural characteristics of RBPs, especially for RBP2 and RBP4, and how these may affect the physiologic actions of RBP2 and RBP4 that contribute to metabolic disease development. Our consideration of RBP2 and RBP4 will not specifically focus on their roles in ATRA formation or degradation but rather on their roles in other signaling pathways which are known to be important in metabolic disease causation and progression.

2. Recent advances in understanding the biochemistry of retinoid-binding proteins (RBPs)

Retinoids, owing to their lipophilic nature and conjugated double-bond structures, face significant challenges in diffusion within aqueous environments and are highly prone to spontaneous oxidation. To mitigate these physicochemical limitations and enhance their transport and utilization in biological systems, specialized carrier proteins have evolved. These proteins are crucial for sequestering and safeguarding labile retinoids, serving as key components of the nonvesicular lipid transport system. Retinoid-binding proteins are essential for ensuring the specificity of both systemic and intracellular retinoid transport, playing a vital role in integrating the compartmentalized metabolic and signaling pathways associated with retinoids. As such, they sit at the intersection of lipid metabolism and signaling.

The primary class of retinoid-binding proteins belongs to the diverse and widely distributed group of lipid-transport proteins known as lipocalins (Akerstrom, Flower, & Salier, 2000; Flower, 1996; Pervaiz & Brew, 1987). These small globular proteins, typically ranging from 15 to 21 kDa, share a conserved structural architecture characterized by an antiparallel β-barrel composed of 8 to 10 β-strands (Fig. 1A) (Newcomer & Ong, 2000). The interior of the β-barrel forms a single binding pocket designed for small hydrophobic ligands, with binding specificity determined by the amino acid composition lining the cavity. This lipocalin fold functions as a versatile platform, enabling the evolution of highly selective binding properties

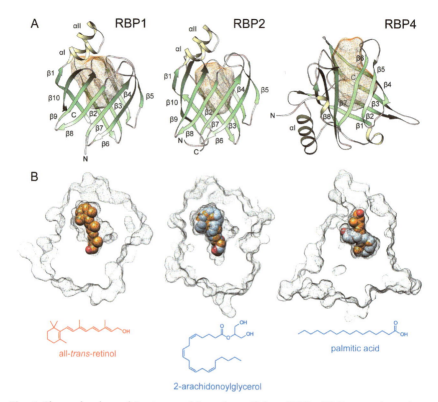

Fig. 1 The molecular architecture and ligand specificity of RBPs. (A) Structural topology comparison of cellular RBP1, RBP2, and serum RBP4. The lipocalin fold, a universal protein scaffold, features a hydrophobic binding cavity within the β-barrel (depicted in orange mesh). This cavity can be adapted to accommodate different lipid ligands. (B) Spatial orientation of endogenous ligands within the RBP binding pockets. RBP1 exhibits binding specificity exclusively for retinoids (shown in orange), while RBP2 and RBP4 have the ability to interact with 2-AG and fatty acids (both depicted in blue), respectively.

and enzymatic activities within the protein family (Achatz, Jarasch, & Skerra, 2022; Skerra, 2000; Urade & Hayaishi, 2000). Notable lipid carrier proteins in the human genome that belong to lipocalin-like families include 5 retinol-binding proteins (RBP1, -2, -4, -5, and -7), 2 retinoic acid-binding proteins (CRABP1 and -2), and 7 fatty acid-binding proteins (FABP1–7) (Flower, North, & Sansom, 2000).

Ligand specificity and diverse functions of intracellular RBPs. Until recently, RBP1 and RBP2 were believed to interact exclusively with all-*trans*-retinol (ROL) and its oxidized form, all-*trans*-retinal (RAL) (Ong, 1984; Ong, 1994; Silvaroli et al., 2016). Consequently, the physiological roles of the intracellular RBPs have been primarily studied in the context of cellular uptake and transport of these two retinoids (Napoli, 2016; Napoli, 2017; Wongsiriroj et al., 2008). Biochemical and *in vivo* studies have provided compelling evidence that RBP1 and RBP2 channel ROL to form retinyl esters and RAL used for ATRA biosynthesis (Boerman & Napoli, 1991; Boerman & Napoli, 1996; Herr & Ong, 1992; Napoli, Posch, Fiorella, & Boerman, 1991). Despite their similarities, these two proteins exhibit distinct tissue expression profiles. RBP1 is ubiquitously expressed, with the highest concentrations found in hepatic stellate cells (HSCs) and retinal pigment epithelium (RPE) cells, while RBP2 expression is predominantly confined to the intestine (Blaner et al., 2020; Huang, Possin, & Saari, 2009; Li et al., 1986; Porter, Fraker, Chytil, & Ong, 1983). This tissue-specific expression underscores the unique functions of these proteins, as evidenced by the different phenotypes observed in *Rbp1* and *Rbp2* knockout mice. Without functional RBP1, mice display an accelerated turnover of hepatic retinoids, reduced retinyl ester content in the liver, and an increased susceptibility to systemic vitamin A-deficiency (Ghyselinck et al., 1999). Additionally, $Rbp1^{-/-}$ mice show delayed regeneration of the visual chromophore after exposure to bright light, leading to slower visual adaptation compared to wild-type (WT) counterparts (Saari et al., 2002). In contrast, RBP2 does not appear to be essential under conditions of dietary retinoid sufficiency. However, $Rbp2^{-/-}$ mice exhibit perinatal lethality when subjected to a vitamin A-deficient diet (E et al., 2002). Moreover, $Rbp2^{-/-}$ mice challenged with a high-fat diet gain more body fat, develop glucose intolerance, and show signs of fatty liver development compared to WT mice (Lee et al., 2020). Interestingly, these metabolic phenotypes do not seem to stem from altered retinoid homeostasis (see below for more details), prompting investigations into potential non-retinoid ligands for RBPs.

To reconcile the diverse functions of RBPs, systematic screening of chemical libraries of bioactive lipids has uncovered unanticipated differences in ligand specificity between RBPs. Unlike RBP1, which demonstrates exclusive selectivity for ROL/RAL as natural ligands, RBP2 exhibits low nanomolar affinity for non-retinoid lipids, including a subset of monoacylglycerols (MAGs) (Fig. 1B) (Silvaroli et al., 2019; Silvaroli et al., 2021). Notably, the binding of MAGs is an intrinsic property of RBP2 that distinguishes it from other intracellular members of the RBP protein subfamily. The ability to interact with MAGs remains selective and results from a defined substitution of two amino acids (T51I and V62M) inside the binding pocket as compared to RBP1 (Silvaroli et al., 2021). Importantly, these gain-of-function modifications in the architecture of the binding cavity broadened but did not eliminate the relatively narrow ligand specificity of RBP2 maintained by a combination of discrete hydrogen bonds formed by the ligands within the binding pocket and hydrophobic and van der Waals contacts with amino acids of the protein's portal region. These interactions collectively define the chemical boundaries for ligands capable of forming a stable complex with RBP2. Altering just one of these interactions can lead to a partial loss of affinity or an inability to stabilize the protein in its ligand-bound closed conformation. The highest affinity (k_d values of 27.1–65.4 nM) was observed for sn-2 MAGs with polyunsaturated fatty acid chains containing 16 to 20 carbons (Silvaroli et al., 2021). Fatty acid chains that are longer do not geometrically fit within the binding pocket, while shorter chains fail to stabilize the portal region of RBP2 in the closed conformation, dramatically increasing the dissociation rate of these ligands. Although RBP2 can also bind sn-1 MAGs, these ligands exhibit approximately two-fold lower affinities compared to the corresponding sn-2 MAGs. Therefore, RBP2 demonstrates the highest binding preference for signaling lipids such as 2-arachidonoylglycerol (2-AG) and 2-oleoylglycerol (2-OG), suggesting a potential role in regulating their concentrations and their biological activities within intestinal tissue (Lee et al., 2020). These topics are considered in greater detail in the text below.

Biochemical basis for the function of RBP4. RBP4, the first discovered and most extensively studied RBP, presents a paradox in that its biological functions and structural characteristics diverge significantly from other RBPs. As the sole carrier of ROL in the bloodstream, RBP4 plays a crucial role in the extrahepatic distribution of ROL mobilized from HSC stores, thereby maintaining systemic retinoid homeostasis (Kanai, Raz, & Goodman, 1968; Quadro et al., 1999). Although it is primarily expressed and secreted by hepatocytes in complex with ROL, RBP4 is also detectable at the protein level in adipocytes and in specialized hormone-secreting enteroendocrine cells (EECs) within the intestine (Calderon et al., 2022; Kloting et al., 2007; Tsutsumi et al., 1992). $Rbp4^{-/-}$ mice are viable and fertile when maintained on a vitamin A-sufficient diet but develop progressive retinal dysfunction due to inadequate vitamin A supplementation of ocular tissues (Quadro et al., 1999; Shen et al., 2016). These mice also exhibit decreased locomotor activity and increased anxiety-like behavior, which have been linked to neuronal loss and gliosis in the cortex and hippocampus (Buxbaum, Roberts, Adame, & Masliah, 2014). In humans, mutations in the RBP4 gene are predominantly associated with eye pathologies. The severity of these phenotypic consequences varies depending on the type of mutation and its impact on ROL binding, interaction with STRA6 (a plasma membrane receptor for RBP4), and the overall vitamin A status of the patient (Chou et al., 2015; Folli, Viglione, Busconi, & Berni, 2005; Kawaguchi et al., 2007; Seeliger et al., 1999). Consequently, mutations can range from causing RPE and retina dystrophy (Biesalski et al., 1999; Cehajic-Kapetanovic, Jasani, Shanks, Clouston, & MacLaren, 2020; Seeliger et al., 1999) to developmental abnormalities, including microphthalmia, anophthalmia, and coloboma (Chou et al., 2015; Cukras et al., 2012). Interestingly, some mutations associated with eye phenotypes have also been linked to systemic conditions like acne vulgaris, osteoarthritis, and hypercholesterolemia, indicating that RBP4 deficiencies may have broader effects on retinoid homeostasis (Cukras et al., 2012; Khan et al., 2017). This notion is further supported by associations between blood RBP4 levels and conditions such as obesity, insulin resistance, and other diseases composing the metabolic syndrome (Codoner-Franch, Carrasco-Luna, Allepuz, Codoner-Alejos, & Guillem, 2016; Kovacs et al., 2007; Wan et al., 2014).

Despite being structurally related to other lipocalin-like proteins, RBP4 exhibits fundamental differences in its mechanism of retinoid binding. These differences arise from alterations in RBP4's tertiary structure, particularly within its β-barrel, which consists of 8 β-strands, two fewer than in other RBPs (Fig. 1A) (Newcomer et al., 1984). Additionally, RBP4 lacks the canonical portal region characteristic of other lipocalins. Consequently, ROL is accommodated in the binding site in an orientation opposite to that seen in RBP1 or RBP2 (Cowan, Newcomer, & Jones, 1990; Silvaroli et al., 2016; Tarter et al., 2008). The β-ionone ring of the retinoid is positioned at the center of the β-barrel, while the isoprene tail extends along the barrel axis toward the entrance of the binding pocket. The interactions between RBP4 and the retinoid moiety are predominantly hydrophobic and van der Waals in nature, with a single hydrogen bond between the ligand's hydroxyl group and the main chain of Gln98. Unlike RBP1 and RBP2, ROL binding does not induce significant conformational changes in RBP4, except in a short loop encompassing amino acids 33–36 (Plau et al., 2023; Silvaroli et al., 2016; Zanotti et al., 2008). However, this minor structural alteration is functionally significant, as it enables the interaction of holo-RBP4 with transthyretin (TTR). Since TTR exists as a tetramer, it can interact with a maximum of two holo-RBP4 molecules, resulting in a 4:2 stoichiometry for the complex. However, due to the limited concentration of RBP4 in plasma, the complex isolated from serum typically contains only one holo-RBP4 per four TTR molecules (Naylor & Newcomer, 1999). The primary function of the TTR interaction is to prolong the serum half-life of holo-RBP4 by forming a higher molecular weight complex that prevents renal filtration (van Bennekum et al., 2001). There is no evidence that RBP4 binding to TTR has a role in coordinating retinoid and thyroid hormone actions.

Biochemical and crystallographic analyses of RBP4 isolated from human or bovine blood have identified ROL as the predominant ligand (Heller, 1975; Perduca, Nicolis, Mannucci, Galliano, & Monaco, 2018). However, the interaction mode with the retinoid moiety does not support selectivity among different retinoid classes. As a result, RBP4 has been shown to interact *in vitro* with RAL, ATRA, and retinyl acetate, with affinities comparable to those for ROL (Berni, Clerici, Malpeli, Cleris, & Formelli, 1993; Zanotti et al., 1994). Although not yet confirmed *in vivo*, it is plausible that interactions with these alternative retinoids may play a role in the physiological functions of extrahepatic RBP4. Notably, RBP4 has also been found to form complexes with endogenous long-chain fatty acids. The protein isolated from human urine or amniotic fluid predominantly carried palmitic acid, along with palmitoleic, oleic, and stearic acids (Fig. 1B) (Perduca et al., 2018). RBP4 is also able to bind with high affinity unesterified fatty acids including palmitic acid, and the long chain unsaturated fatty acids, linolenic acid, γ-linolenic acid, 5Z,8Z,14Z-eucisatrieonic acid and others (Calderon et al., 2022; Perduca et al., 2018). Possibly one or more of these fatty acid ligands influences RBP4 actions intracellularly to affect metabolic disease. Subsequent X-ray crystallography data suggest that these non-retinoid ligands outcompete ROL for the binding site and adopt a similar orientation within the binding pocket, with the carboxyl group near the entrance and the acyl chains buried deeper inside the β-barrel (Perduca et al., 2018). However, fatty acids do not stabilize the complex with TTR, which explains the presence of RBP4 bound to fatty acids in the urine.

The pharmacological manipulation of vitamin A status has long been proposed as a therapeutic strategy for treating diseases associated with impaired ATRA signaling or imbalances in cellular retinoid homeostasis, including certain types of breast cancers, lymphomas, metabolic disorders, hepatic steatosis, and retinal degenerative disorders such as Stargardt disease. In this context, several classes of RBP4 inhibitors have been identified (Chen et al., 2024). The first-in-class was fenretinide, an amide derivative of retinoic acid, which effectively lowered serum RBP4 levels by sterically disrupting complex formation with TTR (Berni & Formelli, 1992).

However, as a retinoid, fenretinide has a complex mechanism of action that extends beyond RBP4 inhibition and includes direct transcriptional regulation of genes (McIlroy et al., 2013). Its use is also accompanied by teratogenic side effects. For these reasons, significant efforts have been directed toward developing specific, non-retinoid inhibitors of RBP4, leading to the identification and optimization of compounds such as A1120 and BPN-14136 (Cioffi et al., 2014; Motani et al., 2009). These compounds exert their biological effects by disrupting ROL-induced RBP4-TTR interactions through a combination of steric hindrance and conformational changes in the TTR binding interface. However, this mode of action raises concerns about potential TTR tetramer destabilization and the formation of amyloid aggregates. To mitigate this risk, bispecific RBP4 inhibitors were designed (Cioffi et al., 2020). These act as both competitive inhibitors of RBP4 and stabilizers of TTR tetramers. Importantly, *in vivo* examination of these inhibitors in mouse models of human diseases revealed their potential in the treatment of age-related macular degeneration and Stargardt disease as well as diet-induced liver steatosis (Cioffi et al., 2014; Cioffi et al., 2019). Despite encouraging results from animal studies, the efficacy of RBP4 inhibitors in treating human conditions remains to be rigorously established. There are ongoing clinical trials to test the efficacy of two commercially developed compounds, tinlarebant, and STG-001 (NCT05949593 and NCT04489511).

3. The unexpected role of RBP2 in regulating both retinoid and neutral lipid signaling and its potential consequences of this for metabolic disease development

We have established that when maintained solely on a control chow diet, 6–7-month-old *Rbp2*-deficient ($Rbp2^{-/-}$) mice accrue significantly more body weight as white adipose tissue (WAT), and they respond significantly less well to a glucose challenge, and possess significantly more hepatic fat than matched WT littermate controls (Lee et al., 2020). These metabolic phenotypes are fully recapitulated by younger 55-day old $Rbp2^{-/-}$ mice fed a high-fat diet for 6–7 weeks (Lee et al., 2020). Thus, RBP2 acts in some manner to regulate energy metabolism and prevent metabolic disease (Lee et al., 2020).

To understand the molecular basis for these phenotypes, we first assessed tissue retinoid (ROL, RAL, and ATRA) concentrations for a number of organs often associated with metabolic disease, including liver, white and brown adipose tissue and pancreas. In agreement with the original characterizations of adult $Rbp2^{-/-}$ mice, we observed lower levels of ROL and retinyl esters in the livers of the mice (E et al., 2002). This has been attributed to inefficient dietary retinoid uptake in the small intestine resulting in lower retinoid stores accumulating in the liver (E et al., 2002). We did not detect differences in hepatic ATRA concentrations for $Rbp2^{-/-}$ versus matched littermate control mice. Nor did we detect differences between $Rbp2^{-/-}$ and matched littermate controls in retinoid concentrations (for ROL, retinyl esters and ATRA) for the two types of adipose tissue and pancreas. We also assessed mRNA expression levels by qRT-PCR for approximately 30 genes involved in retinoid signaling (retinoic acid receptor gene expression) and metabolic regulation (genes encoding proteins involved in retinoid storage and mobilization, the oxidation of retinol to ATRA, and the degradation of ATRA) for knockout and littermate control mice. We did not observe patterns of significant differences in gene expression that might account for the observed metabolic phenotypes we observed. We concluded from these studies that RBP2 is not acting in a manner that affects peripheral tissue ATRA concentrations giving rise to the observed phenotypes. Rather, RBP2 appears to be acting directly in the small intestine.

RBP2 was shown at the time of its discovery in the mid-1980s to bind ROL and not bind a relatively small number of other abundant lipids (Li et al., 1986; Ong, 1984). Based on these findings it was concluded that the protein is a member of the RBP family of proteins. To our knowledge, this has not been reexamined using modern analytical approaches. Consequently, we undertook *in silico* studies to assess the possibility that other low molecular weight lipids might bind this protein. As alluded to above in Section 2, this identified sn-2 MAGs with polyunsaturated fatty acid chains as potentially being able to bind RBP2. Subsequently, we screened a chemical library consisting of more than 900 different lipids for their ability to bind to recombinant human RBP2. Indeed, sn-2 MAGs with polyunsaturated fatty acid chains containing 16 to 20 carbons were found to bind RBP2 with high affinity nearly equivalent to that of ROL (Lee et al., 2020; Silvaroli et al., 2021). Thus, given the relatively large concentrations of 2-MAGs that will be present in the intestine following consumption of a fat-rich meal, several orders of magnitude greater than ROL, RBP2 is a physiologically relevant binding protein for long chain unsaturated 2-monoacylglycerols (2-MAGs), in addition to being a binding protein for ROL and RAL (Blaner et al., 2020; Calderon et al., 2022; Lee et al., 2020; Silvaroli et al., 2021).

The crystal structure of human RBP2 co-crystallized with 2-arachidonoylglycerol (2-AG) was determined at 1.35 Å resolution (Lee et al., 2020; Silvaroli et al., 2021). The determined K^d for 2-AG binding to RBP2 was 27.1 ± 2.4 nM. Moreover, 2-AG was found to displace ROL from RBP2. The K_d values for 2-linoleoylglycerol (2-LG) and 2-oleoylglycerol (2-OG) binding to RBP2 are similar to that determined for 2-AG (Lee et al., 2020; Silvaroli et al., 2021). For comparison, the reported K_d for ROL binding to RBP2 is 10 ± 3 nM (Kane, Bright, & Napoli, 2011). As seen in Fig. 1, the crystal structure shows that 2-AG occupies the same binding site within RBP2 as ROL [see (Silvaroli et al., 2021) for details]. This finding was totally unanticipated by the literature. RBP2 had been extensively studied as a protein present in absorptive cells within the small intestine where it facilitates dietary retinoid packaging into nascent chylomicrons for uptake into the body (E et al., 2002; Lee et al., 2020; Li et al., 1986; Ong, 1984; Wongsiriroj et al., 2008). It is highly expressed in small intestine (Li & Tso, 2003; Li et al., 1986; Ong, 1984; Ong, 1994) and, as discussed below, we have recently shown that is also present in the colon, albeit at lower levels (Blaner et al., 2020), but not in other adult tissues.

2-AG and N-arachidonoylethanolamine (AEA), which unlike 2-AG does not bind well to RBP2, are canonical endocannabinoids that bind to and activate the cell surface cannabinoid receptors 1 and 2 (Cb1 and Cb2) (Cuddihey, MacNaughton, & Sharkey, 2022; Guccio, Gribble, & Reimann, 2022; Rakotoarivelo, Sihag, & Flamand, 2021; Roque-Bravo et al., 2023; Ruiz de Azua & Lutz, 2019). Other long-chain 2-MAGs, referred to as endocannabinoid-like substances, including 2-LG and 2-OG, are also potent regulators of cell activities. The endocannabinoids and endocannabinoid-like substances modulate appetite, inflammation, glucose and lipid metabolism, cell growth and proliferation, smooth muscle contractility, mood regulation, motivation and reward, cognitive functions, nociception, neurogenesis, and neurodegeneration (Cuddihey et al., 2022; Guccio et al., 2022; Rakotoarivelo et al., 2021; Roque-Bravo et al., 2023; Ruiz de Azua & Lutz, 2019). However, for the endocannabinoid-like substances, this involves signaling mediated by other cell surface G-protein coupled receptors, including GPR119 (Cuddihey et al., 2022; Guccio et al., 2022; Rakotoarivelo et al., 2021; Roque-Bravo et al., 2023; Ruiz de Azua & Lutz, 2019).

Our biochemical data establish that RBP2 plays a role in 2-MAG trafficking (Blaner et al., 2020; Calderon et al., 2022; Lee et al., 2020; Silvaroli et al., 2021). We observed that RBP2 acts in a manner that modulates the release of the incretin hormone glucose-dependent insulinotropic polypeptide

(GIP) into the circulation from its site of synthesis in the enteroendocrine K-cells located in the proximal small intestine (Lee et al., 2020). Blood levels of GIP in $Rbp2^{-/-}$ mice following an oral fat challenge are significantly greater within 30 min, by approximately 2-fold, than those observed in matched littermate control mice (Lee et al., 2020). This is accompanied by significant elevations in 2-AG and 2-LG concentrations measured in intestinal scrapings from treated mice (Lee et al., 2020). Elevated blood GIP levels are also observed in both the fasted and fed states (Lee et al., 2020). Importantly, young $Rbp2^{-/-}$ mice (70 days of age) maintain elevated blood GIP levels compared to WT controls (Lee et al., 2020). This is well before any differences in body weights are observed (Lee et al., 2020). We take this to indicate that the elevated GIP levels do not arise as a consequence of the metabolic deficits seen in $Rbp2^{-/-}$ mice but rather precede metabolic disease development. The elevated GIP levels are not seen if the mice are challenged with an oral dose of glucose, only oral fat. This leads us to propose that the chronically elevated GIP levels experienced by $Rbp2^{-/-}$ mice upon dietary fat consumption throughout life contribute to the obesity and metabolic disease phenotypes observed in these mice. We do not yet understand whether the elevated blood GIP levels arise due to differences in K-cell rates of GIP synthesis, GIP secretion, or both. This remains to be established.

GIP is a hormone that is released from the K-cells present in the upper small intestine in response to food intake and contributes to the postprandial control of nutrient deposition, including sugars and fats, by augmenting insulin secretion from the pancreas (Guccio et al., 2022). GIP also exerts an array of other effects on peripheral tissues including effects on adipose tissue that lead to increased lipid accumulation in adipocytes. This has been shown to involve GIP direct effects that stimulate lipoprotein lipase (LPL) hydrolysis of postprandial fat (Eckel, Fujimoto, & Brunzell, 1979; Kim, Nian, & McIntosh, 2007; Kim, Nian, & McIntosh, 2010), and increased adipocyte free fatty acid re-esterification (Eckel et al., 1979; Kim et al., 2007; Kim et al., 2010). A potential correlation between obesity and elevated circulating levels of GIP was first observed almost 40 years ago, when *ob/ob* and *db/db* mice were shown to have 15-fold and 6-fold higher plasma GIP levels compared to lean controls (Flatt, Bailey, Kwasowski, Swanston-Flatt, & Marks, 1983). Because of the known actions of GIP to increase adipose tissue lipid accrual, we propose that the obesity phenotype we observe in $Rbp2^{-/-}$ mice is caused by the elevated blood GIP levels that the mice experience over the course of their lifetime. We wondered whether elevated blood GIP levels arose due to a paracrine action of

RBP2, possibly involving 2-MAG interactions with cell surface receptors, or due to a direct autocrine effect, possibly involving ATRA-RAR regulation of transcription. RBP2 was known to be highly expressed in enterocytes and not in the secretory Goblet cells (Li & Norris, 1996; Li & Tso, 2003; Ong, 1994; Roque-Bravo et al., 2023; Ruiz de Azua & Lutz, 2019). However, its presence in other secretory lineages, specifically Paneth and EECs, had not been investigated. Consequently, we undertook immunohistochemical studies of RBP2 expression in the proximal mouse small intestine (Calderon et al., 2022). We found that RBP2 is highly expressed in nearly all EECs, including both K- and L-cells, but not in Paneth cells. RBP2 expression completely overlaps with that of GIP in cells present in crypts and along the villus. We also observed high levels of expression of RARs and aldehyde dehydrogenase 1 family member A1 (ALDH1A1), an enzyme needed for ATRA synthesis. Paradoxically, GIP-positive EEC numbers in the proximal 10 cm of the small intestine were significantly lower in $Rbp2^{-/-}$ mice. However, no differences in the numbers of other secretory cells including Goblet, Paneth, and L-cells were observed; nor did we observe any loss in intestinal barrier integrity as assessed by FITC-dextran uptake (Calderon et al., 2022). Based on these published findings, we propose that RBP2 acts in an autocrine manner within EECs to modulate K-cell numbers. Thus, we hypothesize, given our findings regarding RAR and ALDH1A1 expression in EECs, that at least some of these effects are mediated by ATRA-RAR signaling, whereas others including the release of peptide hormone from EECs involve 2-MAG actions.

Glucagon-like polypeptide-1 (GLP-1), like GIP, is an incretin hormone that acts as a sensor of nutrient intake, signaling pancreatic secretion of insulin. GLP-1 is synthesized and secreted by enteroendocrine L-cells and, unlike the proximally located (jejunal) K-cells, are found in the more distal small intestine and colon (Coskun et al., 2018; Dahl et al., 2022; Frias et al., 2021; Guccio et al., 2022; Holst, 2022; Huang et al., 2024; Jastreboff, Aronne, & Stefanski, 2022; Kumar & D'Alessio, 2022; Plau, Golczak, Paik, Calderon, & Blaner, 2022; Williams, Staff, Bain, & Min, 2022). L-cells express both *Rbp2* mRNA and protein (Calderon et al., 2022). However, unlike K-cells and GIP, when $Rbp2^{-/-}$ mice are challenged with an oral fat load, we do not see differences in the GLP-1 concentrations present in blood obtained from matched $Rbp2^{-/-}$ and littermate control mice (Lee et al., 2020). Thus, although RBP2 is expressed in both K- and L-cells, there are apparent differences in its actions in the different EEC types.

The paradoxical finding that GIP increases fat deposition in adipocytes has limited the development of GIP-based therapies for the treatment of metabolic disease and obesity (Guccio et al., 2022; Kumar & D'Alessio, 2022; Plau et al., 2022; Williams et al., 2022). This is unlike the other incretin GLP-1, which is synthesized and secreted by L-cells located in the more distal small intestine and colon. GLP-1 receptor agonists have found widespread clinical usage for the treatment of type II diabetes and obesity (Guccio et al., 2022; Holst, 2022; Kumar & D'Alessio, 2022; Williams et al., 2022). Recently, however, the therapeutic benefits of GIP usage in treatments aimed at obesity and type II diabetes has been established (Coskun et al., 2018; Dahl et al., 2022; Frias et al., 2021; Huang et al., 2024; Jastreboff, Ahmad et al., 2022). Published in mid-2022 (Jastreboff, Aronne, & Stefanski, 2022), findings from a large phase 3 double-blind, randomized, controlled trial of tirzepatide, involving 2539 obese non-diabetic patients, showed that many patients receiving one of the 2 higher doses of the drug (from a total of 3 doses) showed weight losses of 20 % or more (average weight reduction of 35–52 pounds) over the 72-week treatment period. Tirzepatide, now prescribed and marketed as Mounjaro, is an engineered peptide which contains the native GIP sequence and displays agonist activity for both the GIP and GLP-1 cell surface receptors (Coskun et al., 2018). In preclinical studies, GIP receptor activation was reported to be synergistic with GLP-1 receptor activation, allowing for greater weight reduction in mice than that achieved with GLP-1 receptor monoagonism (Coskun et al., 2018). Thus, GIP is now finding usage in the clinic for the treatment of obesity (Coskun et al., 2018; Dahl et al., 2022; Frias et al., 2021; Huang et al., 2024; Jastreboff, Ahmad et al., 2022). This raises a very important question regarding how dietary retinoids and fat acting through RBP2 binding of ROL, RAL and 2-MAGs may adversely affect GIP actions leading to excessive fat accrual and metabolic disease development.

RBP2 expression outside of the small intestine. As noted above, the most extensively studied role of RBP2 in retinoid metabolism and actions involves vitamin A absorption, and retinyl ester formation, within enterocytes of the small intestine. However, RBP2's role in retinoic acid synthesis in the intestine is less well studied. We and many others previously reported that enzymes involved in retinoid metabolism are differentially expressed in small intestine and colon. The capacity for ATRA synthesis in the mouse small intestine decreases from proximal to distal (Plau et al., 2022), similar to the reported expression pattern of RBP2

(Crow & Ong, 1985). Using qRT-PCR, we found that colon highly expresses one of the major retinoic acid synthesizing enzymes, ALDH1A1, but expresses only very low levels of lecithin:retinol acyltransferase (LRAT) which is needed for retinyl ester synthesis and dietary vitamin A uptake. Thus, the colonic mucosa shows a similar capacity for retinoic acid synthesis to that of the proximal small intestine (Plau et al., 2022). Interestingly, while undertaking our studies of *Rpb2* expression in EEC (Calderon et al., 2022), we observed that the colon also expresses *Rbp2* mRNA (Plau et al., 2022). These observations when considered collectively suggest that RBP2 contributes to ATRA synthesis in murine colon. Our preliminary immunolocalization studies show that RBP2 is localized mainly in colonocytes and not overlapping significantly with GLP1 expression in enteroendocrine L-cells in the colon (Fig. 2). This is unlike the expression patterns seen in small intestine (Calderon et al., 2022). However, it is known that L-cells are expressed along the crypt-villus axis and thus, it is possible that L-cells near colonocytes may express both GLP-1 and RBP2. More detailed co-localization studies will be needed to determine if both proteins are also expressed in L-cells in colon similar to small intestine. Nevertheless, the role of GLP-1 in colon is likely different from that of GLP-1 in small intestine as the colon is not involved in nutrient absorption. The GLP-1 response seen in blood after food intake (Orskov, Wettergren, & Holst, 1996) is unlikely to originate from the colonic L-cells. Regulation of GLP-1 secretion in L-cells of colon may come from bile acids, short-chain fatty acids or other microbial metabolites (Kuhre, Deacon, Holst, & Petersen, 2021). Distal intestinal L-cells may be involved in maintaining basal levels of GLP-1 postprandially and may regulate gastric emptying to ensure the amount of food leaving the stomach matches the digestion and absorption capacity of small intestine (Gribble & Reimann, 2016; Lewis et al., 2020).

Summary. Collectively, these findings lead us to propose that 2-MAG binding to RBP2 is important for the maintenance of EEC actions in the small intestine, especially for EECs that synthesize and secrete incretin hormones, GIP by K-cells, and glucagon-like peptide-1 (GLP-1) by intestinal L-cells. Moreover, our finding that both 2-MAGs and ROL bind the same location within the RBP2 molecule and that each can displace the binding of the other (Blaner et al., 2020; Calderon et al., 2022; Lee et al., 2020; Silvaroli et al., 2021), indicates that RBP2 is a nexus where the very potent retinoid and endocannabinoid signaling pathways converge (Fig. 2). This role involves RBP2's actions in the synthesis/release of the incretin

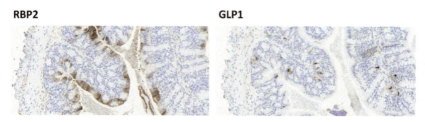

Fig. 2 RBP2 is expressed mainly in colonocytes (A) while L-cells appear deeper in the mucosa and adjacent to goblet cells (B). Anti-RBP2 was obtained from Abcam (ab180494, 1:500) and anti-GLP-1 from Invitrogen (ma5–42868, 1:100).

hormone GIP from intestinal EECs. Regulation of hormone secretion by RBP2 is associated with its ability to bind, with a near equal affinity, low molecular weight ligands that are critical components of two distinct cell signaling pathways, ligands responsible for retinoid signaling and ligands responsible for endocannabinoids/endocannabinoid-like substance signaling (Blaner et al., 2020; Calderon et al., 2022; Lee et al., 2020; Silvaroli et al., 2021). As depicted in Fig. 3, we propose that RBP2 is a node where these two very potent signaling pathways converge within the intestine. It will be important to identify how binding of each class of ligand to RBP2 affects metabolism and metabolic disease development through effects on GIP synthesizing/secreting K-cells. Specifically, we wish to understand which of RBP2's effects on metabolism arise due to its actions as a retinoid-binding protein and which arise due to its actions in binding 2-mono-acylglycerols (2-MAGs).

4. Diverse and unanticipated actions of retinol-binding protein 4 (RBP4) in retinoid signaling, metabolism and metabolic disease

RBP4 was first identified in the late 1960s (Kanai et al., 1968) and was studied for more than 35 years solely from the perspective of facilitating the delivery of retinoids stored in the liver to peripheral tissues where it was used to maintain retinoid-dependent functions. To prevent confusion on the part of the reader, we note that retinol-binding protein (RBP) and retinol-binding protein 4 (RBP4) are alternative names for the same protein/gene. RBP was used until the early 2000s when the gene nomenclature RBP4 was adopted. This singular research interest focused on RBP4 as a circulating transport protein for ROL changed in 2005 when Kahn and colleagues

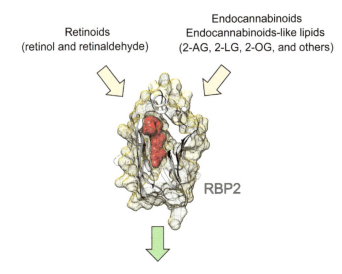

Fig. 3 RBP2 is a point of convergence inking two potent signaling pathways: retinoid signaling and endocannabinoid or endocannabinoid-like substance signaling.

reported both human and mouse model studies which established that relatively high levels of RBP4 in the circulation and in tissues were associated with obesity, insulin resistance, fatty liver, cardiovascular disease, as well as other metabolic diseases (Graham et al., 2006; Yang et al., 2005). This led to much research exploring the molecular bases for this relationship (Blaner, 2019; Lee, Yuen, Jiang, Kahn, & Blaner, 2016; Moraes-Vieira et al., 2020; Steinhoff et al., 2022). It is now thought that overexpression of *Rbp4* in adipose tissue brings about local inflammation, increased cytokine/chemokine production, and increased triglyceride hydrolysis. This propagates to other tissues, adversely affecting their health (Blaner, 2019; Lee et al., 2016; Moraes-Vieira et al., 2020; Steinhoff et al., 2022). But what is responsible for this? Is this a retinoid-dependent or retinoid-independent action of RBP4? Is it an intracellular or extracellular action of RBP4? Answers to these questions are needed but not yet fully available.

There have been no previous reports of RBP4 expression in small intestine. But while undertaking our studies of RBP2 distribution and actions in the small intestine (see Section 3 above), our recently published single-cell transcriptomics and immunohistochemical studies revealed that RBP4 is highly expressed within intestinal EECs. However, it is not expressed in other secretory or absorptive cell lineages (Calderon et al., 2022). Observing RBP4 expression was surprising since the small intestine was not thought to express *Rbp4*. However, intestinal EECs represent <1 % of the total cells present in the small intestine (Ahlman & Nilsson, 2001; Gribble & Reimann, 2019) so *Rbp4* expression might have been overlooked in early studies.

Our immunohistochemical studies established that RBP4 colocalizes with GIP within intracellular secretory vesicles present within K-cells (see Fig. 4 in Calderon et al., 2022). Moreover, GIP and RBP4 are co-secreted when primary murine intestinal cultures are stimulated with a cocktail of forskolin/3-isobutyl-1-methylxanthine/10 mM glucose (F/I/10 G) (Calderon et al., 2022). This observation is intriguing. RBP4, like RBP2, is a retinol-binding protein, but unlike RBP2 which never leaves the cell, RBP4 transports ROL in the circulation. The most often encountered explanation for why extrahepatic cells synthesize and secrete RBP4 is that this provides a mechanism for recycling ROL back to the liver for storage and secretion back into the circulation (Blomhoff, Green, Berg, & Norum, 1990). But this rationale is counterintuitive for the small intestine since the small intestine accounts for the uptake of all of the dietary retinoid that is found in the body. Why would RBP4 synthesized by intestinal K-cells be needed for this purpose given the specific physiological role of this organ in absorbing all dietary retinoids taken into the body?

Since the incretin hormones act to stimulate pancreatic islet hormone production, this raises intriguing questions as to whether *Rbp4* expressed in K-cells in the small intestine has actions which influence systemic metabolism and metabolic regulation through direct effects on pancreatic hormone production. Along the same lines, more than 40 years ago RBP4 expression was immunolocalized to the glucagon-secreting alpha-cells present in pancreatic islets (Kato, Kato, Blaner, Chertow, & Goodman, 1985).

What is vitamin A's role in alpha cells? Data from Gudas and colleagues (Trasino, Benoit, & Gudas, 2015) suggest that systemic vitamin A deficiency increased pancreatic alpha-cell mass and hyperglucagonemia in wild-type mice (Blaner et al., 2020). Other earlier studies from Chertow and colleagues (Chertow et al., 1994; Chertow, Driscoll, Primerano, Cordle, & Matthews, 1996), who fed a vitamin A-deficient diet to rats, showed impairments in arginine-stimulated glucagon secretion from islets isolated from the pancreases of the vitamin A-deficient animals. These authors concluded that vitamin A was needed to assure normal glucagon secretion from alpha cells (Chertow et al., 1994; Chertow et al., 1996).

However, since vitamin A-deficiency affects all tissues and cells in the body, it is unclear whether the reported effects on alpha-cells are direct primary ones or secondary to other changes.

What is RBP4's role in pancreatic alpha-cells? We have carried out preliminary studies addressing this question. Using standard protocols, we isolated primary pancreatic islets from healthy chow-fed adult male wild-type mice (Brun et al., 2015). We then treated the cultured islets with arginine to stimulate glucagon release into the culture medium. As expected from the extensive literature, we observed arginine-stimulated glucagon release from the islet cultures (Brun et al., 2015). When we assessed RBP4 concentrations in the same islet culture media samples, we observed arginine-stimulated RBP4 release into the medium. This is analogous to what we observed for the co-secretion of GIP and RBP4 from primary cultures of isolated intestinal cells (Calderon et al., 2022). Treatments that stimulate the release of the hormone also stimulate RBP4 release. Very surprisingly, when we treated the isolated islets with ROL alone and without arginine, we observed ROL-dependent stimulation of glucagon release into the medium. As can be seen in Fig. 4, treatment of the cultured islets with ATRA does not give rise to the same degree of stimulation of glucagon release as we see for ROL treatment.

Summary. Our data establish that RBP4 has a previously unanticipated role in the secretory processes resulting in GIP release from intestinal K-cells and glucagon from alpha cells. Both GIP and glucagon are hormones that

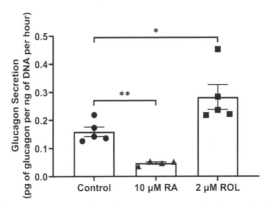

Fig. 4 Effects of treatment of cultured mouse pancreatic islets obtained from chow fed wild type mice with either 2 μM retinol (ROL) or 10 μM all-*trans*-retinoic acid (RA) on glucagon release into the medium. The cultured islets were treated for 2 h after isolation/overnight culture in control medium. Significance: * p < 0.05; ** p < 0.01.

regulate whole body metabolism and that can directly act in metabolic disease causation. These effects do not seem to significantly involve ATRA or other retinoid effects on transcription. Rather, they all directly involve RBP4 and possibly one of its retinoid or fatty acid ligands. We suggest that when considering the associations of RBP4 with metabolic disease one needs to consider novel indirect mechanisms that may still not be fully understood.

Acknowledgments

The authors wish to acknowledge grant support from National Institutes of Health grants R01DK068437 (WSB and MG), R01DK122071 (WSB, MG, JP), and R01EY023948 (MG). This study was also supported by the Columbia University Digestive and Liver Disease Research Center funded by NIH grant 5P30DK132710.

We thank Dr. Jessica M. Snyder (University of Washington, Comparative Medicine and Histology and Imaging Core) for help with the immunohistochemistry.

References

Achatz, S., Jarasch, A., & Skerra, A. (2022). *Journal of Structural Biology: X, 6*, 100054.
Ahlman, H., & Nilsson, O. (2001). *Annals of Oncology, 12*, S63–S68.
Akerstrom, B., Flower, D. R., & Salier, J. P. (2000). *Biochimica et Biophysica Acta, 1482*, 1–8.
Berni, R., Clerici, M., Malpeli, G., Cleris, L., & Formelli, F. (1993). *The FASEB Journal, 7*, 1179–1184.
Berni, R., & Formelli, F. (1992). *FEBS Letters, 308*, 43–45.
Biesalski, H. K., Frank, J., Beck, S. C., Heinrich, F., Illek, B., Reifen, R., ... Zrenner, E. (1999). *The American Journal of Clinical Nutrition, 69*, 931–936.
Blaner, W. S. (2019). *Pharmacology & Therapeutics, 197*, 153–178.
Blaner, W. S., Brun, P. J., Calderon, R. M., & Golczak, M. (2020). *Critical Reviews in Biochemistry and Molecular Biology, 55*, 197–218.
Blomhoff, R., Green, M. H., Berg, T., & Norum, K. R. (1990). *Science (New York, N. Y.), 250*, 399–404.
Boerman, M. H., & Napoli, J. L. (1991). *The Journal of Biological Chemistry, 266*, 22273–22278.
Boerman, M. H., & Napoli, J. L. (1996). *The Journal of Biological Chemistry, 271*, 5610–5616.
Brun, P. J., Grijalva, A., Rausch, R., Watson, E., Yuen, J. J., Das, B. C., ... Blaner, W. S. (2015). *The FASEB Journal, 29*, 671–683.
Buxbaum, J. N., Roberts, A. J., Adame, A., & Masliah, E. (2014). *Neuroscience, 275*, 352–364.
Calderon, R. M., Smith, C. A., Miedzybrodzka, E. L., Silvaroli, J. A., Golczak, M., Gribble, F. M., ... Blaner, W. S. (2022). *Endocrinology, 163*.
Cehajic-Kapetanovic, J., Jasani, K. M., Shanks, M., Clouston, P., & MacLaren, R. E. (2020). *Ophthalmic Genetics, 41*, 288–292.
Chen, S., Pan, Z., Liu, M., Guo, L., Jiang, X., & He, G. (2024). *Journal of Medicinal Chemistry, 67*, 5144–5167.
Chertow, B. S., Driscoll, H. K., Blaner, W. S., Meda, P., Cordle, M. B., & Matthews, K. A. (1994). *Pancreas, 9*, 475–484.
Chertow, B. S., Driscoll, H. K., Primerano, D. A., Cordle, M. B., & Matthews, K. A. (1996). *Metabolism: Clinical and Experimental, 45*, 300–305.
Chou, C. M., Nelson, C., Tarle, S. A., Pribila, J. T., Bardakjian, T., Woods, S., ... Glaser, T. (2015). *Cell, 161*, 634–646.

Cioffi, C. L., Dobri, N., Freeman, E. E., Conlon, M. P., Chen, P., Stafford, D. G., ... Petrukhin, K. (2014). *Journal of Medicinal Chemistry, 57*, 7731-7757.
Cioffi, C. L., Muthuraman, P., Raja, A., Varadi, A., Racz, B., & Petrukhin, K. (2020). *Journal of Medicinal Chemistry, 63*, 11054-11084.
Cioffi, C. L., Racz, B., Varadi, A., Freeman, E. E., Conlon, M. P., Chen, P., ... Petrukhin, K. (2019). *Journal of Medicinal Chemistry, 62*, 5470-5500.
Codoner-Franch, P., Carrasco-Luna, J., Allepuz, P., Codoner-Alejos, A., & Guillem, V. (2016). *Pediatric Diabetes, 17*, 576-583.
Coskun, T., Sloop, K. W., Loghin, C., Alsina-Fernandez, J., Urva, S., Bokvist, K. B., ... Haupt, A. (2018). *Molecular Metabolism, 18*, 3-14.
Cowan, S. W., Newcomer, M. E., & Jones, T. A. (1990). *Proteins, 8*, 44-61.
Crow, J. A., & Ong, D. E. (1985). *Federation Proceedings, 44*, 773-773.
Cuddihey, H., MacNaughton, W. K., & Sharkey, K. A. (2022). *Cellular and Molecular Gastroenterology and Hepatology, 14*, 947-963.
Cukras, C., Gaasterland, T., Lee, P., Gudiseva, H. V., Chavali, V. R., Pullakhandam, R., ... Ayyagari, R. (2012). *PLoS One, 7*, e50205.
Dahl, D., Onishi, Y., Norwood, P., Huh, R., Bray, R., Patel, H., & Rodriguez, A. (2022). *JAMA: The Journal of the American Medical Association, 327*, 534-545.
Demmer, L. A., Levin, M. S., Elovson, J., Reuben, M. A., Lusis, A. J., & Gordon, J. I. (1986). *Proceedings of the National Academy of Sciences of the United States of America, 83*, 8102-8106.
E, X., Zhang, L., Lu, J., Tso, P., Blaner, W. S., Levin, M. S., & Li, E. (2002). *The Journal of Biological Chemistry, 277*, 36617-36623.
Eckel, R. H., Fujimoto, W. Y., & Brunzell, J. D. (1979). *Diabetes, 28*, 1141-1142.
Flatt, P. R., Bailey, C. J., Kwasowski, P., Swanston-Flatt, S. K., & Marks, V. (1983). *Diabetes, 32*(5), 433.
Flower, D. R. (1996). *The Biochemical Journal, 318*(Pt 1), 1-14.
Flower, D. R., North, A. C., & Sansom, C. E. (2000). *Biochimica et Biophysica Acta, 1482*, 9-24.
Folli, C., Viglione, S., Busconi, M., & Berni, R. (2005). *Biochemical and Biophysical Research Communications, 336*, 1017-1022.
Frias, J. P., Davies, M. J., Rosenstock, J., Perez Manghi, F. C., Fernandez Lando, L., Bergman, B. K., ... Brown, K. (2021). and Investigators, S. *The New England Journal of Medicine, 385*, 503-515.
Ghyselinck, N. B., Bavik, C., Sapin, V., Mark, M., Bonnier, D., Hindelang, C., ... Chambon, P. (1999). *The EMBO Journal, 18*, 4903-4914.
Graham, T. E., Yang, Q., Bluher, M., Hammarstedt, A., Ciaraldi, T. P., Henry, R. R., ... Kahn, B. B. (2006). *The New England Journal of Medicine, 354*, 2552-2563.
Gribble, F. M., & Reimann, F. (2016). *Annual Review of Physiology, 78*, 277-299.
Gribble, F. M., & Reimann, F. (2019). *Nature Reviews Endocrinology, 15*, 226-237.
Guccio, N., Gribble, F. M., & Reimann, F. (2022). *Annual Review of Nutrition, 42*, 21-44.
Heller, J. (1975). *The Journal of Biological Chemistry, 250*, 6549-6554.
Herr, F. M., & Ong, D. E. (1992). *Biochemistry, 31*, 6748-6755.
Holst, J. J. (2022). *Digestive Diseases and Sciences, 67*, 2716-2720.
Huang, J., Possin, D. E., & Saari, J. C. (2009). *Molecular Vision, 15*, 223-234.
Huang, X., Liu, J., Peng, G., Lu, M., Zhou, Z., Jiang, N., & Yan, Z. (2024). *The Journal of Endocrinology, 262*.
Jastreboff, A. M., Aronne, L. J., Ahmad, N. N., Wharton, S., Connery, L., Alves, B., ... Investigators, S. (2022). *The New England Journal of Medicine, 387*, 205-216.
Jastreboff, A. M., Aronne, L. J., & Stefanski, A. (2022). *The New England Journal of Medicine, 387*, 1434-1435.
Kanai, M., Raz, A., & Goodman, D. S. (1968). *The Journal of Clinical Investigation, 47*, 2025-2044.

Kane, M. A., Bright, F. V., & Napoli, J. L. (2011). *Biochimica et Biophysica Acta, 1810*, 514–518.

Kato, M., Kato, K., Blaner, W. S., Chertow, B. S., & Goodman, D. S. (1985). *Proceedings of the National Academy of Sciences of the United States of America, 82*, 2488–2492.

Kawaguchi, R., Yu, J., Honda, J., Hu, J., Whitelegge, J., Ping, P., ... Sun, H. (2007). *Science (New York, N. Y.), 315*, 820–825.

Khan, K. N., Carss, K., Raymond, F. L., Islam, F., Nihr BioResource-Rare Diseases, C., Moore, A. T., ... Arno, G. (2017). *Ophthalmic Genetics, 38*, 465–466.

Kim, S. J., Nian, C., & McIntosh, C. H. (2007). *The Journal of Biological Chemistry, 282*, 8557–8567.

Kim, S. J., Nian, C., & McIntosh, C. H. (2010). *Journal of Lipid Research, 51*, 3145–3157.

Kloting, N., Graham, T. E., Berndt, J., Kralisch, S., Kovacs, P., Wason, C. J., ... Kahn, B. B. (2007). *Cell Metabolism, 6*, 79–87.

Kovacs, P., Geyer, M., Berndt, J., Kloting, N., Graham, T. E., Bottcher, Y., ... Stumvoll, M. (2007). *Diabetes, 56*, 3095–3100.

Kuhre, R. E., Deacon, C. F., Holst, J. J., & Petersen, N. (2021). *Frontiers in Endocrinology (Lausanne), 12*, 694284.

Kumar, N., & D'Alessio, D. A. (2022). *The Journal of Clinical Endocrinology and Metabolism, 107*, 2148–2153.

Lee, S. A., Yang, K. J. Z., Brun, P. J., Silvaroli, J. A., Yuen, J. J., Shmarakov, I., ... Blaner, W. S. (2020). *Science Advances, 6*, eaay8937.

Lee, S. A., Yuen, J. J., Jiang, H., Kahn, B. B., & Blaner, W. S. (2016). *Hepatology (Baltimore, Md.), 64*, 1534–1546.

Lewis, J. E., Miedzybrodzka, E. L., Foreman, R. E., Woodward, O. R. M., Kay, R. G., Goldspink, D. A., ... Reimann, F. (2020). *Diabetologia, 63*, 1396–1407.

Li, E., Demmer, L. A., Sweetser, D. A., Ong, D. E., & Gordon, J. I. (1986). *Proceedings of the National Academy of Sciences of the United States of America, 83*, 5779–5783.

Li, E., & Norris, A. W. (1996). *Annual Review of Nutrition, 16*, 205–234.

Li, E., & Tso, P. (2003). *Current Opinion in Lipidology, 14*, 241–247.

McIlroy, G. D., Delibegovic, M., Owen, C., Stoney, P. N., Shearer, K. D., McCaffery, P. J., & Mody, N. (2013). *Diabetes, 62*, 825–836.

Moraes-Vieira, P. M., Yore, M. M., Sontheimer-Phelps, A., Castoldi, A., Norseen, J., Aryal, P., ... Kahn, B. B. (2020). *Proceedings of the National Academy of Sciences of the United States of America, 117*, 31309–31318.

Motani, A., Wang, Z., Conn, M., Siegler, K., Zhang, Y., Liu, Q., ... Coward, P. (2009). *The Journal of Biological Chemistry, 284*, 7673–7680.

Napoli, J. L. (2016). *Subcellular Biochemistry, 81*, 21–76.

Napoli, J. L. (2017). *Pharmacology & Therapeutics, 173*, 19–33.

Napoli, J. L., Posch, K. P., Fiorella, P. D., & Boerman, M. H. (1991). *Biomedicine & Pharmacotherapy = Biomedecine & Pharmacotherapie, 45*, 131–143.

Naylor, H. M., & Newcomer, M. E. (1999). *Biochemistry, 38*, 2647–2653.

Newcomer, M. E., Jones, T. A., Aqvist, J., Sundelin, J., Eriksson, U., Rask, L., & Peterson, P. A. (1984). *The EMBO Journal, 3*, 1451–1454.

Newcomer, M. E., & Ong, D. E. (2000). *Biochimica et Biophysica Acta, 1482*, 57–64.

Ong, D. E. (1984). *The Journal of Biological Chemistry, 259*, 1476–1482.

Ong, D. E. (1994). *Nutrition Reviews, 52*, S24–S31.

Orskov, C., Wettergren, A., & Holst, J. J. (1996). *Scandinavian Journal of Gastroenterology, 31*, 665–670.

Perduca, M., Nicolis, S., Mannucci, B., Galliano, M., & Monaco, H. L. (2018). *Biochimica et Biophysica Acta (BBA—Molecular and Cell Biology of Lipids), 1863*, 458–466.

Pervaiz, S., & Brew, K. (1987). *The FASEB Journal, 1*, 209–214.

Plau, J., Golczak, M., Paik, J., Calderon, R. M., & Blaner, W. S. (2022). *Biochimica et Biophysica Acta BBA—Molecular and Cell Biology of Lipids, 1867*, 159179.
Plau, J., Morgan, C. E., Fedorov, Y., Banerjee, S., Adams, D. J., Blaner, W. S., ... Golczak, M. (2023). *ACS Chemical Biology, 18*, 2309–2323.
Porter, S. B., Fraker, L. D., Chytil, F., & Ong, D. E. (1983). *Proceedings of the National Academy of Sciences of the United States of America, 80*, 6586–6590.
Quadro, L., Blaner, W. S., Salchow, D. J., Vogel, S., Piantedosi, R., Gouras, P., ... Gottesman, M. E. (1999). *The EMBO Journal, 18*, 4633–4644.
Rakotoarivelo, V., Sihag, J., & Flamand, N. (2021). *Cells, 10*.
Roque-Bravo, R., Silva, R. S., Malheiro, R. F., Carmo, H., Carvalho, F., da Silva, D. D., & Silva, J. P. (2023). *Annual Review of Pharmacology and Toxicology, 63*, 187–209.
Ruiz de Azua, I., & Lutz, B. (2019). *Cellular and Molecular Life Sciences: CMLS, 76*, 1341–1363.
Saari, J. C., Nawrot, M., Garwin, G. G., Kennedy, M. J., Hurley, J. B., Ghyselinck, N. B., & Chambon, P. (2002). *Investigative Ophthalmology & Visual Science, 43*, 1730–1735.
Seeliger, M. W., Biesalski, H. K., Wissinger, B., Gollnick, H., Gielen, S., Frank, J., ... Zrenner, E. (1999). *Investigative Ophthalmology & Visual Science, 40*, 3–11.
Shen, J., Shi, D., Suzuki, T., Xia, Z., Zhang, H., Araki, K., ... Li, Z. (2016). *Laboratory Investigation; A Journal of Technical Methods and Pathology, 96*, 680–691.
Silvaroli, J. A., Arne, J. M., Chelstowska, S., Kiser, P. D., Banerjee, S., & Golczak, M. (2016). *The Journal of Biological Chemistry, 291*, 8528–8540.
Silvaroli, J. A., Plau, J., Adams, C. H., Banerjee, S., Widjaja-Adhi, M. A. K., Blaner, W. S., & Golczak, M. (2021). *Journal of Lipid Research, 62*, 100054.
Silvaroli, J. A., Widjaja-Adhi, M. A. K., Trischman, T., Chelstowska, S., Horwitz, S., Banerjee, S., ... Golczak, M. (2019). *ACS Chemical Biology, 14*, 434–448.
Skerra, A. (2000). *Biochimica et Biophysica Acta, 1482*, 337–350.
Steinhoff, J. S., Lass, A., & Schupp, M. (2022). *Nutrients, 14*.
Tarter, M., Capaldi, S., Carrizo, M. E., Ambrosi, E., Perduca, M., & Monaco, H. L. (2008). *Proteins, 70*, 1626–1630.
Trasino, S. E., Benoit, Y. D., & Gudas, L. J. (2015). *The Journal of Biological Chemistry, 290*, 1456–1473.
Tsutsumi, C., Okuno, M., Tannous, L., Piantedosi, R., Allan, M., Goodman, D. S., & Blaner, W. S. (1992). *The Journal of Biological Chemistry, 267*, 1805–1810.
Urade, Y., & Hayaishi, O. (2000). *Biochimica et Biophysica Acta, 1482*, 259–271.
van Bennekum, A. M., Wei, S., Gamble, M. V., Vogel, S., Piantedosi, R., Gottesman, M., ... Blaner, W. S. (2001). *The Journal of Biological Chemistry, 276*, 1107–1113.
Wan, K., Zhao, J., Deng, Y., Chen, X., Zhang, Q., Zeng, Z., ... Chen, Y. (2014). *International Journal of Molecular Sciences, 15*, 22309–22319.
Williams, D. M., Staff, M., Bain, S. C., & Min, T. (2022). *touchREV Endocrinol, 18*, 43–48.
Wongsiriroj, N., Piantedosi, R., Palczewski, K., Goldberg, I. J., Johnston, T. P., Li, E., & Blaner, W. S. (2008). *The Journal of Biological Chemistry, 283*, 13510–13519.
Yang, Q., Graham, T. E., Mody, N., Preitner, F., Peroni, O. D., Zabolotny, J. M., ... Kahn, B. B. (2005). *Nature, 436*, 356–362.
Zanotti, G., Folli, C., Cendron, L., Alfieri, B., Nishida, S. K., Gliubich, F., ... Berni, R. (2008). *The FEBS Journal, 275*, 5841–5854.
Zanotti, G., Marcello, M., Malpeli, G., Folli, C., Sartori, G., & Berni, R. (1994). *The Journal of Biological Chemistry, 269*, 29613–29620.

CHAPTER FIVE

Rethinking retinoic acid self-regulation: A signaling robustness network approach

Abraham Fainsod[a,*] and Rajanikanth Vadigepalli[b,*]
[a]Department of Developmental Biology and Cancer Research, Institute for Medical Research Israel-Canada, Faculty of Medicine, The Hebrew University of Jerusalem, Jerusalem, Israel
[b]Daniel Baugh Institute for Functional Genomics and Computational Biology, Department of Pathology and Genomic Medicine, Sidney Kimmel Medical College, Thomas Jefferson University, Philadelphia, PA, United States
*Corresponding authors. e-mail address: abraham.fainsod@mail.huji.ac.il;
Rajanikanth.Vadigepalli@jefferson.edu

Contents

1. Retinoic acid is a major regulator of developmental processes and tissue homeostasis	114
2. Environmental dependence of ATRA biosynthesis	116
3. Polymorphisms in ATRA network components and the induction of disease	119
4. Robustness of the ATRA metabolic and signaling network	120
5. The ATRA robustness response exhibits Pareto optimality	122
Acknowledgments	128
Funding	128
References	129

Abstract

All-*trans* retinoic acid (ATRA) signaling is a major pathway regulating numerous differentiation, proliferation, and patterning processes throughout life. ATRA biosynthesis depends on the nutritional availability of vitamin A and other retinoids and carotenoids, while it is sensitive to dietary and environmental toxicants. This nutritional and environmental influence requires a robustness response that constantly fine-tunes the ATRA metabolism to maintain a context-specific, physiological range of signaling levels. The ATRA metabolic and signaling network is characterized by the existence of multiple enzymes, transcription factors, and binding proteins capable of performing the same activity. The partial spatio-temporal expression overlap of these enzymes and proteins yields different network compositions in the cells and tissues where this pathway is active. Genetic polymorphisms affecting the activity of individual network components further impact the network composition variability and the self-regulatory feedback response to ATRA fluctuations. Experiments directly challenging the robustness response uncovered a Pareto optimality in the ATRA network, such that some genetic backgrounds efficiently deal with excess ATRA but are very limited in their robustness response to reduced ATRA and vice versa. We discuss a network-focused framework to describe the robustness response and the Pareto optimality of the ATRA metabolic and signaling network.

1. Retinoic acid is a major regulator of developmental processes and tissue homeostasis

The precise regulation of developmental processes followed by tissue homeostasis is essential for forming and maintaining complex and functional body plans. In vertebrates, one of the main signaling pathways controlling numerous of these processes is the all-*trans* retinoic acid (ATRA) signaling pathway. ATRA is a major, biologically active derivative of vitamin A (retinol, ROL) (Kedishvili, 2016; Metzler & Sandell, 2016; Shabtai & Fainsod, 2018). ATRA signaling is initially activated during early gastrulation in vertebrate embryogenesis and continues to perform its gene-regulatory functions throughout life (Asson-Batres, 2020; Berenguer & Duester, 2022; Gur, Bendelac-Kapon, Shabtai, Pillemer, & Fainsod, 2022). During embryogenesis, ATRA regulates numerous key processes like pattern formation, cell differentiation, cell proliferation, and organ development (Dubey, Rose, Jones, & Saint-Jeannet, 2018; Ghyselinck & Duester, 2019; Gur, Edri, Moody, & Fainsod, 2022; Parihar et al., 2021; Petrelli et al., 2023; Summerbell & Maden, 1990). Many of these regulatory roles will continue postnatally, and new ones will be added for the adult maintenance of tissue homeostasis (Blaner, 2019; Chai et al., 2021; DiKun & Gudas, 2023; Gao et al., 2013; Owusu & Ross, 2016; Timoneda et al., 2018; Wu, Adams et al., 2019; Yang et al., 2018). Multiple disease conditions have been attributed to abnormal ATRA signaling, including developmental syndromes (Clagett-Dame & Knutson, 2011; Coberly, Lammer, & Alashari, 1996; Collins & Mao, 1999; Shenefelt, 1972), several types of cancer (Brown, 2023; Marcato et al., 2015; Penny et al., 2016; Surakhy et al., 2020; Wu et al., 2023; Yan et al., 2018), and even some mental disabilities (Fainsod, Abbou, Bendelac-Kapon, Edri, & Pillemer, 2022; Hu, van Dam, Wang, Lucassen, & Zhou, 2020; Moreno-Ramos, Olivares, Haider, de Autismo, & Lattig, 2015; Stoney & McCaffery, 2016).

The processes regulated by ATRA signaling are very sensitive to the level of ATRA. Numerous genes exhibit concentration-dependent responses to ATRA, for instance, *Hox* genes and the early embryonic organizer genes (Gur, Bendelac-Kapon et al., 2022; Neijts & Deschamps, 2017; Nolte, De Kumar, & Krumlauf, 2019). Excess or impaired ATRA signaling induces a wide array of embryonic malformations (Corcoran, So, & Maden, 2002; Durston et al., 1989; Kessel & Gruss, 1991; Marshall et al., 1992; Papalopulu et al., 1991). These malformations can be phenocopied by manipulating specific ATRA network enzymes (Hollemann, Chen, Grunz, & Pieler, 1998; Kam et al., 2013; Kot-Leibovich & Fainsod, 2009). Severe ATRA deficiency by mutating the gene encoding retinaldehyde dehydrogenase 2 (ALDH1A2), the central ATRA-producing enzyme in the early embryo, results in lethality, supporting its significant role during early embryogenesis (Begemann, Schilling, Rauch, Geisler, & Ingham, 2001; Grandel et al., 2002; Niederreither, Subbarayan, Dollé, & Chambon, 1999). These observations and the role of ATRA signaling on the patterning of the anterior-posterior axis led efforts to identify ATRA signaling gradients in support of a role as a morphogen (Dubey et al., 2018; Schilling, Nie, & Lander, 2012; Schilling, Sosnik, & Nie, 2016; Shimozono, Iimura, Kitaguchi, Higashijima, & Miyawaki, 2013). The concentration-dependent gene responses and the creation of morphogen signaling gradients require tight regulation of ATRA metabolism to achieve its important regulatory roles and function despite changing sources of retinoids such as vitamin A or carotenoids (ß-carotene) for ATRA production (Asson-Batres, 2020; Blaner, 2019; de Souza Mesquita, Mennitti, de Rosso, & Pisani, 2021).

The regulation of gene expression by ATRA involves nuclear receptors and, in particular, the retinoic acid receptor (RAR) subfamily (di Masi et al., 2015; Evans & Mangelsdorf, 2014) that heterodimerize with the retinoid X nuclear receptors (RXRs) (Fattori et al., 2015; Levin et al., 1992). The absence or presence of ATRA that binds as a ligand to the RAR/RXR heterodimer determines its activity. The RAR/RXR heterodimer binds to the ATRA-responsive elements (RAREs) in ATRA-regulated target genes (le Maire et al., 2010; Mangelsdorf, Ong, Dyck, & Evans, 1990; Umesono, Murakami, Thompson, & Evans, 1991; Yu et al., 1991). In the absence of ATRA, the RAR/RXR heterodimer interacts with co-repressors such as NCOR1 or NCOR2 to promote gene silencing or down-regulation (Chen, Umesono, & Evans, 1996; Janesick et al., 2014; Koide, Downes, Chandraratna, Blumberg, & Umesono, 2001; Linney, Donerly, Mackey, & Dobbs-McAuliffe, 2011; Weston, Blumberg, & Underhill, 2003). In the presence of ATRA, conformational changes occur, promoting co-repressor displacement and co-activator binding followed by up-regulation of target gene expression (le Maire & Bourguet, 2014; Liu, Hu, & Rosenfeld, 2014). There are some exceptions to this scenario, where the RAR/RXR heterodimer complex binds the NCOR co-repressors in the presence of ATRA (Kumar, Cunningham, & Duester, 2016) or binds additional co-repressors (LCOR) (Fernandes et al., 2003) or proteins that prevent ATRA binding to the heterodimer (Epping et al., 2005; Gurevich & Aneskievich, 2009; Miro Estruch et al., 2017).

2. Environmental dependence of ATRA biosynthesis

Classically, ATRA biosynthesis involves two sequential oxidation reactions, first from ROL to retinaldehyde (RAL) and subsequently from RAL to ATRA (Fig. 1) (Gudas, 2022; Kedishvili, 2016; Metzler & Sandell, 2016). ATRA can also be produced from ß-carotene or storage forms, i.e., retinyl esters (Blaner, 2019; Napoli, 2020; Sirbu, Chiş, & Moise, 2020). The oxidation of RAL is performed by enzymes of the aldehyde dehydrogenase (ALDH) family. The retinaldehyde dehydrogenase (ALDH1A) activity is a unidirectional reaction that produces ATRA (Mic, Molotkov, Fan, Cuenca, & Duester, 2000; Molotkov & Duester, 2003; Shabtai, Jubran, Nassar, Hirschberg, & Fainsod, 2016; Zhao et al., 1996). The

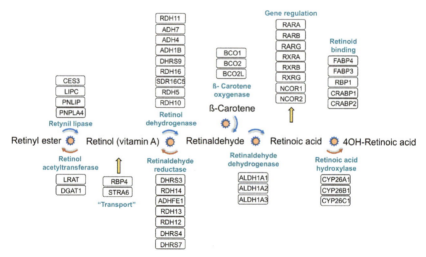

Fig. 1 The ATRA metabolic and signaling network during early embryogenesis. The general composition of the ATRA metabolic and signaling network highlights multiple enzymes that can independently catalyze any given reaction in vitamin A metabolism. There is transcriptomic evidence of embryonic expression for all the genes in the network, albeit in different combinations at multiple developmental stages and spatial domains. The enzymes were placed in the network schematic based on their preferred enzymatic activity.

ATRA produced will participate in gene regulation; if in excess, it will be degraded. The CYP26 enzyme family terminates the ATRA signaling activity by chemically modifying the ATRA and targeting it for degradation (Marikar et al., 1998; Niederreither, Abu-Abed et al., 2002; White et al., 1997). These observations suggest that the production of RAL and its availability as a substrate for enzymes with ALDH1A activity is a major node for regulating ATRA signaling (Fig. 1) (Belyaeva, Adams, Popov, & Kedishvili, 2019).

Most ATRA metabolic reactions leading to the production of RAL are bidirectional, allowing for the regulation of RAL levels. Even RAL produced from ß-carotene can be reduced back to ROL by RAL reductases to limit the substrate availability for ALDH1As and control ATRA production (Fig. 1) (Adams et al., 2021; Billings et al., 2013; Feng, Hernandez, Waxman, Yelon, & Moens, 2010; Porté et al., 2013; Shabtai & Fainsod, 2018). ATRA signaling is further controlled by ROL, RAL, and ATRA-binding proteins that limit or chaperone ATRA or its precursor(s) in their translocation within the cell and across cells (Napoli, 2017). When a significant retinoid fluctuation occurs, a response is triggered involving the ATRA metabolic enzymes already present and extensive network-wide transcriptional changes elicited by the ATRA-dependent autoregulatory feedback regulation of genes encoding ATRA metabolic and signaling network components. Evidence of this response to excess ATRA was obtained when embryos were shown to efficiently metabolize excess retinoids to restore physiological levels (Kraft & Juchau, 1992). To elucidate this response, multiple studies have shown that following an increase in ATRA levels, genes encoding enzymes that reduce ATRA levels (DHRS3 and CYP26A1) are up-regulated, and enzymes that produce ATRA (RDH10 and ALDH1A2) are down-regulated (Chen, Pollet, Niehrs, & Pieler, 2001; de Roos et al., 1999; Hollemann et al., 1998; Kam et al., 2013; Roberts, 2020; Shabtai, Bendelac, Jubran, Hirschberg, & Fainsod, 2018; Strate, Min, Iliev, & Pera, 2009; Zhong et al., 2019). ATRA signaling autoregulation was also speculated in studies with reduced ATRA, where the developmental malformations induced were relatively mild, suggesting the presence of compensatory mechanisms (Blumberg et al., 1997; Gur, Bendelac-Kapon et al., 2022; Hollemann et al., 1998; Janesick et al., 2014; Koide et al., 2001; Kot-Leibovich & Fainsod, 2009; Shabtai et al., 2018; Sharpe & Goldstone, 1997; Shukrun, Shabtai, Pillemer, & Fainsod, 2019). ATRA levels are tightly regulated throughout life, relying on quantitative, spatial, and temporal regulation of the expression and activity of ATRA biosynthetic or metabolizing enzymes, and signaling factors (Blaner, 2019; Ghyselinck & Duester, 2019; Kedishvili, 2016).

Vitamin A Deficiency syndrome (VAD) arises due to nutritional deficits, and it is the leading cause of preventable child blindness worldwide (Bastos Maia et al., 2019; Wiseman, Bar-El Dadon, & Reifen, 2017). VAD can be rescued by vitamin A supplementation with the caveat that excess ROL can be extremely teratogenic (Clagett-Dame & Knutson, 2011; Coberly et al., 1996; Collins & Mao, 1999; Shenefelt, 1972). Multiple developmental syndromes and adult diseases have also been linked to reduced ATRA signaling, including DiGeorge/VeloCardioFacial syndrome (DG/VCF) (Coberly et al., 1996; See, Kaiser, White, & Clagett-Dame, 2008), Fetal Alcohol Spectrum Disorder (FASD) (Fainsod et al., 2022), Congenital Heart Disease (CHD) (Pavan et al., 2009), neural tube defects (NTDs) (Edri, Cohen, Shabtai, & Fainsod, 2023; Li et al., 2018; Pangilinan et al., 2014; Urbizu et al., 2013), and multiple types of cancer (Kim et al., 2005; Vermot, Niederreither, Garnier, Chambon, & Dollé, 2003). The ATRA signaling pathway can also be hampered by toxicants in food sources or the environment (Chen & Reese, 2013; Shmarakov, 2015). The induction of Fetal Alcohol Spectrum Disorder (FASD) by ingestion of alcohol (ethanol) during pregnancy has been shown to involve a reduction in ATRA biosynthesis through its clearance intermediate, acetaldehyde (Fainsod et al., 2022). Another ATRA inhibitor in food is citral, a natural chemical contributing to the citrus aroma and taste in fruits, which is a known inhibitor of ALDH1A activity (Kikonyogo, Abriola, Dryjanski, & Pietruszko, 1999; Laughton, Skakum, & Levi, 1962). In addition, multiple environmental toxicants have been described that disrupt the activity of the ATRA signaling pathway (Chen & Reese, 2013; Estrada-Ortiz et al., 2022; Lie et al., 2019; Paganelli, Gnazzo, Acosta, López, & Carrasco, 2010; Papis, Bernardini, Gornati, Menegola, & Prati, 2007).

As nutritional sources of ATRA precursors continuously change with diet and ATRA-inhibitory toxicant exposures can be sporadic or chronic, the ATRA metabolic and signaling pathway has a built-in self-regulatory mechanism to maintain normal ATRA signaling levels (Reijntjes, Blentic, Gale, & Maden, 2005; Sosnik et al., 2016). This complex self-regulatory response involves epigenetic, transcriptomic, proteomic, and metabolomic changes aimed at restoring normal ATRA signaling levels (Abed, Foucher, Bernard, Tancrède-Bohin, & Cavusoglu, 2023; Berenguer, Meyer, Yin, & Duester, 2020; Bohn et al., 2023; Jiao, Li, & Chiu, 2012; Rossetti & Sacchi, 2020;

Samrani et al., 2023; Wahl, Wyatt, Turner, & Dickinson, 2018). Most of our understanding of the self-regulatory response of the ATRA metabolic and signaling network comes from the analysis of individual genes and few network-wide studies (Blaner, 2019; D'Aniello, Rydeen, Anderson, Mandal, & Waxman, 2013; Parihar et al., 2021; Rydeen et al., 2015).

3. Polymorphisms in ATRA network components and the induction of disease

The response to fluctuations in retinoid and carotenoid availability or exposure to nutritional or environmental ATRA-affecting toxicants is further complicated by naturally occurring genetic polymorphisms in ATRA metabolic and signaling network components. Genetic linkage studies of multiple conditions and diseases have identified the possible involvement of single nucleotide or copy number variants in genes encoding ATRA network components. Several types of adult and children tumors (Mazul et al., 2019; Niu et al., 2023; Pérez-Sayáns et al., 2023; Surakhy et al., 2020), osteoarthritis (Khosasih et al., 2023; Zhu et al., 2022), VAD susceptibility (Suzuki & Tomita, 2022), skeletal problems (Nilsson et al., 2016), Crohn's disease (Fransén et al., 2013), and neural tube defects (Deak et al., 2005; Li et al., 2018; Pangilinan et al., 2014; Rat et al., 2006; Tran et al., 2011; Wolujewicz et al., 2021; Zou et al., 2020), among others, have been linked to polymorphisms in ATRA network components (Tran et al., 2011). However, the relatively low incidence of developmental syndromes attributed to polymorphisms in ATRA network component might be due to spontaneous abortions or medically terminated pregnancies due to severe malformations, and only those mild enough that proceed to full-term are available for further study (Abdelmaksoud, Wollina, Lotti, & Temiz, 2023; Altıntaş Aykan & Ergün, 2020; Cha et al., 2022).

These genetic variants (single nucleotide or copy number) can map to regulatory elements, resulting in changes in gene expression levels or affect the spatial or temporal expression pattern of the gene in question. Alternatively, the polymorphism could affect the activity or abundance of the encoded protein. Some of the missense, frameshift, or regulatory variants can result in reduced activity. Alternatively, missense or regulatory polymorphisms could have the opposite effect and result in overexpression or enhanced activity, but for many of them, the outcome is not known (D'Aniello & Waxman, 2015; Guida et al., 2021; Khosasih et al., 2023; Lee et al., 2017; Li et al., 2018;

Merino et al., 2019; Zou et al., 2020). In the ATRA metabolic and signaling network, where some biosynthetic steps can be reversed or regulated, the increase or decrease in the expression and activity of a specific component can either promote or hamper ATRA signaling. Commonly observed disease-linked variants involve members of the CYP26 family of cytochrome P450 that target the ATRA for degradation (Khosasih et al., 2023; Niu et al., 2023; Rat et al., 2006). Increased CYP26 activity would reduce ATRA signaling, while decreased activity would allow higher signaling levels and the spreading of the signal from the site of ATRA production. This and multiple other scenarios, depending on the affected enzyme or gene in the network, will affect the dynamic response of the ATRA metabolic and signaling network to fluctuations in retinoid or carotenoid availability or exposure to ATRA-inhibiting toxicants.

4. Robustness of the ATRA metabolic and signaling network

Regulatory pathway consistency by adapting and maintaining normal signaling levels despite environmental changes, insults, and genetic polymorphisms is termed robustness (Eldar, Shilo, & Barkai, 2004; Nijhout, Best, & Reed, 2019). Multiple reports support the activation of an ATRA-driven self-regulatory mechanism, i.e., robustness response, to restore normal signaling levels (D'Aniello & Waxman, 2015; Reijntjes et al., 2005; Sosnik et al., 2016). Early on, retinoid-injected rat embryos were shown to efficiently metabolize the excess retinoids to try and restore physiological levels (Kraft & Juchau, 1992). Since then, the expression of multiple ATRA network components has been shown to respond to changes in ATRA signaling levels (Chen et al., 2001; de Roos et al., 1999; Hollemann et al., 1998; Kam et al., 2013; Ribes et al., 2007; Roberts, 2020; Shabtai et al., 2018; Soref et al., 2001; Strate et al., 2009; Zhong et al., 2019). Although these transcriptional changes in response to ATRA fluctuations agree with a robustness response, most studies focused on individual ATRA metabolic and signaling network genes. Few studies have taken a more expansive view by analyzing multiple ATRA network components or performing transcriptomic analyses (Chai et al., 2021; Owusu & Ross, 2016; Parihar et al., 2021; Steinhoff, Lass, & Schupp, 2022).

Functional overlap between multiple enzymes and proteins in the ATRA network has been described taking advantage of numerous experimental genetic models and *in vitro* kinetic analysis (Abbou, Bendelac-Kapon, Sebag, & Fainsod, 2022; Adams et al., 2021; Belyaeva et al., 2019; Kedishvili, 2016; Kumar, Sandell, Trainor, Koentgen, & Duester, 2012; Metzler & Sandell, 2016; Napoli, 2020; O'Connor, Varshosaz, & Moise, 2022; Shannon, Moise, & Trainor, 2017; Wu, Adams et al., 2019; Wu, Kedishvili, & Belyaeva, 2019). These observations established groups of enzymes that preferentially perform the same reaction in the metabolism of ATRA (Fig. 1), although these enzymes are encoded by different genes, and the enzymes exhibit different kinetic characteristics and partially overlapping and dynamic spatiotemporal expression patterns (Blentic, Gale, & Maden, 2003; Chen et al., 2001; Hollemann et al., 1998; Niederreither Fraulob et al. 1999; Niederreither, Fraulob, Garnier, Chambon, & Dollé, 2002; Romand et al., 2004). It is important to consider that the structural redundancy in the ATRA network with multiple components capable of performing the same reaction (Fig. 1) and their partially overlapping spatiotemporal expression patterns will result in different network compositions. Even in the same tissue or organ, the composition can vary in different regions or at different times, as in the case of the developing retina and the heart (Lupo et al., 2005; Molotkov, Molotkova, & Duester, 2006; Perl & Waxman, 2019). These different compositions, enzymatic kinetic characteristics, and self-regulatory responses will likely result in different robustness responses. These expression and intrinsic kinetic differences between enzymes performing the same ATRA metabolic reaction will be differentially affected by the dietary availability of retinoids or carotenoids, which, in turn, will differentially affect the amount of RAL produced (Blaner et al., 2016; Fainsod & Kot-Leibovich, 2018; Ghyselinck & Duester, 2019; Kedishvili, 2016).

In the case of the ATRA metabolic and signaling network, simultaneously regulating multiple embryonic processes requires delicate fine-tuning of signal timing, localization, and strength. Then, the robustness response has evolved to overcome or at least lessen the teratogenic effects of the ATRA fluctuations in different cell-types or regions. The ATRA-dependent transcriptional regulation of ATRA network components comprises a central mechanism in the ATRA robustness response. The complexity and redundancy in the ATRA metabolic and signaling network require coordination of the individual transcriptional responses to achieve a coherent reaction dependent on the direction of the ATRA fluctuation. Several studies have described conditions where the ATRA robustness response cannot overcome the change in ATRA signaling (D'Aniello & Waxman, 2015; Lee et al., 2017; Saili et al., 2020). Upregulation of the

CYP26 enzymes is an important and common outcome of the robustness response to excess ATRA (Hollemann et al., 1998; Pennimpede et al., 2010; Roberts, 2020; Topletz et al., 2015; Zhong et al., 2019). As part of the response to excess ATRA, the expression of *Aldh1a2* is down-regulated to prevent further ATRA biosynthesis, in addition to the CYP26-dependent ATRA degradation (Chen et al., 2001; Niederreither, McCaffery, Dräger, Chambon, & Dollé, 1997). It has been suggested that an excessive self-regulatory feedback response to increased ATRA can ultimately result in reduced ATRA signaling, partly explaining the similarity in the developmental malformations induced by excess and reduced ATRA (Lee et al., 2012; Rydeen et al., 2015). These observations support the activity of an efficient robustness response that requires constant fine-tuning, whose regulatory rules remain to be elucidated.

To directly challenge the autoregulation of ATRA signaling and study the robustness response, we established an experimental system based on transient (pulse-chase) ATRA manipulations in *Xenopus* embryos (Parihar et al., 2021). In this study, embryos treated for 2 h with ATRA (excess) or with 4-di-ethyl-amino-benzaldehyde (DEAB, inhibition of ATRA biosynthesis) were studied to understand the kinetics of restoration of the normal transcriptome by RNAseq analysis. At the transcriptomic level, the transient ATRA perturbations led to limited gene expression changes over an extended period. These observations support the induction of an effective ATRA robustness response, in which the experimental and control samples could not be significantly discriminated in the principal component analysis (Parihar et al., 2021). Further analysis of the genes responding to the ATRA manipulation identified several groups of genes. Among the responsive genes, the majority (417 genes) only responded to one type of ATRA fluctuation: excess or reduction. A smaller group of genes (136) responded to both excess and reduced ATRA. Among the ATRA-responsive genes, some were down-regulated, while others were up-regulated. These observations led us to propose a complex gene regulatory response to ATRA fluctuation to achieve an efficient robustness response (Parihar et al., 2021).

5. The ATRA robustness response exhibits Pareto optimality

While the early embryo showed a relatively small number of statistically significant gene expression changes to transient ATRA perturbations, there was a wide range of gene expression responses between individual clutches. Notably, the variability from clutch to clutch in the ATRA target gene group (e.g., *Hox* genes) manifested as the differences in the extent to which these genes were down or up-regulated. In other words, the progression along a response trajectory was differentially affected clutch-to-clutch. These experiments were performed in *Xenopus laevis* embryos, which is an outbred experimental model exhibiting genetic polymorphisms much like humans (Leibovich, Moody, Klein, & Fainsod, 2022; Savova et al., 2017; Session et al., 2016). The clutch-to-clutch differences in the gene expression responses to ATRA manipulation likely depend on genetic variations that affect the ATRA network and its robustness response. Guided by this intuition, we pursued an analysis approach that first visualized the dynamic transcriptomic changes in three principal components and then fit a curve to connect all the biological repeats at all time points in ATRA-manipulated and control samples in the 3-dimensional principal component space. A principal curve was then fit to the projected gene expression data, representing the trajectory in which the system evolves over time for all clutches under the combined effects of the developmental program and the robustness response to ATRA manipulation (Fig. 2). Comparing the distance between ATRA-manipulated and control conditions along the trajectory enabled an integrated quantitative assessment of the difference of a set of multiple target gene expression changes affected by the ATRA robustness response. We derived a distance score as the sum of the normalized absolute distance between ATRA manipulated and control embryos at each time point. The clutches with closing gaps between ATRA-manipulated samples and untreated controls over time demonstrate high robustness to ATRA perturbations (Fig. 2B and C). By contrast, clutches with poor robustness responses show extensive, and in some cases even widening, separation along the trajectory over time between ATRA-manipulated and untreated control embryos (Fig. 2A and D). This multi-gene integrative approach to quantitatively measure the robustness response allows the ranking of individual clutches for the ability to rebalance the ATRA biosynthetic activities towards normalizing ATRA levels and signaling. We observed that the clutches differ in their capacity to mount effective regulatory feedback responses by adjusting the expression of multiple genes in the ATRA metabolic network, yielding variable robustness responses.

Most surprisingly, our results uncovered an asymmetry in the ATRA robustness response. The clutches that displayed a very high degree of robustness to increased ATRA levels were comparatively poor in

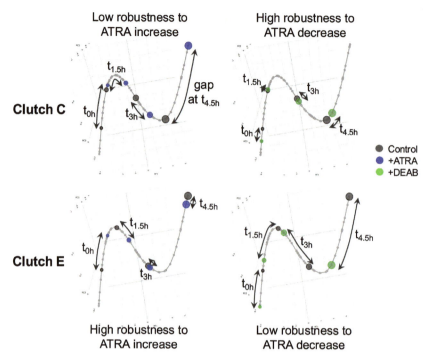

Fig. 2 Trajectory analysis of the ATRA robustness response as a function of time. Three-dimensional principal curves based on expression profiles of *Hox* genes, showing projections of the sample points on the curve for ATRA (left) and DEAB (right) treatments. In this approach, the samples of all biological repeats at all time points are projected onto the first three principal components based on the expression profiles of *Hox* genes (Parihar et al., 2021). A principal curve is fit to the projected data, which represents a trajectory along which the developmental process characterized by the expression of *Hox* genes progresses over time. The origin and direction of the curve are constrained by specifying the centroid of the control samples at the t_{0h} time point and the centroid of all the remaining samples. As each clutch is visualized separately along the trajectory, this approach allows a comparison of the distance between ATRA-manipulated conditions and controls along the trajectory over time. The arc distance between ATRA manipulations and controls (indicated by double-headed arrows) decreased over time, with variability across clutches in the time taken to close the gap. Clutches C and E are highlighted on the principal curves for their opposite patterns of robustness to increased versus decreased ATRA levels. The ranking of clutches is based on the total (absolute) normalized distance of treatment samples from the corresponding control samples for each time point, representing a multi-gene measure of the robustness of each clutch. The t_{0h} time point corresponds to the wash time after a 2-hour treatment starting at developmental stage 9.5 (late blastula). Gene expression profiling was performed at time points t_{0h}, $t_{1.5h}$, t_{3h}, and t_{4h} (Parihar et al., 2021).

counteracting a reduction in ATRA levels and vice versa (Fig. 3). Most clutches exhibited average robustness to both increased and reduced ATRA levels. Put together, these clutches were distributed along a line where increasing robustness to excess ATRA in any given clutch (i.e., genotype) comes at the cost of reduced capacity to counteract decreases in ATRA, and vice versa (Fig. 3). Such a 'trade-off' has been termed Pareto

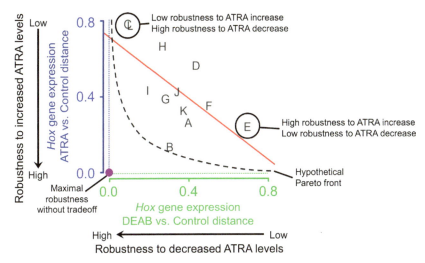

Fig. 3 **Pareto optimality is evident from a relative ranking of different polymorphic backgrounds based on robustness to ATRA manipulation.** The 12 clutches (A to L) are plotted based on the sum of net (absolute) normalized distances of ATRA manipulated vs. control samples across all time points in the *Hox* gene expression response trajectory (see Fig. 2, and Parihar et al., 2021). Clutches C and L show the highest deviation between ATRA and Control conditions along the principal curve (i.e., low robustness to ATRA addition) while showing the least deviation between DEAB and Control conditions (i.e., high robustness to ALDH1A inhibition). Clutch E shows an opposite pattern of robustness response and is located diagonally farthest apart from clutches C an L. The negative slope of the best-fit line (in red) indicates a robustness trade-off in clutch distribution such that an increase in robustness to DEAB exposure comes at the cost of decreasing robustness to ATRA addition, suggesting a Pareto optimality in response to ATRA fluctuations. Several clutches show varying levels of intermediate robustness and are distributed along the best-fit line. The dotted lines represent axes for independently modulating the robustness response, with the (0,0) point of origin representing a theoretical maximal robustness possible without any trade-off. The dashed curve (in black) indicates a hypothetical Pareto front representing a boundary for the distribution of polymorphic backgrounds with differently balanced levels of the robustness response. The "archetypal" genotypes representing the extreme differential robustness to increased vs. decreased ATRA are highlighted in circles (clutches C, L, and E).

optimality, which typically arises from a competing allocation of a constant resource across disparate variables (Pallasdies, Norton, Schleimer, & Schreiber, 2021; Shoval et al., 2012). Pareto optimality has been observed in many biological scenarios, including morphological features (Kavanagh, Evans, & Jernvall, 2007), bacterial adaptation to changing metabolic conditions (Shoval et al., 2012), the topology of neural arbors (Chandrasekhar & Navlakha, 2019), epithelial-mesenchymal plasticity in cancer (Cook & Wrana, 2022), among others. Pareto optimality arises from a multitask or multi-objective optimization of structure or function. Under these requirements, the underlying variables are constrained and hence cannot be adjusted to achieve the most optimal levels for all the objectives or tasks simultaneously. The optimal solution to a given objective or task leads to an "archetype" representing the best possible scenario when focused on a sole purpose (Shoval et al., 2012). Typically, biological systems are distributed between multiple "archetypes" such that any given system can approach one of the archetypes in its structure/function only at the cost of distancing itself from other archetypes, i.e., the Pareto trade-off. For the ATRA signaling network, the two archetypes appear to be most robust to only an excess or reduced ATRA signaling. The clutches distribute along a restricted set of phenotypes between the two extremities of robustness such that any given clutch is partly optimized for a robust response to both increased and decreased ATRA signaling. This set of partially optimal phenotypes represents a "Pareto front" that indicates the range of robustness of ATRA metabolic and signaling networks arising from varying combinations of enzymes and genetic polymorphisms affecting those components.

Multiple questions remain unresolved in understanding the mechanistic underpinnings of the robustness response, its asymmetry, and the underlying structural and regulatory constraints that result in a trade-off of intrinsic capacity to counteract perturbations to ATRA levels. It is difficult to explain this trade-off at the level of ATRA clearance. For instance, it is not readily apparent why an individual with a specific variant genotype that can efficiently up-regulate *Cyp26a1* expression in response to increased ATRA levels is comparatively inefficient at reducing CYP26A1 levels and, hence, its activity in response to ATRA reduction. Why is the enhancer strength for *Cyp26a1* induction in a competing balance with the efficiency of *Cyp26a1* repression and CYP26A1 removal? Similar puzzles arise when considering any individual enzyme in the ATRA production and the upstream biochemical fluxes leading to RAL production. The asymmetric robustness response raises an intriguing possibility of correlative constraints between enzymatic activities in the ATRA metabolic network.

One possible partial explanation for the Pareto optimality observed in the ATRA metabolic and signaling network could involve enzymatic complexes like the DHRS3/RDH10 retinoid oxidoreductase complex (ROC) (Adams et al., 2021). Its initial description showed that DHRS3 and RDH10 require each other's presence in the ROC to achieve their full enzymatic activity (Adams et al., 2014). Importantly, catalytically inactive forms of DHRS3 or RDH10 still had the activity-enhancing effect on the other enzyme. Then, a situation could come about where a genetic variant in either DHRS3 or RDH10 could drive the system to one of the robustness archetypes. In this scenario, either DHRS3 or RDH10 could have reduced activity due to a polymorphism while still enhancing the activity of its counterpart. This would result in an amplified effect in the same direction, i.e., the relative loss of the intrinsic enzymatic activity and, at the same time, the promotion of the opposite enzymatic activity. Although the DHRS3 and RDH10 ROC has been characterized in a number of studies, recent proteomic evidence suggested the interaction between other components of the ATRA network. Some examples include the possible interaction between DHRS3 and DHRS4, or between DHRS7 and ALDH1A3, in addition to interactions between ALDH1A1, ALDH1A2, and ALDH1A3 (Huttlin et al., 2021; Oughtred et al., 2021). These uncharacterized protein-protein interactions among ATRA network components raise the possibility of additional enzymatic complexes where variants can strongly affect the activity of other complex components and the overall ATRA robustness response. Multiple studies have identified abnormal gene expression patterns or polymorphisms in ATRA network components, linking them to developmental malformations or disease conditions (D'Aniello & Waxman, 2015; Guida et al., 2021; Khosasih et al., 2023; Lee et al., 2017; Merino et al., 2019; Zou et al., 2020). Some polymorphisms map to protein-coding regions, possibly affecting enzymatic activity, while others map to upstream regions, raising the possibility of regulatory effects and gene expression changes (Li et al., 2018). Detailed, in-depth proteomic and transcriptomic analysis is required to understand the effect of these ATRA network gene variants on the robustness response to ATRA fluctuations.

What are the limits of the robustness response of the ATRA metabolic and signaling network? The ATRA robustness response will be initially undertaken by the ATRA metabolic and signaling network components already expressed in the cells, region, or organ being studied. A key observation is that the self-regulatory feedback to perturbations of ATRA levels changes the expression of ATRA network component genes whose chromatin structure is accessible for regulation, focusing mainly on genes that are already expressed at some level. The robustness response involves limited *de novo* expression of network constituents, as some of them might not be accessible to transcriptional regulation. This observation is supported by the limited transcriptomic changes following ATRA manipulation (Abbou et al., 2022; Gur, Bendelac-Kapon et al., 2022; Parihar et al., 2021). It follows that the composition of the ATRA metabolic and signaling network in a specific cell-type will constrain the set of self-regulatory feedback responses available to respond to ATRA fluctuations and hence limit the hierarchical regulatory decisions implemented to achieve robust maintenance of ATRA levels. For a given ATRA metabolic and signaling network composition, the strength of enhancers and other regulatory elements can control the speed and magnitude of the gene regulatory feedback response (Lee et al., 2012; Rydeen et al., 2015; Zolfaghari et al., 2019) to determine the overall robustness to perturbations in ATRA levels. Viewed through this lens, developing a mechanistic understanding of how robustness is achieved in ATRA signaling requires a structure-function strategy that includes (1) determination of tissue/cell type-specific ATRA network composition, (2) targeted perturbations that elicit feedback response and illuminate the regulatory layers of control, and (3) genotype-wide assessment of the robustness response to increases and decreases in ATRA signaling to characterize the relative ability to counteract excess versus deficiency in ATRA levels. The combination of enzymes expressed in the ATRA network in a given cell and the polymorphisms affecting the corresponding gene regulatory elements limit the robustness response, yielding a spatiotemporal tissue context-specific capacity for robustness in ATRA signaling.

Acknowledgments

We wish to thank Sally Moody and Martin Blum for critically reading the manuscript, having lengthy discussions, and providing extensive comments. We also wish to thank Liat Bendelac-Kapon, Tali Abbou-Levi, Yehuda Shabtai, Michal Gur, and Graciela Pillemer for discussions on the regulation of retinoic acid signaling.

Funding

This work was funded in part by grants from the United States-Israel Binational Science Foundation (2017199), The Israel Science Foundation (668/17), the Manitoba Liquor and

Lotteries (RG-003-21), and the Wolfson Family Chair in Genetics to AF; National Institute on Alcohol Abuse and Alcoholism R01 AA018873 to RV; and, a Pilot Funding grant from Thomas Jefferson University and The Hebrew University of Jerusalem collaborative research program to AF and RV.

References

Abbou, T., Bendelac-Kapon, L., Sebag, A., & Fainsod, A. (2022). Enhanced loss of retinoic acid network genes in xenopus laevis achieves a tighter signal regulation. *Cells, 11*. https://doi.org/10.3390/cells11030327.

Abdelmaksoud, A., Wollina, U., Lotti, T., & Temiz, S. A. (2023). Isotretinoin and pregnancy termination: An overview. *International Journal of Dermatology, 62*, e255–e256. https://doi.org/10.1111/ijd.16570.

Abed, K., Foucher, A., Bernard, D., Tancrède-Bohin, E., & Cavusoglu, N. (2023). One-year longitudinal study of the stratum corneum proteome of retinol and all-trans-retinoic acid treated human skin: An orchestrated molecular event. *Scientific Reports, 13*, 11196. https://doi.org/10.1038/s41598-023-37750-5.

Adams, M. K., Belyaeva, O. V., Wu, L., Chaple, I. F., Dunigan-Russell, K., Popov, K. M., & Kedishvili, N. Y. (2021). Characterization of subunit interactions in the hetero-oligomeric retinoid oxidoreductase complex. *The Biochemical Journal, 478*, 3597–3611. https://doi.org/10.1042/BCJ20210589.

Altıntaş Aykan, D., & Ergün, Y. (2020). Isotretinoin: Still the cause of anxiety for teratogenicity. *Dermatologic Therapy, 33*, e13192. https://doi.org/10.1111/dth.13192.

Asson-Batres, M. A. (2020). How dietary deficiency studies have illuminated the many roles of vitamin A during development and postnatal life. *Sub-Cellular Biochemistry, 95*, 1–26. https://doi.org/10.1007/978-3-030-42282-0_1.

Bastos Maia, S., Rolland Souza, A. S., Costa Caminha, M., de, F., Lins da Silva, S., Callou Cruz, R. de S. B. L., ... Batista Filho, M. (2019). Vitamin A and pregnancy: A narrative review. *Nutrients, 11*. https://doi.org/10.3390/nu11030681.

Begemann, G., Schilling, T. F., Rauch, G. J., Geisler, R., & Ingham, P. W. (2001). The zebrafish neckless mutation reveals a requirement for raldh2 in mesodermal signals that pattern the hindbrain. *Development (Cambridge, England), 128*, 3081–3094.

Belyaeva, O. V., Adams, M. K., Popov, K. M., & Kedishvili, N. Y. (2019). Generation of retinaldehyde for retinoic acid biosynthesis. *Biomolecules, 10*. https://doi.org/10.3390/biom10010005.

Berenguer, M., & Duester, G. (2022). Retinoic acid, RARs and early development. *Journal of Molecular Endocrinology, 69*, T59–T67. https://doi.org/10.1530/JME-22-0041.

Berenguer, M., Meyer, K. F., Yin, J., & Duester, G. (2020). Discovery of genes required for body axis and limb formation by global identification of retinoic acid-regulated epigenetic marks. *PLoS Biology, 18*, e3000719. https://doi.org/10.1371/journal.pbio.3000719.

Billings, S. E., Pierzchalski, K., Butler Tjaden, N. E., Pang, X.-Y., Trainor, P. A., Kane, M. A., & Moise, A. R. (2013). The retinaldehyde reductase DHRS3 is essential for preventing the formation of excess retinoic acid during embryonic development. *The FASEB Journal, 27*, 4877–4889. https://doi.org/10.1096/fj.13-227967.

Blaner, W. S., Li, Y., Brun, P.-J., Yuen, J. J., Lee, S.-A., & Clugston, R. D. (2016). Vitamin A absorption, storage and mobilization. *Sub-cellular Biochemistry, 81*, 95–125. https://doi.org/10.1007/978-94-024-0945-1_4.

Blaner, W. S. (2019). Vitamin A signaling and homeostasis in obesity, diabetes, and metabolic disorders. *Pharmacology & Therapeutics, 197*, 153–178. https://doi.org/10.1016/j.pharmthera.2019.01.006.

Blentic, A., Gale, E., & Maden, M. (2003). Retinoic acid signalling centres in the avian embryo identified by sites of expression of synthesising and catabolising enzymes. *Developmental Dynamics: An Official Publication of the American Association of Anatomists, 227*, 114–127. https://doi.org/10.1002/dvdy.10292.

Blumberg, B., Bolado, J., Moreno, T. A., Kintner, C., Evans, R. M., & Papalopulu, N. (1997). An essential role for retinoid signaling in anteroposterior neural patterning. *Development (Cambridge, England), 124*, 373–379.

Bohn, T., Balbuena, E., Ulus, H., Iddir, M., Wang, G., Crook, N., & Eroglu, A. (2023). Carotenoids in health as studied by omics-related endpoints. *Advances in Nutrition, 14*, 1538–1578. https://doi.org/10.1016/j.advnut.2023.09.002.

Brown, G. (2023). Deregulation of all-trans retinoic acid signaling and development in cancer. *International Journal of Molecular Sciences, 24*. https://doi.org/10.3390/ijms241512089.

Chai, Z., Lyu, Y., Chen, Q., Wei, C.-H., Snyder, L. M., Weaver, V., ... Ross, A. C. (2021). RNAseq studies reveal distinct transcriptional response to vitamin A deficiency in small intestine versus colon, uncovering novel vitamin A-regulated genes. *The Journal of Nutritional Biochemistry, 98*, 108814. https://doi.org/10.1016/j.jnutbio.2021.108814.

Chandrasekhar, A., & Navlakha, S. (2019). Neural arbors are Pareto optimal. *Proceedings. Biological Sciences/The Royal Society, 286*, 20182727. https://doi.org/10.1098/rspb.2018.2727.

Cha, E.-H., Kim, N., Kwak, H.-S., Han, H. J., Joo, S. H., Choi, J.-S., ... Han, J. Y. (2022). Pregnancy and neonatal outcomes after periconceptional exposure to isotretinoin in Koreans. *Obstetrics & Gynecology Science, 65*, 166–175. https://doi.org/10.5468/ogs.21354.

Chen, J. D., Umesono, K., & Evans, R. M. (1996). SMRT isoforms mediate repression and anti-repression of nuclear receptor heterodimers. *Proceedings of the National Academy of Sciences of the United States of America, 93*, 7567–7571. https://doi.org/10.1073/pnas.93.15.7567.

Chen, Y., Pollet, N., Niehrs, C., & Pieler, T. (2001). Increased XRALDH2 activity has a posteriorizing effect on the central nervous system of Xenopus embryos. *Mechanisms of Development, 101*, 91–103. https://doi.org/10.1016/S0925-4773(00)00558-X.

Chen, Y., & Reese, D. H. (2013). A screen for disruptors of the retinol (vitamin A) signaling pathway. *Birth Defects Research. Part B, Developmental and Reproductive Toxicology, 98*, 276–282. https://doi.org/10.1002/bdrb.21062.

Clagett-Dame, M., & Knutson, D. (2011). Vitamin A in reproduction and development. *Nutrients, 3*, 385–428. https://doi.org/10.3390/nu3040385.

Coberly, S., Lammer, E., & Alashari, M. (1996). Retinoic acid embryopathy: Case report and review of literature. *Pediatric Pathology & Laboratory Medicine, 16*, 823–836. https://doi.org/10.1080/15513819609169308.

Collins, M. D., & Mao, G. E. (1999). Teratology of retinoids. *Annual Review of Pharmacology and Toxicology, 39*, 399–430. https://doi.org/10.1146/annurev.pharmtox.39.1.399.

Cook, D. P., & Wrana, J. L. (2022). A specialist-generalist framework for epithelial-mesenchymal plasticity in cancer. *Trends Cancer, 8*, 358–368. https://doi.org/10.1016/j.trecan.2022.01.014.

Corcoran, J., So, P. L., & Maden, M. (2002). Absence of retinoids can induce motoneuron disease in the adult rat and a retinoid defect is present in motoneuron disease patients. *Journal of Cell Science, 115*, 4735–4741.

D'Aniello, E., Rydeen, A. B., Anderson, J. L., Mandal, A., & Waxman, J. S. (2013). Depletion of retinoic acid receptors initiates a novel positive feedback mechanism that promotes teratogenic increases in retinoic acid. *PLoS Genetics, 9*, e1003689. https://doi.org/10.1371/journal.pgen.1003689.

D'Aniello, E., & Waxman, J. S. (2015). Input overload: Contributions of retinoic acid signaling feedback mechanisms to heart development and teratogenesis. *Developmental Dynamics: An Official Publication of the American Association of Anatomists, 244*, 513–523. https://doi.org/10.1002/dvdy.24232.

Deak, K. L., Dickerson, M. E., Linney, E., Enterline, D. S., George, T. M., Melvin, E. C., ... NTD Collaborative Group (2005). Analysis of ALDH1A2, CYP26A1, CYP26B1, CRABP1, and CRABP2 in human neural tube defects suggests a possible association with alleles in ALDH1A2. *Birth Defects Research Part A: Clinical and Molecular Teratology, 73*, 868–875. https://doi.org/10.1002/bdra.20183.

de Roos, K., Sonneveld, E., Compaan, B., ten Berge, D., Durston, A. J., & van der Saag, P. T. (1999). Expression of retinoic acid 4-hydroxylase (CYP26) during mouse and Xenopus laevis embryogenesis. *Mechanisms of Development, 82*, 205–211. https://doi.org/10.1016/s0925-4773(99)00016-7.

de Souza Mesquita, L. M., Mennitti, L. V., de Rosso, V. V., & Pisani, L. P. (2021). The role of vitamin A and its pro-vitamin carotenoids in fetal and neonatal programming: gaps in knowledge and metabolic pathways. *Nutrition Reviews, 79*, 76–87. https://doi.org/10.1093/nutrit/nuaa075.

DiKun, K. M., & Gudas, L. J. (2023). Vitamin A and retinoid signaling in the kidneys. *Pharmacology & Therapeutics, 248*, 108481. https://doi.org/10.1016/j.pharmthera.2023.108481.

di Masi, A., Leboffe, L., De Marinis, E., Pagano, F., Cicconi, L., Rochette-Egly, C., ... Nervi, C. (2015). Retinoic acid receptors: From molecular mechanisms to cancer therapy. *Molecular Aspects of Medicine, 41*, 1–115. https://doi.org/10.1016/j.mam.2014.12.003.

Dubey, A., Rose, R. E., Jones, D. R., & Saint-Jeannet, J.-P. (2018). Generating retinoic acid gradients by local degradation during craniofacial development: One cell's cue is another cell's poison. *Genesis (New York, N. Y.: 2000), 56*. https://doi.org/10.1002/dvg.23091.

Durston, A. J., Timmermans, J. P., Hage, W. J., Hendriks, H. F., de Vries, N. J., Heideveld, M., & Nieuwkoop, P. D. (1989). Retinoic acid causes an anteroposterior transformation in the developing central nervous system. *Nature, 340*, 140–144. https://doi.org/10.1038/340140a0.

Edri, T., Cohen, D., Shabtai, Y., & Fainsod, A. (2023). Alcohol induces neural tube defects by reducing retinoic acid signaling and promoting neural plate expansion. *Frontiers in Cell and Developmental Biology, 11*, 1282273. https://doi.org/10.3389/fcell.2023.1282273.

Eldar, A., Shilo, B.-Z., & Barkai, N. (2004). Elucidating mechanisms underlying robustness of morphogen gradients. *Current Opinion in Genetics & Development, 14*, 435–439. https://doi.org/10.1016/j.gde.2004.06.009.

Epping, M. T., Wang, L., Edel, M. J., Carlée, L., Hernandez, M., & Bernards, R. (2005). The human tumor antigen PRAME is a dominant repressor of retinoic acid receptor signaling. *Cell, 122*, 835–847. https://doi.org/10.1016/j.cell.2005.07.003.

Estrada-Ortiz, N., Starokozhko, V., van Steenwijk, H., van der Heide, C., Permentier, H., van Heemskerk, L., ... de Graaf, I. A. M. (2022). Disruption of vitamin A homeostasis by the biocide tetrakis(hydroxymethyl) phosphonium sulphate in pregnant rabbits. *Journal of Applied Toxicology: JAT, 42*, 1921–1936. https://doi.org/10.1002/jat.4364.

Evans, R. M., & Mangelsdorf, D. J. (2014). Nuclear receptors, RXR, and the big bang. *Cell, 157*, 255–266. https://doi.org/10.1016/j.cell.2014.03.012.

Fainsod, A., Abbou, T., Bendelac-Kapon, L., Edri, T., & Pillemer, G. (2022). Fetal alcohol spectrum disorder as a retinoic acid deficiency syndrome. In A. E. Chudley, & G. G. Hicks (Eds.), *Fetal alcohol spectrum disorder: Advances in research and practice, neuromethods* (pp. 49–76). New York, NY: Springer US. https://doi.org/10.1007/978-1-0716-2613-9_4.

Fainsod, A., & Kot-Leibovich, H. (2018). Xenopus embryos to study fetal alcohol syndrome, a model for environmental teratogenesis. *Biochemistry and Cell Biology = Biochimie et Biologie Cellulaire, 96*, 77–87. https://doi.org/10.1139/bcb-2017-0219.

Fattori, J., Campos, J. L. O., Doratioto, T. R., Assis, L. M., Vitorino, M. T., Polikarpov, I., ... Figueira, A. C. M. (2015). RXR agonist modulates TR: Corepressor dissociation upon 9-cis retinoic acid treatment. *Molecular Endocrinology (Baltimore, Md.), 29*, 258–273. https://doi.org/10.1210/me.2014-1251.

Feng, L., Hernandez, R. E., Waxman, J. S., Yelon, D., & Moens, C. B. (2010). Dhrs3a regulates retinoic acid biosynthesis through a feedback inhibition mechanism. *Developmental Biology, 338*, 1–14. https://doi.org/10.1016/j.ydbio.2009.10.029.

Fernandes, I., Bastien, Y., Wai, T., Nygard, K., Lin, R., Cormier, O., ... White, J. H. (2003). Ligand-dependent nuclear receptor corepressor LCoR functions by histone deacetylase-dependent and -independent mechanisms. *Molecular Cell, 11*, 139–150. https://doi.org/10.1016/s1097-2765(03)00014-5.

Fransén, K., Franzén, P., Magnuson, A., Elmabsout, A. A., Nyhlin, N., Wickbom, A., ... Halfvarson, J. (2013). Polymorphism in the retinoic acid metabolizing enzyme CYP26B1 and the development of Crohn's Disease. *PLoS One, 8*, e72739. https://doi.org/10.1371/journal.pone.0072739.

Gao, T., He, B., Pan, Y., Li, R., Xu, Y., Chen, L., ... Wang, S. (2013). The association of retinoic acid receptor beta2(RARβ2) methylation status and prostate cancer risk: A systematic review and meta-analysis. *PLoS One, 8*, e62950. https://doi.org/10.1371/journal.pone.0062950.

Ghyselinck, N. B., & Duester, G. (2019). Retinoic acid signaling pathways. *Development (Cambridge, England), 146*. https://doi.org/10.1242/dev.167502.

Grandel, H., Lun, K., Rauch, G.-J., Rhinn, M., Piotrowski, T., Houart, C., ... Brand, M. (2002). Retinoic acid signalling in the zebrafish embryo is necessary during pre-segmentation stages to pattern the anterior-posterior axis of the CNS and to induce a pectoral fin bud. *Development (Cambridge, England), 129*, 2851–2865.

Gudas, L. J. (2022). Retinoid metabolism: New insights. *Journal of Molecular Endocrinology, 69*, T37–T49. https://doi.org/10.1530/JME-22-0082.

Guida, V., Sparascio, F. P., Bernardini, L., Pancheri, F., Melis, D., Cocciadiferro, D., ... De Luca, A. (2021). Copy number variation analysis implicates novel pathways in patients with oculo-auriculo-vertebral-spectrum and congenital heart defects. *Clinical Genetics, 100*, 268–279. https://doi.org/10.1111/cge.13994.

Gurevich, I., & Aneskievich, B. J. (2009). Liganded RARalpha and RARgamma interact with but are repressed by TNIP1. *Biochemical and Biophysical Research Communications, 389*, 409–414. https://doi.org/10.1016/j.bbrc.2009.08.159.

Gur, M., Bendelac-Kapon, L., Shabtai, Y., Pillemer, G., & Fainsod, A. (2022). Reduced retinoic acid signaling during gastrulation induces developmental microcephaly. *Frontiers in Cell and Developmental Biology, 10*, 844619. https://doi.org/10.3389/fcell.2022.844619.

Gur, M., Edri, T., Moody, S. A., & Fainsod, A. (2022). Retinoic acid is required for normal morphogenetic movements during gastrulation. *Frontiers in Cell and Developmental Biology, 10*, 857230. https://doi.org/10.3389/fcell.2022.857230.

Hollemann, T., Chen, Y., Grunz, H., & Pieler, T. (1998). Regionalized metabolic activity establishes boundaries of retinoic acid signalling. *The EMBO Journal, 17*, 7361–7372. https://doi.org/10.1093/emboj/17.24.7361.

Huttlin, E. L., Bruckner, R. J., Navarrete-Perea, J., Cannon, J. R., Baltier, K., Gebreab, F., ... Gygi, S. P. (2021). Dual proteome-scale networks reveal cell-specific remodeling of the human interactome. *Cell, 184*, 3022–3040.e28. https://doi.org/10.1016/j.cell.2021.04.011.

Hu, P., van Dam, A.-M., Wang, Y., Lucassen, P. J., & Zhou, J.-N. (2020). Retinoic acid and depressive disorders: Evidence and possible neurobiological mechanisms. *Neuroscience and Biobehavioral Reviews, 112*, 376–391. https://doi.org/10.1016/j.neubiorev.2020.02.013.

Janesick, A., Nguyen, T. T. L., Aisaki, K., Igarashi, K., Kitajima, S., Chandraratna, R. A. S., ... Blumberg, B. (2014). Active repression by RARγ signaling is required for vertebrate axial elongation. *Development (Cambridge, England), 141*, 2260–2270. https://doi.org/10.1242/dev.103705.

Jiao, R.-Q., Li, G., & Chiu, J.-F. (2012). Comparative proteomic analysis of differentiation of mouse F9 embryonic carcinoma cells induced by retinoic acid. *Journal of Cellular Biochemistry, 113*, 1811–1819. https://doi.org/10.1002/jcb.24091.

Kam, R. K. T., Shi, W., Chan, S. O., Chen, Y., Xu, G., Lau, C. B.-S., ... Zhao, H. (2013). Dhrs3 protein attenuates retinoic acid signaling and is required for early embryonic patterning. *The Journal of Biological Chemistry, 288*, 31477–31487. https://doi.org/10.1074/jbc.M113.514984.

Kavanagh, K. D., Evans, A. R., & Jernvall, J. (2007). Predicting evolutionary patterns of mammalian teeth from development. *Nature, 449*, 427–432. https://doi.org/10.1038/nature06153.

Kedishvili, N. Y. (2016). Retinoic acid synthesis and degradation. *Sub-cellular Biochemistry, 81*, 127–161. https://doi.org/10.1007/978-94-024-0945-1_5.

Kessel, M., & Gruss, P. (1991). Homeotic transformations of murine vertebrae and concomitant alteration of Hox codes induced by retinoic acid. *Cell, 67*, 89–104. https://doi.org/10.1016/0092-8674(91)90574-i.

Khosasih, V., Liu, K.-M., Huang, C.-M., Liou, L.-B., Hsieh, M.-S., Lee, C.-H., ... Wu, J.-Y. (2023). A functional polymorphism downstream of vitamin A regulator gene CYP26B1 is associated with hand osteoarthritis. *International Journal of Molecular Sciences, 24*. https://doi.org/10.3390/ijms24033021.

Kikonyogo, A., Abriola, D. P., Dryjanski, M., & Pietruszko, R. (1999). Mechanism of inhibition of aldehyde dehydrogenase by citral, a retinoid antagonist. *European Journal of Biochemistry/FEBS, 262*, 704–712. https://doi.org/10.1046/j.1432-1327.1999.00415.x.

Kim, H., Lapointe, J., Kaygusuz, G., Ong, D. E., Li, C., van de Rijn, M., ... Pollack, J. R. (2005). The retinoic acid synthesis gene ALDH1a2 is a candidate tumor suppressor in prostate cancer. *Cancer Research, 65*, 8118–8124. https://doi.org/10.1158/0008-5472.CAN-04-4562.

Koide, T., Downes, M., Chandraratna, R. A., Blumberg, B., & Umesono, K. (2001). Active repression of RAR signaling is required for head formation. *Genes & Development, 15*, 2111–2121. https://doi.org/10.1101/gad.908801.

Kot-Leibovich, H., & Fainsod, A. (2009). Ethanol induces embryonic malformations by competing for retinaldehyde dehydrogenase activity during vertebrate gastrulation. *Disease Models & Mechanisms, 2*, 295–305. https://doi.org/10.1242/dmm.001420.

Kraft, J. C., & Juchau, M. R. (1992). Correlations between conceptal concentrations of all-trans-retinoic acid and dysmorphogenesis after microinjections of all-trans-retinoic acid, 13-cis-retinoic acid, all-trans-retinoyl-beta-glucuronide, or retinol in cultured whole rat embryos. *Drug Metabolism and Disposition: The Biological Fate of Chemicals, 20*, 218–225.

Kumar, S., Cunningham, T. J., & Duester, G. (2016). Nuclear receptor corepressors Ncor1 and Ncor2 (Smrt) are required for retinoic acid-dependent repression of Fgf8 during somitogenesis. *Developmental Biology, 418*, 204–215. https://doi.org/10.1016/j.ydbio.2016.08.005.

Kumar, S., Sandell, L. L., Trainor, P. A., Koentgen, F., & Duester, G. (2012). Alcohol and aldehyde dehydrogenases: Retinoid metabolic effects in mouse knockout models. *Biochimica et Biophysica Acta, 1821*, 198–205. https://doi.org/10.1016/j.bbalip.2011.04.004.

Laughton, P. M., Skakum, W., & Levi, L. (1962). Citral in flavoring extracts, determination of citral in food and drug products in the barbituric acid condensation method. *Journal of Agricultural and Food Chemistry, 10*, 49–51. https://doi.org/10.1021/jf60119a016.

Lee, L. M. Y., Leung, C.-Y., Tang, W. W. C., Choi, H.-L., Leung, Y.-C., McCaffery, P. J., ... Shum, A. S. W. (2012). A paradoxical teratogenic mechanism for retinoic acid. *Proceedings of the National Academy of Sciences of the United States of America, 109*, 13668–13673. https://doi.org/10.1073/pnas.1200872109.

Lee, L. M. Y., Leung, M. B. W., Kwok, R. C. Y., Leung, Y. C., Wang, C. C., McCaffery, P. J., ... Shum, A. S. W. (2017). Perturbation of retinoid homeostasis increases malformation risk in embryos exposed to pregestational diabetes. *Diabetes, 66*, 1041–1051. https://doi.org/10.2337/db15-1570.

Leibovich, A., Moody, S. A., Klein, S. L., & Fainsod, A. (2022). Xenopus: A model to study natural genetic variation and its disease implications. In A. Fainsod, & S. A. Moody (Eds.). *Xenopus: From basic biology to disease models in the genomic era* (pp. 313–324). Boca Raton: CRC Press. https://doi.org/10.1201/9781003050230-25.

Levin, A. A., Sturzenbecker, L. J., Kazmer, S., Bosakowski, T., Huselton, C., Allenby, G., ... Lovey, A. (1992). 9-cis retinoic acid stereoisomer binds and activates the nuclear receptor RXR alpha. *Nature, 355*, 359–361. https://doi.org/10.1038/355359a0.

le Maire, A., & Bourguet, W. (2014). Retinoic acid receptors: Structural basis for coregulator interaction and exchange. *Sub-cellular Biochemistry, 70*, 37–54. https://doi.org/10.1007/978-94-017-9050-5_3.

le Maire, A., Teyssier, C., Erb, C., Grimaldi, M., Alvarez, S., de Lera, A. R., ... Bourguet, W. (2010). A unique secondary-structure switch controls constitutive gene repression by retinoic acid receptor. *Nature Structural & Molecular Biology, 17*, 801–807. https://doi.org/10.1038/nsmb.1855.

Linney, E., Donerly, S., Mackey, L., & Dobbs-McAuliffe, B. (2011). The negative side of retinoic acid receptors. *Neurotoxicology and Teratology, 33*, 631–640. https://doi.org/10.1016/j.ntt.2011.06.006.

Liu, Z., Hu, Q., & Rosenfeld, M. G. (2014). Complexity of the RAR-mediated transcriptional regulatory programs. *Sub-cellular Biochemistry, 70*, 203–225. https://doi.org/10.1007/978-94-017-9050-5_10.

Li, H., Zhang, J., Chen, S., Wang, F., Zhang, T., & Niswander, L. (2018). Genetic contribution of retinoid-related genes to neural tube defects. *Human Mutation, 39*, 550–562. https://doi.org/10.1002/humu.23397.

Lie, K. K., Meier, S., Sørhus, E., Edvardsen, R. B., Karlsen, Ø., & Olsvik, P. A. (2019). Offshore crude oil disrupts retinoid signaling and eye development in larval atlantic haddock. *Frontiers in Marine Science, 6*. https://doi.org/10.3389/fmars.2019.00368.

Lupo, G., Liu, Y., Qiu, R., Chandraratna, R. A. S., Barsacchi, G., He, R.-Q., & Harris, W. A. (2005). Dorsoventral patterning of the Xenopus eye: A collaboration of Retinoid, Hedgehog and FGF receptor signaling. *Development (Cambridge, England), 132*, 1737–1748. https://doi.org/10.1242/dev.01726.

Mangelsdorf, D. J., Ong, E. S., Dyck, J. A., & Evans, R. M. (1990). Nuclear receptor that identifies a novel retinoic acid response pathway. *Nature, 345*, 224–229. https://doi.org/10.1038/345224a0.

Marcato, P., Dean, C. A., Liu, R.-Z., Coyle, K. M., Bydoun, M., Wallace, M., ... Lee, P. W. K. (2015). Aldehyde dehydrogenase 1A3 influences breast cancer progression via differential retinoic acid signaling. *Molecular Oncology, 9*, 17–31. https://doi.org/10.1016/j.molonc.2014.07.010.

Marikar, Y., Wang, Z., Duell, E. A., Petkovich, M., Voorhees, J. J., & Fisher, G. J. (1998). Retinoic acid receptors regulate expression of retinoic acid 4-hydroxylase that specifically inactivates all-trans retinoic acid in human keratinocyte HaCaT cells. *The Journal of Investigative Dermatology, 111*, 434–439. https://doi.org/10.1046/j.1523-1747.1998.00297.x.

Marshall, H., Nonchev, S., Sham, M. H., Muchamore, I., Lumsden, A., & Krumlauf, R. (1992). Retinoic acid alters hindbrain Hox code and induces transformation of rhombomeres 2/3 into a 4/5 identity. *Nature, 360*, 737–741. https://doi.org/10.1038/360737a0.

Mazul, A. L., Weinberg, C. R., Engel, S. M., Siega-Riz, A. M., Zou, F., Carrier, K. S., ... Olshan, A. F. (2019). Neuroblastoma in relation to joint effects of vitamin A and maternal and offspring variants in vitamin A-related genes: A report of the Children's Oncology Group. *Cancer Epidemiol, 61*, 165–171. https://doi.org/10.1016/j.canep.2019.06.009.

Merino, J., Dashti, H. S., Li, S. X., Sarnowski, C., Justice, A. E., Graff, M., ... Tanaka, T. (2019). Genome-wide meta-analysis of macronutrient intake of 91,114 European ancestry participants from the cohorts for heart and aging research in genomic epidemiology consortium. *Molecular Psychiatry, 24*, 1920–1932. https://doi.org/10.1038/s41380-018-0079-4.

Metzler, M. A., & Sandell, L. L. (2016). Enzymatic metabolism of vitamin A in developing vertebrate embryos. *Nutrients, 8*, pii: E812. https://doi.org/10.3390/nu8120812.

Mic, F. A., Molotkov, A., Fan, X., Cuenca, A. E., & Duester, G. (2000). RALDH3, a retinaldehyde dehydrogenase that generates retinoic acid, is expressed in the ventral retina, otic vesicle and olfactory pit during mouse development. *Mechanisms of Development, 97*, 227–230. https://doi.org/10.1016/S0925-4773(00)00434-2.

Miro Estruch, I., Melchers, D., Houtman, R., de Haan, L. H. J., Groten, J. P., Louisse, J., & Rietjens, I. M. C. M. (2017). Characterization of the differential coregulator binding signatures of the Retinoic Acid Receptor subtypes upon (ant)agonist action. *Biochimica et Biophysica Acta (BBA) – Proteins and Proteomics, 1865*, 1195–1206. https://doi.org/10.1016/j.bbapap.2017.06.011.

Molotkov, A., & Duester, G. (2003). Genetic evidence that retinaldehyde dehydrogenase Raldh1 (Aldh1a1) functions downstream of alcohol dehydrogenase Adh1 in metabolism of retinol to retinoic acid. *The Journal of Biological Chemistry, 278*, 36085–36090. https://doi.org/10.1074/jbc.M303709200.

Molotkov, A., Molotkova, N., & Duester, G. (2006). Retinoic acid guides eye morphogenetic movements via paracrine signaling but is unnecessary for retinal dorsoventral patterning. *Development (Cambridge, England), 133*, 1901–1910. https://doi.org/10.1242/dev.02328.

Moreno-Ramos, O. A., Olivares, A. M., Haider, N. B., de Autismo, L. C., & Lattig, M. C. (2015). Whole-exome sequencing in a South American cohort links ALDH1A3, FOXN1 and retinoic acid regulation pathways to autism spectrum disorders. *PLoS One, 10*, e0135927. https://doi.org/10.1371/journal.pone.0135927.

Napoli, J. L. (2020). Post-natal all-trans-retinoic acid biosynthesis. *Methods in Enzymology, 637*, 27–54. https://doi.org/10.1016/bs.mie.2020.02.003.

Napoli, J. L. (2017). Cellular retinoid binding-proteins, CRBP, CRABP, FABP5: Effects on retinoid metabolism, function and related diseases. *Pharmacology & Therapeutics, 173*, 19–33. https://doi.org/10.1016/j.pharmthera.2017.01.004.

Neijts, R., & Deschamps, J. (2017). At the base of colinear Hox gene expression: Cis-features and trans-factors orchestrating the initial phase of Hox cluster activation. *Developmental Biology, 428*, 293–299. https://doi.org/10.1016/j.ydbio.2017.02.009.

Niederreither, K., Abu-Abed, S., Schuhbaur, B., Petkovich, M., Chambon, P., & Dollé, P. (2002). Genetic evidence that oxidative derivatives of retinoic acid are not involved in retinoid signaling during mouse development. *Nature Genetics, 31*, 84–88. https://doi.org/10.1038/ng876.

Niederreither, K., Fraulob, V., Garnier, J.-M., Chambon, P., & Dollé, P. (2002). Differential expression of retinoic acid-synthesizing (RALDH) enzymes during fetal development and organ differentiation in the mouse. *Mechanisms of Development, 110*, 165–171. https://doi.org/10.1016/S0925-4773(01)00561-5.

Niederreither, K., McCaffery, P., Dräger, U. C., Chambon, P., & Dollé, P. (1997). Restricted expression and retinoic acid-induced downregulation of the retinaldehyde dehydrogenase type 2 (RALDH-2) gene during mouse development. *Mechanisms of Development, 62*, 67–78. https://doi.org/10.1016/S0925-4773(96)00653-3.

Niederreither, K., Subbarayan, V., Dollé, P., & Chambon, P. (1999). Embryonic retinoic acid synthesis is essential for early mouse post-implantation development. *Nature Genetics, 21*, 444–448. https://doi.org/10.1038/7788.

Nijhout, H. F., Best, J. A., & Reed, M. C. (2019). Systems biology of robustness and homeostatic mechanisms. *Wiley Interdisciplinary Reviews: Systems Biology and Medicine, 11*, e1440. https://doi.org/10.1002/wsbm.1440.

Nilsson, O., Isoherranen, N., Guo, M. H., Lui, J. C., Jee, Y. H., Guttmann-Bauman, I., ... Baron, J. (2016). Accelerated skeletal maturation in disorders of retinoic acid metabolism: A case report and focused review of the literature. *Hormone and Metabolic Research. Hormon- und Stoffwechselforschung. Hormones et Metabolisme, 48*, 737–744. https://doi.org/10.1055/s-0042-114038.

Niu, S., Shi, K., Yue, X., Pan, M., Song, L., Gu, L., ... Chang, J. (2023). A variant in CYP26B1 associated with esophageal squamous cell carcinoma risk by affecting retinoic acid metabolism. *Molecular Carcinogenesis, 62*, 991–1000. https://doi.org/10.1002/mc.23540.

Nolte, C., De Kumar, B., & Krumlauf, R. (2019). Hox genes: Downstream "effectors" of retinoic acid signaling in vertebrate embryogenesis. *Genesis (New York, N. Y.: 2000), 57*, e23306. https://doi.org/10.1002/dvg.23306.

O'Connor, C., Varshosaz, P., & Moise, A. R. (2022). Mechanisms of feedback regulation of vitamin A metabolism. *Nutrients, 14*, 1312. https://doi.org/10.3390/nu14061312.

Oughtred, R., Rust, J., Chang, C., Breitkreutz, B.-J., Stark, C., Willems, A., ... Tyers, M. (2021). The BioGRID database: A comprehensive biomedical resource of curated protein, genetic, and chemical interactions. *Protein Science: A Publication of the Protein Society, 30*, 187–200. https://doi.org/10.1002/pro.3978.

Owusu, S. A., & Ross, A. C. (2016). Retinoid homeostatic gene expression in liver, lung and kidney: Ontogeny and response to vitamin A-retinoic acid (VARA) supplementation from birth to adult age. *PLoS One, 11*, e0145924. https://doi.org/10.1371/journal.pone.0145924.

Paganelli, A., Gnazzo, V., Acosta, H., López, S. L., & Carrasco, A. E. (2010). Glyphosate-based herbicides produce teratogenic effects on vertebrates by impairing retinoic acid signaling. *Chemical Research in Toxicology, 23*, 1586–1595. https://doi.org/10.1021/tx1001749.

Pallasdies, F., Norton, P., Schleimer, J.-H., & Schreiber, S. (2021). Neural optimization: Understanding trade-offs with Pareto theory. *Current Opinion in Neurobiology, 71*, 84–91. https://doi.org/10.1016/j.conb.2021.08.008.

Pangilinan, F., Molloy, A. M., Mills, J. L., Troendle, J. F., Parle-McDermott, A., Kay, D. M., ... Brody, L. C. (2014). Replication and exploratory analysis of 24 candidate risk polymorphisms for neural tube defects. *BMC Medical Genetics, 15*, 102. https://doi.org/10.1186/s12881-014-0102-9.

Papalopulu, N., Clarke, J. D., Bradley, L., Wilkinson, D., Krumlauf, R., & Holder, N. (1991). Retinoic acid causes abnormal development and segmental patterning of the anterior hindbrain in Xenopus embryos. *Development (Cambridge, England), 113*, 1145–1158.

Papis, E., Bernardini, G., Gornati, R., Menegola, E., & Prati, M. (2007). Gene expression in Xenopus laevis embryos after Triadimefon exposure. *Gene Expression Patterns: GEP, 7*, 137–142. https://doi.org/10.1016/j.modgep.2006.06.003.

Parihar, M., Bendelac-Kapon, L., Gur, M., Abbou, T., Belorkar, A., Achanta, S., ... Fainsod, A. (2021). Retinoic acid fluctuation activates an uneven, direction-dependent network-wide robustness response in early embryogenesis. *Frontiers in Cell and Developmental Biology, 9*, 747969. https://doi.org/10.3389/fcell.2021.747969.

Pavan, M., Ruiz, V. F., Silva, F. A., Sobreira, T. J., Cravo, R. M., Vasconcelos, M., ... Xavier-Neto, J. (2009). ALDH1A2 (RALDH2) genetic variation in human congenital heart disease. *BMC Medical Genetics, 10*, 113. https://doi.org/10.1186/1471-2350-10-113.

Pennimpede, T., Cameron, D. A., MacLean, G. A., Li, H., Abu-Abed, S., & Petkovich, M. (2010). The role of CYP26 enzymes in defining appropriate retinoic acid exposure during embryogenesis. *Birth Defects Research Part A: Clinical and Molecular Teratology, 88*, 883–894. https://doi.org/10.1002/bdra.20709.

Penny, H. L., Prestwood, T. R., Bhattacharya, N., Sun, F., Kenkel, J. A., Davidson, M. G., ... Engleman, E. G. (2016). Restoring retinoic acid attenuates intestinal inflammation and tumorigenesis in apcmin/+ mice. *Cancer Immunology Research, 4*, 917–926. https://doi.org/10.1158/2326-6066.CIR-15-0038.

Pérez-Sayáns, M., Chamorro-Petronacci, C. M., Bravo, S. B., Padín-Iruegas, M. E., Guitián-Fernández, E., Barros-Angueira, F., ... García-García, A. (2023). Genetic linkage analysis of head and neck cancer in a Spanish family. *Oral Diseases*. https://doi.org/10.1111/odi.14572.

Perl, E., & Waxman, J. S. (2019). Reiterative mechanisms of retinoic acid signaling during vertebrate heart development. *Journal of Developmental Biology, 7*. https://doi.org/10.3390/jdb7020011.

Petrelli, B., Öztürk, A., Pind, M., Ayele, H., Fainsod, A., & Hicks, G. G. (2023). Genetically programmed retinoic acid deficiency during gastrulation phenocopies most known developmental defects due to acute prenatal alcohol exposure in FASD. *Frontiers in Cell and Developmental Biology, 11*, 1208279. https://doi.org/10.3389/fcell.2023.1208279.

Porté, S., Xavier Ruiz, F., Giménez, J., Molist, I., Alvarez, S., Domínguez, M., ... Farrés, J. (2013). Aldo-keto reductases in retinoid metabolism: Search for substrate specificity and inhibitor selectivity. *Chemico-Biological Interactions, 202*, 186–194. https://doi.org/10.1016/j.cbi.2012.11.014.

Rat, E., Billaut-Laden, I., Allorge, D., Lo-Guidice, J.-M., Tellier, M., Cauffiez, C., ... Broly, F. (2006). Evidence for a functional genetic polymorphism of the human retinoic acid-metabolizing enzyme CYP26A1, an enzyme that may be involved in spina bifida. *Birth Defects Research Part A: Clinical and Molecular Teratology, 76*, 491–498. https://doi.org/10.1002/bdra.20275.

Reijntjes, S., Blentic, A., Gale, E., & Maden, M. (2005). The control of morphogen signalling: Regulation of the synthesis and catabolism of retinoic acid in the developing embryo. *Developmental Biology, 285*, 224–237. https://doi.org/10.1016/j.ydbio.2005.06.019.

Ribes, V., Otto, D. M. E., Dickmann, L., Schmidt, K., Schuhbaur, B., Henderson, C., ... Dollé, P. (2007). Rescue of cytochrome P450 oxidoreductase (Por) mouse mutants reveals functions in vasculogenesis, brain and limb patterning linked to retinoic acid homeostasis. *Developmental Biology, 303*, 66–81. https://doi.org/10.1016/j.ydbio.2006.10.032.

Roberts, C. (2020). Regulating retinoic acid availability during development and regeneration: The role of the CYP26 enzymes. *Journal of Developmental Biology, 8*. https://doi.org/10.3390/jdb8010006.

Romand, R., Niederreither, K., Abu-Abed, S., Petkovich, M., Fraulob, V., Hashino, E., & Dollé, P. (2004). Complementary expression patterns of retinoid acid-synthesizing and -metabolizing enzymes in pre-natal mouse inner ear structures. *Gene Expression Patterns: GEP, 4*, 123–133. https://doi.org/10.1016/j.modgep.2003.09.006.

Rossetti, S., & Sacchi, N. (2020). Emerging cancer epigenetic mechanisms regulated by all-trans retinoic acid. *Cancers (Basel), 12*. https://doi.org/10.3390/cancers12082275.

Rydeen, A., Voisin, N., D'Aniello, E., Ravisankar, P., Devignes, C.-S., & Waxman, J. S. (2015). Excessive feedback of Cyp26a1 promotes cell non-autonomous loss of retinoic acid signaling. *Developmental Biology, 405*, 47–55. https://doi.org/10.1016/j.ydbio.2015.06.008.

Saili, K. S., Antonijevic, T., Zurlinden, T. J., Shah, I., Deisenroth, C., & Knudsen, T. B. (2020). Molecular characterization of a toxicological tipping point during human stem cell differentiation. *Reproductive Toxicology (Elmsford, N. Y.), 91*, 1–13. https://doi.org/10.1016/j.reprotox.2019.10.001.

Samrani, L. M. M., Dumont, F., Hallmark, N., Bars, R., Tinwell, H., Pallardy, M., & Piersma, A. H. (2023). Retinoic acid signaling pathway perturbation impacts mesodermal-tissue development in the zebrafish embryo: biomarker candidate identification using transcriptomics. *Reproductive Toxicology (Elmsford, N. Y.), 119*, 108404. https://doi.org/10.1016/j.reprotox.2023.108404.

Savova, V., Pearl, E. J., Boke, E., Nag, A., Adzhubei, I., Horb, M. E., & Peshkin, L. (2017). Transcriptomic insights into genetic diversity of protein-coding genes in X. laevis. *Developmental Biology, 424*, 181–188. https://doi.org/10.1016/j.ydbio.2017.02.019.

Schilling, T. F., Nie, Q., & Lander, A. D. (2012). Dynamics and precision in retinoic acid morphogen gradients. *Current Opinion in Genetics & Development, 22*, 562–569. https://doi.org/10.1016/j.gde.2012.11.012.

Schilling, T. F., Sosnik, J., & Nie, Q. (2016). Visualizing retinoic acid morphogen gradients. *Methods in Cell Biology, 133*, 139–163. https://doi.org/10.1016/bs.mcb.2016.03.003.

See, A. W.-M., Kaiser, M. E., White, J. C., & Clagett-Dame, M. (2008). A nutritional model of late embryonic vitamin A deficiency produces defects in organogenesis at a high penetrance and reveals new roles for the vitamin in skeletal development. *Developmental Biology, 316*, 171–190. https://doi.org/10.1016/j.ydbio.2007.10.018.

Session, A. M., Uno, Y., Kwon, T., Chapman, J. A., Toyoda, A., Takahashi, S., ... Rokhsar, D. S. (2016). Genome evolution in the allotetraploid frog Xenopus laevis. *Nature, 538*, 336–343. https://doi.org/10.1038/nature19840.

Shabtai, Y., Bendelac, L., Jubran, H., Hirschberg, J., & Fainsod, A. (2018). Acetaldehyde inhibits retinoic acid biosynthesis to mediate alcohol teratogenicity. *Scientific Reports, 8*, 347. https://doi.org/10.1038/s41598-017-18719-7.

Shabtai, Y., & Fainsod, A. (2018). Competition between ethanol clearance and retinoic acid biosynthesis in the induction of fetal alcohol syndrome. *Biochemistry and Cell Biology = Biochimie et Biologie Cellulaire, 96*, 148–160. https://doi.org/10.1139/bcb-2017-0132.

Shabtai, Y., Jubran, H., Nassar, T., Hirschberg, J., & Fainsod, A. (2016). Kinetic characterization and regulation of the human retinaldehyde dehydrogenase 2 enzyme during production of retinoic acid. *The Biochemical Journal, 473*, 1423–1431. https://doi.org/10.1042/BCJ20160101.

Shannon, S. R., Moise, A. R., & Trainor, P. A. (2017). New insights and changing paradigms in the regulation of vitamin A metabolism in development. *Wiley Interdisciplinary Reviews: Developmental Biology, 6*. https://doi.org/10.1002/wdev.264.

Sharpe, C. R., & Goldstone, K. (1997). Retinoid receptors promote primary neurogenesis in Xenopus. *Development (Cambridge, England), 124*, 515–523.

Shenefelt, R. E. (1972). Gross congenital malformations. Animal model: treatment of various species with a large dose of vitamin A at known stages in pregnancy. *The American Journal of Pathology, 66*, 589–592.

Shimozono, S., Iimura, T., Kitaguchi, T., Higashijima, S.-I., & Miyawaki, A. (2013). Visualization of an endogenous retinoic acid gradient across embryonic development. *Nature, 496*, 363–366. https://doi.org/10.1038/nature12037.

Shmarakov, I. O. (2015). Retinoid-xenobiotic interactions: The Ying and the Yang. *Hepatobiliary Surgery and Nutrition, 4*, 243–267. https://doi.org/10.3978/j.issn.2304-3881.2015.05.05.

Shoval, O., Sheftel, H., Shinar, G., Hart, Y., Ramote, O., Mayo, A., ... Alon, U. (2012). Evolutionary trade-offs, Pareto optimality, and the geometry of phenotype space. *Science (New York, N. Y.), 336*, 1157–1160. https://doi.org/10.1126/science.1217405.

Shukrun, N., Shabtai, Y., Pillemer, G., & Fainsod, A. (2019). Retinoic acid signaling reduction recapitulates the effects of alcohol on embryo size. *Genesis (New York, N. Y.: 2000), 57*, e23284. https://doi.org/10.1002/dvg.23284.

Sirbu, I. O., Chiş, A. R., & Moise, A. R. (2020). Role of carotenoids and retinoids during heart development. *Biochimica et Biophysica Acta (BBA) – Molecular and Cell Biology of Lipids, 1865*, 158636. https://doi.org/10.1016/j.bbalip.2020.158636.

Soref, C. M., Di, Y. P., Hayden, L., Zhao, Y. H., Satre, M. A., & Wu, R. (2001). Characterization of a novel airway epithelial cell-specific short chain alcohol dehydrogenase/reductase gene whose expression is up-regulated by retinoids and is involved in the metabolism of retinol. *The Journal of Biological Chemistry, 276*, 24194–24202. https://doi.org/10.1074/jbc.M100332200.

Sosnik, J., Zheng, L., Rackauckas, C. V., Digman, M., Gratton, E., Nie, Q., & Schilling, T. F. (2016). Noise modulation in retinoic acid signaling sharpens segmental boundaries of gene expression in the embryonic zebrafish hindbrain. *eLife, 5*, e14034. https://doi.org/10.7554/eLife.14034.

Steinhoff, J. S., Lass, A., & Schupp, M. (2022). Retinoid homeostasis and beyond: how retinol binding protein 4 contributes to health and disease. *Nutrients, 14*. https://doi.org/10.3390/nu14061236.

Stoney, P. N., & McCaffery, P. (2016). A vitamin on the mind: New discoveries on control of the brain by vitamin A. *World Review of Nutrition and Dietetics, 115*, 98–108. https://doi.org/10.1159/000442076.

Strate, I., Min, T. H., Iliev, D., & Pera, E. M. (2009). Retinol dehydrogenase 10 is a feedback regulator of retinoic acid signalling during axis formation and patterning of the central nervous system. *Development (Cambridge, England), 136*, 461–472. https://doi.org/10.1242/dev.024901.

Summerbell, D., & Maden, M. (1990). Retinoic acid, a developmental signalling molecule. *Trends in Neurosciences, 13*, 142–147.

Surakhy, M., Wallace, M., Bond, E., Grochola, L. F., Perez, H., Di Giovannantonio, M., … Bond, G. L. (2020). A common polymorphism in the retinoic acid pathway modifies adrenocortical carcinoma age-dependent incidence. *British Journal of Cancer, 122*, 1231–1241. https://doi.org/10.1038/s41416-020-0764-3.

Suzuki, M., & Tomita, M. (2022). Genetic variations of vitamin A-absorption and storage-related genes, and their potential contribution to vitamin A deficiency risks among different ethnic groups. *Frontiers in Nutrition, 9*, 861619. https://doi.org/10.3389/fnut.2022.861619.

Timoneda, J., Rodríguez-Fernández, L., Zaragozá, R., Marín, M. P., Cabezuelo, M. T., Torres, L., … Barber, T. (2018). Vitamin A deficiency and the lung. *Nutrients, 10*. https://doi.org/10.3390/nu10091132.

Topletz, A. R., Tripathy, S., Foti, R. S., Shimshoni, J. A., Nelson, W. L., & Isoherranen, N. (2015). Induction of CYP26A1 by metabolites of retinoic acid: Evidence that CYP26A1 is an important enzyme in the elimination of active retinoids. *Molecular Pharmacology, 87*, 430–441. https://doi.org/10.1124/mol.114.096784.

Tran, P. X., Au, K. S., Morrison, A. C., Fletcher, J. M., Ostermaier, K. K., Tyerman, G. H., & Northrup, H. (2011). Association of retinoic acid receptor genes with meningomyelocele. *Birth Defects Research Part A: Clinical and Molecular Teratology, 91*, 39–43. https://doi.org/10.1002/bdra.20744.

Umesono, K., Murakami, K. K., Thompson, C. C., & Evans, R. M. (1991). Direct repeats as selective response elements for the thyroid hormone, retinoic acid, and vitamin D3 receptors. *Cell, 65*, 1255–1266. https://doi.org/10.1016/0092-8674(91)90020-y.

Urbizu, A., Toma, C., Poca, M. A., Sahuquillo, J., Cuenca-León, E., Cormand, B., & Macaya, A. (2013). Chiari malformation type I: A case-control association study of 58 developmental genes. *PLoS One, 8*, e57241. https://doi.org/10.1371/journal.pone.0057241.

Vermot, J., Niederreither, K., Garnier, J.-M., Chambon, P., & Dollé, P. (2003). Decreased embryonic retinoic acid synthesis results in a DiGeorge syndrome phenotype in newborn mice. *Proceedings of the National Academy of Sciences of the United States of America, 100*, 1763–1768. https://doi.org/10.1073/pnas.0437920100.

Wahl, S. E., Wyatt, B. H., Turner, S. D., & Dickinson, A. J. G. (2018). Transcriptome analysis of Xenopus orofacial tissues deficient in retinoic acid receptor function. *BMC Genomics, 19*, 795. https://doi.org/10.1186/s12864-018-5186-8.

Weston, A. D., Blumberg, B., & Underhill, T. M. (2003). Active repression by unliganded retinoid receptors in development: Less is sometimes more. *The Journal of Cell Biology, 161*, 223–228. https://doi.org/10.1083/jcb.200211117.

White, J. A., Beckett-Jones, B., Guo, Y.-D., Dilworth, F. J., Bonasoro, J., Jones, G., & Petkovich, M. (1997). cDNA cloning of human retinoic acid-metabolizing enzyme (hP450RAI) identifies a novel family of cytochromes P450 (CYP26). *Journal of Biological Chemistry, 272*, 18538–18541. https://doi.org/10.1074/jbc.272.30.18538.

Wiseman, E. M., Bar-El Dadon, S., & Reifen, R. (2017). The vicious cycle of vitamin a deficiency: A review. *Critical Reviews in Food Science and Nutrition, 57*, 3703–3714. https://doi.org/10.1080/10408398.2016.1160362.

Wolujewicz, P., Aguiar-Pulido, V., AbdelAleem, A., Nair, V., Thareja, G., Suhre, K., ... Ross, M. E. (2021). Genome-wide investigation identifies a rare copy-number variant burden associated with human spina bifida. *Genetics in Medicine: Official Journal of the American College of Medical Genetics, 23*, 1211–1218. https://doi.org/10.1038/s41436-021-01126-9.

Wu, L., Belyaeva, O. V., Adams, M. K., Klyuyeva, A. V., Lee, S.-A., Goggans, K. R., ... Kedishvili, N. Y. (2019). Mice lacking the epidermal retinol dehydrogenases SDR16C5 and SDR16C6 display accelerated hair growth and enlarged meibomian glands. *The Journal of Biological Chemistry, 294*, 17060–17074. https://doi.org/10.1074/jbc.RA119.010835.

Wu, L., Kedishvili, N. Y., & Belyaeva, O. V. (2019). Retinyl esters are elevated in progeny of retinol dehydrogenase 11 deficient dams. *Chemico-Biological Interactions, 302*, 117–122. https://doi.org/10.1016/j.cbi.2019.01.041.

Wu, Z., Zhang, X., An, Y., Ma, K., Xue, R., Ye, G., ... Bu, P. (2023). CLMP is a tumor suppressor that determines all-trans retinoic acid response in colorectal cancer. *Developmental Cell, 58*, 2684–2699.e6. https://doi.org/10.1016/j.devcel.2023.10.006.

Yang, D., Vuckovic, M. G., Smullin, C. P., Kim, M., Lo, C. P.-S., Devericks, E., ... Napoli, J. L. (2018). Modest decreases in endogenous all-trans-retinoic acid produced by a mouse Rdh10 heterozygote provoke major abnormalities in adipogenesis and lipid metabolism. *Diabetes, 67*, 662–673. https://doi.org/10.2337/db17-0946.

Yan, W., Wu, K., Herman, J. G., Xu, X., Yang, Y., Dai, G., & Guo, M. (2018). Retinoic acid-induced 2 (RAI2) is a novel tumor suppressor, and promoter region methylation of RAI2 is a poor prognostic marker in colorectal cancer. *Clinical Epigenetics, 10*, 69. https://doi.org/10.1186/s13148-018-0501-4.

Yu, V. C., Delsert, C., Andersen, B., Holloway, J. M., Devary, O. V., Näär, A. M., ... Rosenfeld, M. G. (1991). RXRβ: A coregulator that enhances binding of retinoic acid, thyroid hormone, and vitamin D receptors to their cognate response elements. *Cell, 67*, 1251–1266. https://doi.org/10.1016/0092-8674(91)90301-E.

Zhao, D., McCaffery, P., Ivins, K. J., Neve, R. L., Hogan, P., Chin, W. W., & Dräger, U. C. (1996). Molecular identification of a major retinoic-acid-synthesizing enzyme, a retinaldehyde-specific dehydrogenase. *European Journal of Biochemistry / FEBS, 240*, 15–22. https://doi.org/10.1111/j.1432-1033.1996.0015h.x.

Zhong, G., Hogarth, C., Snyder, J. M., Palau, L., Topping, T., Huang, W., ... Isoherranen, N. (2019). The retinoic acid hydroxylase Cyp26a1 has minor effects on postnatal vitamin A homeostasis, but is required for exogenous atRA clearance. *The Journal of Biological Chemistry, 294*, 11166–11179. https://doi.org/10.1074/jbc.RA119.009023.

Zhu, L., Kamalathevan, P., Koneva, L. A., Zarebska, J. M., Chanalaris, A., Ismail, H., ... Shirley, R. (2022). Variants in ALDH1A2 reveal an anti-inflammatory role for retinoic acid and a new class of disease-modifying drugs in osteoarthritis. *Science Translational Medicine, 14*, eabm4054. https://doi.org/10.1126/scitranslmed.abm4054.

Zolfaghari, R., Mattie, F. J., Wei, C.-H., Chisholm, D. R., Whiting, A., & Ross, A. C. (2019). CYP26A1 gene promoter is a useful tool for reporting RAR-mediated retinoid activity. *Analytical Biochemistry, 577*, 98–109. https://doi.org/10.1016/j.ab.2019.04.022.

Zou, J., Wang, F., Yang, X., Wang, H., Niswander, L., Zhang, T., & Li, H. (2020). Association between rare variants in specific functional pathways and human neural tube defects multiple subphenotypes. *Neural Dev, 15*, 8. https://doi.org/10.1186/s13064-020-00145-7.

CHAPTER SIX

The action of retinoic acid on spermatogonia in the testis

Shelby L. Havel and Michael D. Griswold*
School of Molecular Biosciences, Washington State University, Pullman, Washington, United States
*Corresponding author. e-mail address: mgriswold@wsu.edu

Contents

1. Male germ cell development in the mouse	144
2. The essential role of sertoli cells in spermatogenesis	146
3. The cycle of the seminiferous epithelium	147
4. Retinoic acid synthesis in the testis	148
5. Elucidating testicular *at*RA synthesis activity through genetic knockout studies	151
5.1 Rdh10 expression by sertoli cells is required for juvenile spermatogenesis	151
5.2 RAL dehydrogenase activity in Sertoli cells, but not germ cells, is essential for the initial A to A1 transition	152
5.3 Loss of Cyp26b1, but not Cyp26a1, affects spermatogenesis	154
5.4 Gene expression changes induced by atRA support spermatogonial differentiation	154
6. Synchronizing spermatogenesis using the drug WIN 18,446	156
7. Conclusions	159
References	160

Abstract

For mammalian spermatogenesis to proceed normally, it is essential that the population of testicular progenitor cells, A undifferentiated spermatogonia (A$_{undiff}$), undergoes differentiation during the A to A1 transition that occurs at the onset of spermatogenesis. The commitment of the A$_{undiff}$ population to differentiation and leaving a quiescent, stem-like state gives rise to all the spermatozoa produced across the lifespan of an individual, and ultimately determines male fertility. The action of *all-trans* retinoic acid (*at*RA) on the A$_{undiff}$ population is the determining factor that induces this change. Sertoli cells, omnipresent, nurse cells within the mammalian testis are responsible for synthesizing the *at*RA that prompts this change in the neonatal testicular environment. The mechanism of *at*RA synthesis and signaling has been robustly explored and, in this review, we have summarized what is currently known about the action of testicular *at*RA at the onset of spermatogenesis. We have combined this with evidence gained from prominent genetic studies that have further elucidated the function of genes critical to *at*RA synthesis. We have additionally described the effects of the first pulse of *at*RA delivered to the germ cells of the testis, which has been investigated using WIN 18,446 treatment which prevents *at*RA synthesis and induces spermatogenic synchrony. This method provides unparalleled resolution into cell and stage specific testicular changes, and combined with transgenic animal models, has allowed researchers to elucidate much regarding the onset of spermatogenesis.

1. Male germ cell development in the mouse

The efficient, asynchronous production of male germ cells is essential to a species' survival, and the process by which all the mature spermatozoa are made is called spermatogenesis. Research focused on male, mammalian reproductive biology is more essential than ever, as male infertility currently contributes to more than half of all diagnosed cases of infertility, and, unlike many cases of female infertility, is without an identifiable cause in more than 80 % of all cases (Agarwal et al., 2015; Hamada et al., 2012; Kothandaraman et al., 2016). *Mus musculus* is one of the most broadly used model organisms to study spermatogenesis, and as such our focus in this review will remain in the context of in vivo work utilizing mouse models. For a review of *at*RA action within humans, see (Schleif et al., 2022).

Within the testis, all the male germ cells produced start as a pool of gonocytes, also called pre-/pro-spermatogonia, which develop from primordial germ cells (PGCs) formed during early embryogenesis (Culty, 2009). The specialization of the germline in the mouse initiates at embryonic day 6.5 (E6.5), and the PGCs formed begin migrating to the gonadal ridge as early as E7.5. This journey carries the developing PGCs from the posterior primitive streak to the endoderm, the hindgut endoderm, and mesentery, before finally arriving at the gonadal ridge via bilateral migration where the movement of these cells completes by E13.5 (Alberts et al., 2002; Nikolic et al., 2016). PGCs in male embryos then enter a state of mitotic arrest, where they wait in a quiescent state until birth to resume proliferating and differentiating. This is in contrast to female PGCs, which continue their development embryonically, undergoing meiosis and providing the female with all of her oogonia by birth (Alberts et al., 2002; Hu et al., 2015; Nikolic et al., 2016).

Postnatally, the male PGCs, now referred to as gonocytes (or pre-/prospermatogonia), migrate to the edge of the basement membrane of the seminiferous tubules of the testis by approximately 3 days post-partum (DPP) forming a heterogenous pool of progenitor cells. Spermatogenesis occurs across three developmental stages that can be broadly classified as proliferative, meiotic, and spermiogenic. The proliferative phase initiates when the gonocytes resume dividing postnatally, and these cells are responsible for both (1) dividing mitotically to maintain a population of stem spermatogonia cells (SSCs) through self-renewal, as well as (2) producing daughter cells referred to as A undifferentiated spermatogonia (A_{undiff}).

The A_{undiff} population starts as single cells (A_s or A_{single}) which divide once mitotically to form pairs of cells (A_{pr} or A_{paired}), then subsequent divisions produce aligned rows of four, eight, and ultimately 16-cell-chains of undifferentiated spermatogonia (A_{al} or $A_{aligned}$). This growth phase encompasses three rounds of mitotic divisions before the A_{undiff} cells undergo an irreversible commitment to differentiation and meiotic initiation termed the A to A1 transition (Griswold, 2016). This is an essential step in the germ cells' development, and the first cells produced from this commitment step are the A1 differentiating spermatogonia (A1). The A1 cells produced during the A to A1 transition continue to divide mitotically, undergoing another six rounds of mitotic division which produces sequential cohorts of differentiating spermatogonia [A2, A3, A4, Intermediate (In), and B], before ultimately forming preleptotene spermatocytes. Meiosis begins within the preleptotene spermatocytes, which migrate into the adluminal compartment by crossing the blood-testis barrier (BTB). The spermatocytes can be separated into two groups depending on their progress through meiosis: early (preleptotene, leptotene, and zygotene) and late (pachytene, diplotene, and secondary). The secondary spermatocytes formed at the end of this first round of meiosis begin undergoing a second round of meiotic division that forms spermatids.

During this third phase of spermatogenesis, known as spermiogenesis, spermatids undergo multiple phases of development: Golgi phase, cap phase, tail formation, and final maturation. During the Golgi phase, the round spermatids develop cellular polarity, transitioning away from the previously maintained radial symmetry as what will eventually become the head and tail of the spermatozoa begins to form. Additionally, during the Golgi phase, the genetic material stored inside of the spermatid nuclei also goes through a condensation event where histones are exchanged for protamines. This process inactivates the chromatin to make the cells transcriptionally inactive and allows the paternal genome to be condensed so it may fit inside the head of the forming spermatozoa. The histone-protamine swap additionally protects the paternal genetic material from damage or mutation. During the following cap phase, the Golgi apparatus surrounds the newly condensed spermatid nucleus and forms the acrosome, a unique organelle that secretes hydrolytic enzymes that will eventually aid the spermatozoa in penetrating the outer layers of the ovum. The spermatids then begin elongating, as the tail of the spermatid continues growing and the cell begins orienting itself so that the newly forming tail is pointed towards the lumen space and away from the tubule epithelium. Last, during the maturation phase, the elongated spermatids leave the seminiferous tubule environment after being freed from the seminiferous epithelium during spermiation and undergo final morphologic and chemical changes within the epididymis to become mature spermatozoa capable of fertilization (Bellvé et al., 1977; Hess and de Franca, 2008; Mruk and Cheng, 2004).

2. The essential role of sertoli cells in spermatogenesis

Aside from the various types of germ cells produced across the complete process of spermatogenesis, the only other cell type present within the seminiferous tubules of the testis are the Sertoli cells. Sertoli cells are omnipresent, somatic cells within the testis that carry out many functional roles to create the testicular microenvironment required to support the ongoing production of spermatozoa. It is because of the multitude of supporting functions the Sertoli cells perform that they are known as nurse cells. These include the regulation of hormone synthesis and the immune response, nutrient delivery to the developing male germ cells, and forming the BTB which is critical to germ cell growth as this barrier formation compartmentalizes the seminiferous tubule interior into a basal and adluminal compartment (França et al., 2016; Griswold, 1988; Kaur et al., 2014; Petersen and Söder, 2006). The BTB, formed by tight junctions between neighboring Sertoli cells, separates the pre-meiotic germ cells kept in the basal compartment from the spermatocytes and spermatids formed later in the adluminal compartment. This compartmentalization of the seminiferous epithelium supports germ cell development as each of these compartments provide separate environmental stimulus that supports the development of the different germ cell populations. This is the same barrier that preleptotene spermatocytes are required to cross in order to continue their development and begin meiosis as the environmental stimulus the germ cells receive in the basal compartment does not support meiotic progression (Cheng and Mruk, 2012; Luaces et al., 2023; Mruk and Cheng, 2015). All the germ cells formed within the seminiferous tubules after the onset of spermatogenesis are completely immotile, and because of this they rely entirely on the Sertoli cells to aid them in progressing from the basal compartment to the adluminal compartment within the seminiferous tubules as they develop. This occurs through the formation of multiple types of adherent cell-cell junctions between the Sertoli cells and the various germ cell subtypes found within the seminiferous tubules. The Sertoli cells are also responsible for freeing the elongated spermatids from the seminiferous epithelium by secreting enzymes that dissolve the ectoplasmic specializations that join the spermatids to the Sertoli cells, and phagocytizing the excess cytoplasm and organelles no longer needed by the spermatozoa (Griswold, 2016; Mruk and Cheng, 2004). This excess cytoplasm is known as the *residual body of Regaud*, or more commonly as the residual body (Firlit and Davis, 1965). Sertoli cells additionally are responsible for secreting the fluid that carries the freed elongated spermatids to the epididymis where final physical and chemical changes make them both motile and capable of fertilization (Richburg et al., 1994).

3. The cycle of the seminiferous epithelium

The cellular associations made between the germ cells and Sertoli cells are identifiable, cyclical, and can be sorted into 12 distinct stages in the mouse. The continual cycling of these stages is most commonly referred to as the cycle of the seminiferous epithelium (Hess and de Franca, 2008; Meistrich and Hess, 2013; Russell et al., 1993). This cycle occurs continuously and asynchronously, allowing for the continual production of readily available, viable spermatozoa. One complete cycle of the seminiferous epithelium in the mouse takes approximately 35 days, starting with the A_{undiff} population, eventually culminating with the release of elongated spermatids. The time between the occurrence of one stage until the next occurrence of that specific stage takes approximately 8.6 days, with the first wave of spermatogenesis being an exception to this rule, and taking closer to 6 days to complete the cycle (Agrimson et al., 2016). This timing was previously speculated to be a result of the absence of a self-renewal step by the SSCs, as the A1 spermatogonia are likely formed directly from the gonocytes, rather than from the A_{undiff} progenitor cells produced mitotically by the SSCs (Evans et al., 2014; Yoshida et al., 2006). Evidence for this phenomenon was compiled by evaluating the expression of SSC marker Neurogenin 3 (*Ngn3*). It was found that the germ cells that differentiate into the first A1 population never express *Ngn3*, suggesting that these cells most likely do not originate from the population of A$_{undiff}$ spermatogonia, but rather from gonocytes which never develop into SSCs prior to the A to A1 transition (Yoshida et al., 2006). Further studies are still required to fully identify the unique patterns of cell differentiation that occur during the first cycle of the seminiferous epithelium, and full elucidation of spermatogonial differentiation at this stage of neonatal testicular development will be essential to future reproductive healthcare.

4. Retinoic acid synthesis in the testis

The mechanism of testicular *at*RA synthesis has been well documented. *All-trans* retinol (ROL) circulates in plasma in a protein complex with transthyretin (TTR) and retinol binding protein 4 (RBP4) which is taken into the Sertoli cell via the cell membrane receptor stimulated by retinoic acid 6 (STRA6) (Amengual et al., 2014; Newcomer and Ong, 2013). Once ROL has entered the cell, it forms a complex with cellular retinoid binding protein (RBP1) where it can either be converted into *at*RA through two oxidation reactions, or esterified by lecithin retinol acyltransferase (LRAT) into retinyl esters for storage. The formed retinyl esters can later be converted into ROL by retinyl ester hydrolases (REH) (Shingleton et al., 1989; O'Byrne and Blaner, 2013) (Fig. 1).

The first oxidative step in the synthesis of *at*RA from ROL is catalyzed by retinol dehydrogenase 10 (RDH10), which requires NAD+ as a cofactor, to form *all-trans* retinaldehyde (RAL). This enzymatic process is known to be the rate-limiting step of *at*RA synthesis, and is reversible (Farjo et al., 2011). The reduction to form ROL from RAL can be catalyzed by multiple enzymes present in Sertoli cells such as the retinol dehydrogenases RDH11 and RDH14, as well as the dehydrogenase reductase enzymes DHRS3 and DHRS4 (Fig. 1). All of these enzymes require NADPH as a cofactor to carry out the conversion of RAL into ROL (Adams et al., 2014; Belyaeva et al., 2018; Hogarth et al., 2015a).

Following the synthesis of RAL from ROL, another oxidation step is performed by the three testicular RAL dehydrogenase enzymes (ALDH1A1, ALDH1A2, and ALDH1A3) to produce *at*RA (Napoli, 1996) (Fig. 1). *Aldh1a1* is the primary RAL dehydrogenase present in Sertoli cells, *Aldh1a2* is primarily located in germ cells, and *Aldh1a3* is only present at very low levels in the testis (Vernet et al., 2006; Kent et al., 2016). *At*RA appears to act in an autocrine and paracrine manner by changing gene expression in the

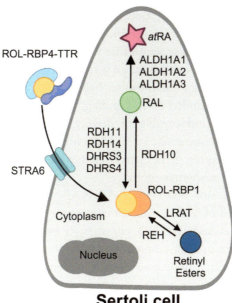

Fig. 1 Synthesis of *at*RA within Sertoli cells. *All-trans* retinol (ROL) is taken into the cell from circulating plasma while bound in a complex with RBP4 and TTR by the transmembrane protein STRA6. Once inside the cell, ROL, now bound to RBP1, can be either oxidized by RDH10 to produced *all-trans* retinaldehyde (RAL) or esterified by LRAT to form retinyl esters. Both of these processes are reversible, with the reduction of RAL to ROL carried out by the enzymes RDH11, RDH14, DHRS3 or DHRS4, and the conversion of retinyl esters to ROL by retinyl ester hydrolases (REH). Once produced within the cell, RAL can be further oxidized by the RAL dehydrogenase enzymes ALDH1A1, ALDH1A2, or ALDH1A3 to ultimately form *at*RA. *Created with Biorender.com.*

Sertoli cells and in the germ cells. The action is initiated by the entrance of *at*RA into the nucleus of the target cell, where it binds to a retinoic-acid-receptor (RAR) heterodimerized with a retinoid-x-receptor (RXR). The RAR:RXR heterodimers are bound on retinoic acid response elements (RAREs) located in target genes (Bastien and Rochette-Egly, 2004; Laudet and Gronemeyer, 2002; Rochette-Egly and Germain, 2009). There are three isotypes of the RARs (RARA, RARB, and RARG), which require either *at*RA or 9-*cis*-retinoic acid isomer (9-*cis*-RA) as binding substrate, and three isotypes of the RXRs (RXRA, RXRB, and RXRG), which use 9-*cis*-RA as the preferred binding substrate. RARs have been shown previously to act as the primary drivers of *at*RA-induced signaling in the testis (Gaemers et al., 1998). RXR action in the testis has been suggested to be primarily due to the necessary involvement of the RXRs in the formation of RAR:RXR heterodimers, rather than activation by 9-*cis*-RA. Previous published work has evaluated the effects of exogenous 9-cis-RA treatment on a vitamin A–deficient testis, and it was found that while 9-*cis*-RA is able to permeate the testis, a much higher concentration is required compared to the action of *at*RA, and 9-*cis*-RA has a much shorter half-life than *at*RA (Gaemers et al., 1998). Later work expanded on this finding, and showed that *at*RA was the only endogenous ligand capable of driving downstream *at*RA signaling within the testis (Allenby et al., 1993; Mouchon et al., 1999). RARA is the primary receptor expressed in Sertoli cells while RARG is the primary receptor expressed in spermatogonia.

When *at*RA is not bound to RAR:RXR heterodimers, a repressive complex comprised of nuclear receptor corepressors (NCOR1 and NCOR2) and histone deacetylase (HDAC) enzymes, is bound to the RAR:RXR heterodimer which, in turn, suppresses transcription (Fig. 2A). When *at*RA is bound, the heterodimer undergoes a conformational change that prompts the release of the corepressive complex, which allows nuclear receptor coactivators (NCOA1, NCOA2 and NCOA3) and histone acetylase enzymes (HAT) to bind. This induces chromatin relaxation, increased promoter activity, and ultimately upregulated *at*RA-driven gene expression (Fig. 2B) (Bastien and Rochette-Egly, 2004; O'Connor et al., 2022). Signaling mediated by *at*RA drives many homeostatic and developmental processes including embryonic body axis formation, neurogenesis, cardiogenesis, tissue

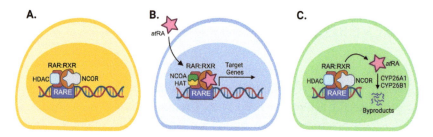

Fig. 2 The action of retinoic acid (*at*RA) within the A$_{undiff}$ spermatogonia leading to the A to A1 transition. (A) Without *at*RA, the RAR:RXR heterodimer is bound to retinoic acid response elements (RARE) and interacts with corepressors (NCOR-HDAC), maintaining a transcriptionally inactive state. (B) *at*RA enters the target cell, prompting the A to A1 transition, induces conformation changes that release NCOR-HDAC from the RAR:RXR heterodimers, and coactivators (NCOA-HAT) are recruited. This, in turn, relaxes the chromatin and enables the transcription of *at*RA-responsive genes. (C) After the A to A1 transition, RAR:RXR releases *at*RA which is degraded into water soluble byproducts by CYP26A1 or CYP26B1, and NCOR-HDAC binds to the RAR:RXR heterodimer preventing further gene transcription. *Created with Biorender.com.*

homeostasis, eye development, and, most importantly to this work, spermatogonial differentiation (Chambon, 1996; Duester, 2008; Ghyselinck and Duester, 2019; Mark et al., 2006; Raverdeau et al., 2012).

Following the response of the targeted cell population, *at*RA is ultimately degraded via a final third oxidation step that is carried out by the cytochrome p450 enzymes CYP26A1, CYP26B1 and CYP26C1, which breakdown *at*RA to produce inactive, water-soluble metabolite byproducts (Fig. 2C) (Isoherranen and Zhong, 2019; Thatcher and Isoherranen, 2009). The turnover of *at*RA in the murine testis occurs relatively quickly, and endogenous *at*RA has been documented to have a half-life of only 1.3 h (Arnold et al., 2015).

5. Elucidating testicular *at*RA synthesis activity through genetic knockout studies

While much of the above information has been identified through many morphologic and quantitative studies assessing how developing male germ cells change through the progression of spermatogenesis, more strides still have been, and actively are being made by utilizing genetically engineered model systems. These models have provided previously unseen resolution into the multiple steps of *at*RA synthesis in the testis.

5.1 Rdh10 expression by sertoli cells is required for juvenile spermatogenesis

The initial conversion of ROL to RAL, as facilitated by RDH10, has been previously evaluated using cell specific *Rdh10*-deficient mouse models (Tong et al., 2013). Previous investigations into *Rdh10* using a knockout model (KO) showed embryonic lethality because *Rdh10* is essential during mid-gestational embryogenesis by synthesizing *at*RA needed for limb, organ, and craniofacial development (Sandell et al. 2007, 2012). It was found that embryonic lethality induced by global knocking out *Rdh10* could be rescued by providing pregnant dams with exogenous RAL, however the overall fitness of the produced animals was still decreased (Rhinn et al., 2011).

Innovating on these results, germ cell-specific, Sertoli cell-specific, or both germ cell- and Sertoli cell-specific *Rdh10* conditional knockout (cKO) animals were viable, and revealed that *Rdh10* expression by Sertoli cells, not germ cells, is necessary for proper spermatogenic entry. Germ cell-specific *Rdh10* cKO mice showed no significant changes in testicular weight, morphology, or TUNEL-positive germ cells from the wild-type (WT) animals. By two weeks of age, however, Sertoli cell-specific as well as germ cell- and Sertoli cell-specific cKO animals showed a significant change from the WT animals in all these metrics. The testes collected and analyzed from animals with a Sertoli cell-specific deletion of *Rdh10* contained only A_{undiff} and Sertoli cells within nearly half of all the tubules analyzed. This phenotype was exacerbated within the Sertoli cell- and germ cell-specific cKO animals, manifesting in approximately 85% of all tubules. This result showed that while Sertoli cells are the primary synthesizers of RAL via RDH10 action, germ cells contribute to RAL production as well. However, a deficit in the ability of the germ cells to produce RAL is not significantly deleterious to the individual as long as Sertoli cells remain functional in their capacity to synthesize RAL. The germ cell loss seen in the Sertoli cell-specific cKO animals, and in the Sertoli cell- and germ cell-specific cKO animals, was found to recover by approximately nine weeks of age, with these animals eventually being capable of forming mature spermatozoa. The mechanism for why this recovery occurs is currently unclear, but it has been speculated that in the adult testis, round and elongating spermatids may be synthesizing RAL from β-Carotene via the enzyme β-Carotene-15,15'-dioxygenase (BCO1) (Vernet et al., 2006).

5.2 RAL dehydrogenase activity in Sertoli cells, but not germ cells, is essential for the initial A to A1 transition

As described above, the oxidation of RAL to *at*RA is known to be carried out by either ALDH1A1, ALDH1A2, or ALDH1A3. The latter is expressed in the interstitial cells of the testis (Teletin et al., 2019; Vernet et al., 2006), and its functional role remains unknown. However *Aldh1a3* KO animals are non-viable and die shortly after birth, which has suggested its importance may lie during embryonic development (Dupé et al., 2003). Similarly, *Aldh1a2* KO animals die during embryonic development (Niederreither et al., 1999). On the other hand, *Aldh1a1* KO animals are viable and fertile (Fan et al., 2003; Matt et al., 2005). Currently, evidence supports that the first germ cells capable of producing *at*RA are spermatocytes, as measured by in situ hybridization targeting *Aldh1a2* (Endo et al., 2015; Raverdeau et al., 2012). Chemical treatments which eliminated populations of spermatocytes led to a marked loss of germ cell expressed *Aldh1a2*, and resulted in a halt in germ cell maturation (Endo et al., 2015; Kirsanov et al., 2023). These combined works have provided valuable insight into the mechanism and importance of *at*RA synthesis in the testis, however the redundancy seen in the expression patterns of *Aldh1a1* and *Aldh1a2* remained a gap in our knowledge of the function of these enzymes.

To address this, all three RAL dehydrogenase enzymes were deleted from Sertoli cells, circumventing the risk of a rescue phenotype that could arise had they not targeted each of the three enzymes. Evaluation of the Sertoli cell-specific cKO animals found that they were unable to produce sufficient *at*RA to clear the initial A to A1 transition, leaving them infertile as spermatogenesis was never initiated (Raverdeau et al., 2012; Teletin et al., 2019). This result demonstrated that *at*RA synthesized by the Sertoli cells was essential to the onset of spermatogenesis, and subsequently the production of all mature male germ cells. When this same experiment was performed on animals deficient in the three RAL dehydrogenase enzymes solely within the germ cells, they found that *at*RA concentrations were decreased by approximately 63% compared to WT animals, there was a significant increase in upstream retinoid concentrations (both ROL and retinyl esters), but it was found that germ cell synthesized *at*RA was ultimately non-essential to spermatogenesis or the progression of the developing germ cells through the cycle of the seminiferous epithelium (Teletin et al., 2019). As long as the Sertoli cells were still able to produce *at*RA, regardless of the ability of the germ cells in synthesizing *at*RA, spermatogenesis was able to successfully continue. Interestingly, *Aldh1a1* KO animals were fertile (Fan et al., 2003; Matt et al., 2005). This suggests that the deficit caused by the loss of Sertoli cell expressed *Aldh1a1* could be rescued by Sertoli cell-expressed *Aldh1a2* in conjunction with germ cell action (Raverdeau et al., 2012; Teletin et al., 2019; Vernet et al., 2006).

The findings of this study were later confirmed by a different approach showing first that *Aldh1a3* activity was not sufficient to rescue spermatogenesis when *Aldh1a1* and *Aldh1a2* expression was deleted from germ cells and Sertoli cells, emphasizing that *Aldh1a3* is dispensable for *at*RA synthesis in the seminiferous epithelium (Topping and Griswold, 2022). It was then additionally shown that ablation of *Aldh1a1* and *Aldh1a2* in Sertoli cells prevented the A_{undiff} spermatogonia to clear the A to A1 transition, although this was not seen in the germ cell-specific cKO animals. These results directly support previous findings (Raverdeau et al., 2012; Teletin et al., 2019), confirming that RAL dehydrogenase activity loss within germ does not hinder spermatogenesis and that Sertoli cell-synthesized *at*RA was essential to the initiation of spermatogenesis and to the commitment of the A_{undiff} spermatogonia population to differentiation. Within a normally cycling murine testis, *at*RA synthesized by germ cells is nevertheless instrumental to the post-meiotic processes of spermiogenesis and spermiation. Loss of pachytene spermatocytes in vivo resulted in a significant decrease in total *at*RA concentrations in adult testes, without affecting the expression profiles of RAL dehydrogenases in the other germ cell populations or within the Sertoli cells. It additionally led to an increase in round spermatid populations, and in aligned elongated spermatids that had not been successfully released from the seminiferous tubules. These defects were found to be directly tied to reduced *at*RA, as they were reversed by an exogenous bolus of *at*RA (Endo et al., 2015). Exogenous *at*RA however was insufficient to rescue long-lasting spermatogenesis in testes in which the RAL dehydrogenases were knocked out in both the germ and Sertoli cell populations. This result showed that although not required for the initiation of spermatogenesis, germ cell synthesized *at*RA is, in fact, required for continuing spermatogenesis when Sertoli cell derived *at*RA is absent (Teletin et al., 2019; Topping and Griswold, 2022).

5.3 Loss of Cyp26b1, but not Cyp26a1, affects spermatogenesis

The action of the cytochrome p450 enzymes CYP26A1 and CYP26B1 has also been evaluated using genetic knockout models, in order to determine whether their action is essential to *at*RA degradation within the seminiferous tubule (Hogarth et al., 2015b). The authors found using germ cell- and Sertoli cell-specific cKO models that *Cyp26a1* is dispensable to spermatogenesis, as knocking out *Cyp26a1* in either cell type showed no morphological impact on the testes of male mice. In contrast, *Cyp26b1* Sertoli cell-specific cKO had an effect, as the testes collected from these animals contained seminiferous tubules completely devoid of differentiating germ cells, supporting that *Cyp26b1* in Sertoli cells was needed for normal spermatogenesis. When *Cyp26b1* was deleted in both germ cells and Sertoli cells, massive defects could be found with an average of 96% of tubules being devoid of differentiating germ cells. Combined, the results of this work ultimately supported that CYP26B1 activity was required in germ cells and Sertoli cells for normal spermatogenesis (Hogarth et al., 2015b).

5.4 Gene expression changes induced by atRA support spermatogonial differentiation

One of the most easily identifiable and observable responses of germ cells to *at*RA is the expression of the transcription factor stimulated by retinoid acid 8 (*Stra8*) which is specific to the differentiating spermatogonia and preleptotene spermatocytes (Anderson et al., 2008; Endo et al., 2015; Mark et al., 2008; Mark et al., 2015; Zhou et al., 2008). STRA8 has been shown to be required for successful meiotic initiation, as spermatocytes from STRA8 KO mice are unable to clear meiotic prophase (Anderson et al., 2008). This is not the only function of STRA8 however, as significant increases in the sizes of testes isolated from STRA8 KO mice were found and this result was coupled with an accumulation of undifferentiated spermatogonia (Endo et al., 2015). It was additionally found that the population of differentiating population was significantly reduced, and this result was visible in animals as young as 10 days DPP. These findings supported that STRA8 has a functional role, prior to meiosis, in supporting spermatogonial differentiation. Later work found that differentiating spermatogonia isolated from STRA8 KO testes immediately following the A to A1 transition had increased levels of *Neurog3, Sox3, Lin28a, Zbtb16, Pou5f1, and Nanos2*, which are all characteristic of the A_{undiff} spermatogonia, and would normally be downregulated in differentiating cells (Gewiss et al., 2021). These changes in gene expression correlated with a decrease in differentiating cells, and an increase in the undifferentiated population, reaffirming the functional role of STRA8 to the commitment of spermatogonia differentiation, and ongoing spermatogenesis.

The KIT proto-oncogene receptor tyrosine kinase (*Kit*) is present in the same population of differentiating spermatogonia that expresses STRA8 (Busada et al., 2015; Schrans-Stassen et al., 1999; Yoshinaga et al., 1991). Evaluations into STRA8 KO mice showed that KIT expression was not dependent on STRA8, but induced in differentiating spermatogonia following treatment with exogenous *at*RA (Busada et al., 2015). Kit ligand (*Kitl*) is expressed on Sertoli cells and is an essential binding partner to KIT (Loveland and Schlatt, 1997). *Kitl* expression has been found localized to the basement membrane of the seminiferous tubules, suggesting a functional role related to the proliferation of some spermatogonia populations (Manova et al., 1993). This model was supported by other findings indicating the importance of KIT to the survival of the differentiating spermatogonia populations. Loss of KIT in the mouse did not impact the self-renewal of the SSCs or the A_{undiff} daughter cells produced mitotically but did result in a depletion of the differentiating spermatogonia populations between 6 and 7 DPP causing a block in spermatogenesis, and prompting the authors to conclude that KIT is essential for ongoing spermatogenesis (Yoshinaga et al., 1991).

6. Synchronizing spermatogenesis using the drug WIN 18,446

In a normal testicular environment, spermatogenesis occurs asynchronously. This is an advantage for the male, as it provides a readily available pool of mature spermatozoa when needed but is a significant hurdle for any researchers interested in evaluating stage-specific, or cell-specific functions in the testis. At any given time during the cycle of the seminiferous epithelium, there are at least four separate germ cell-types present within a single cross-section of a seminiferous tubule (Hess and de Franca, 2008; Meistrich and Hess, 2013; Russell et al., 1993). The action of *at*RA proceeds in a pulsatile manner, where the response to atRA can be visualized occurring periodically in defined patches along the length of the tubule, moving along the length of the tubule like a wave (Evans et al., 2014; Gewiss et al., 2020; Hogarth et al., 2011; Snyder et al., 2010) (Fig. 3A). A cross-section of a normally cycling, adult testis will yield a variety of stages and a multitude of individual cell-types, making attempts at narrowing specific gene expression or function to a single point stage or cell-type challenging.

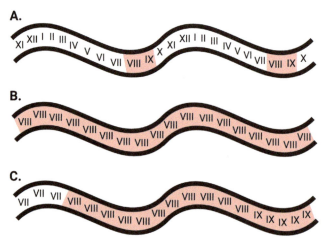

Fig. 3 Stage progression along the seminiferous tubules. Red shading represents retinoic acid (*at*RA) activity during stages VIII-IX that correspond with the A to A1 transition. (A) In a normally cycling murine testis, the stages of the seminiferous epithelium progress sequentially with pulses of *at*RA activity occurring only during stages VIII-IX. (B) In a testis following synchronization by WIN 18,446, and then *at*RA treatments, the A to A1 transition occurs in all A_{undiff} spermatogonia and is observed along the entire length of the seminiferous tubule. (C) Late after the treatment, the seminiferous epithelium gradually returns to the typical sequential arrangement of stages, with 3–5 stages present along the length of the tubule. *Created with Biorender.com.*

Bis-diclorodiacetyl-diamines (BDADs) are a class of drug that inhibit the action of RAL dehydrogenase enzymes throughout the body (Coulston et al., 1960). Disulfiram is one such member of this drug family, which is available as a prescribable alcohol deterrent drug (Stokes and Abdijadid, 2023). By inhibiting the enzyme ALDH1A1 within the liver, the primary method of enzymatic alcohol breakdown is blocked. Alcohol consumption in tandem with disulfiram treatment results in many symptoms including nausea, flushing of the face, and disorientation on the milder end and hepatic failure, cardiac failure, and death in extreme response cases. The side-effects induced by disulfiram treatment is both dependent on the dose of disulfiram the patient is taking, as well as the quantity of alcohol consumed while taking the drug. Critically, this medication does not modify how the brain responds to alcohol or acts to treat alcohol-addiction, but rather functions as an alcohol discouraging drug. N,N'-(octane-1,8-diyl)bis (2,2-dichloroacetamide), also called WIN 18,446, is another BDAD that had been considered a viable candidate for a male contraceptive. Named Fertilysin during clinical trials, WIN 18,446 showed promising results inhibiting the production of sperm in human, as well as in many animal species including rodents, wolves, cats, and shrews, providing what appeared to be a safe, reversible, option for a novel male contraceptive (Coulston et al., 1960; Heller et al., 1961). However, testing in humans revealed that alcohol consumption alongside Fertilysin treatment resulted in similar side-effects as produced during disulfiram treatment by inhibiting alcohol breakdown in the liver, and, following these findings, the drug was never approved for production (Heller et al., 1961).

While Fertilysin never produced the male contraceptive researchers hoped it would, it has provided reproductive biologists with a powerful tool to assess spermatogenesis in a stage specific manner. By inhibiting testicular RAL dehydrogenase enzymes (Amory et al., 2011; Paik et al., 2014), daily WIN 18,446 treatment provided orally at a dosage of 100 mg/kg beginning at 2 DPP in the mouse prevents the initial synthesis of *at*RA in Sertoli cells, and effectively blocks the A to A1 transition from occurring (Hogarth et al., 2011, 2013; Paik et al., 2014). As long as this treatment regimen is maintained within a 24-hour window of the previous dose, so too is the inhibition of the RAL dehydrogenase enzymes and ultimately *at*RA production. During this period, it can be observed that the SSCs continue to divide mitotically but cannot undergo the A to A1 transition (Hogarth et al., 2011; Paik et al., 2014).

Following WIN 18,446 treatment, exogenous *at*RA can be given to the male mouse via intraperitoneal injection which will, in turn, induce differentiation within all the accumulated A_{undiff} spermatogonia at once. Within the seminiferous tubules of treated mice, the response to *at*RA does not occur in a pulsatile manner, rather the initiation of the A to A1 transition can be observed along the entire length of the tubule (Fig. 3B). Amounts of exogenous *at*RA ranging between 50 μg to 200 μg have been found to be sufficient in inducing the A to A1 transition, although in practice 100 μg has been found to be ideal to initiate spermatogenesis, without sacrificing the overall health and fitness of the experimental animal (Agrimson et al., 2016). The A to A1 transition occurs synchronously, and a cross-section of a testis collected from a treated animal shows that all the seminiferous tubules can be sorted into approximately 3–5 distinct stages of the seminiferous epithelium, instead of 12 (Agrimson et al., 2016; Gewiss et al., 2021). Then, spermatogenesis continues in a synchronized manner into adulthood but will gradually return to the typical pattern with successive stages being visible. Evaluations of treated testes 90 days after *at*RA injection, show a sequential arrangement of stages is visible across an entire testis cross-section (Fig. 3C) (Hogarth et al., 2013).

By controlling spermatogenesis initiation, researchers are allowed much more control over their animal models and their experimental designs. WIN 18,446 produces a testicular environment that proceeds with normal timing throughout the cycle of the seminiferous epithelium, and a synchronized testis allows researchers to capture and evaluate individual cell-types and stages with high specificity.

Prior to the A to A1 transition, WIN 18,446 halts A_{undiff} development during the G0 phase of mitosis, and during the first wave of spermatogenesis the A_{undiff} spermatogonia likely skip either the A1 or A2 step, shortening the cycle from 8.6 to 6 days (Agrimson et al., 2016). This finding is consistent with a non-synchronized environment, supporting that WIN 18,446 and *at*RA treatment does not significantly alter normal spermatogenic timing, and supporting also previous results investigating the length of the first cycle of spermatogenesis (Yoshida et al., 2006). Later studies showed that one bolus of exogenous *at*RA was sufficient to promote the A to A1 transition, and additional administered amounts of exogenous *at*RA does not alter the developmental timeline of the differentiating spermatogonia (Johnson et al., 2023).

Previous work evaluating the Sertoli cell transcriptome in adult, non-synchronized testes using single cell RNA-sequencing (scRNA-seq) was able to identify four major groups of Sertoli cells, that could be further separated into nine subtypes. However, this heterogeneity could not be attributed to specific stages, as each subclass of Sertoli cell identified could be found in multiple stages throughout the cycle of the seminiferous epithelium (Green et al., 2018). Later work performed bulk RNA sequencing (RNA-seq) on Sertoli cells purified from a synchronized testis, and paired this analysis with scRNA-seq of adult synchronized testes to confirm cell-specificity of the assessed transcripts (Gewiss et al., 2021). This allowed the researchers to propose that Sertoli cells could be largely grouped into two distinct clusters: one occurring at stages VII-VIII and another encompassing all the other stages across the cycle of the seminiferous epithelium.

Another study using WIN 18,446 treatment allowed to isolate A1 spermatogonia, and all the following germ cell-subtypes formed prior to the elongating spermatids, to perform scRNA-seq. This study identified new markers for the various subtypes of postnatal male germ cells, and provided a comprehensive resource for future studies (Chen et al., 2018).

WIN 18,446 and *at*RA treatment has been additionally utilized to isolate early undifferentiated and differentiating spermatogonia prior to the assay for transposase-accessible chromatin with sequencing (ATAC-seq) to determine if chromatin states are altered within the spermatogonial populations in response to *at*RA (Schleif et al., 2023). This work supported previous results identified via transmission electron microscope (TEM) which suggested that there is a large amount of accessible chromatin prior to spermatogonial differentiation during the A to A1 transition, and showed that *at*RA causes a shift in chromatin structure in the spermatogonia resulting in a higher proportion of closed chromatin following *at*RA exposure (Agrimson et al., 2016; Chiarini-Garcia and Russell, 2001).

7. Conclusions

Within this review, we have summarized the current understanding of the essential role *at*RA plays during the commitment of A_{undiff} spermatogonia to differentiation at the onset of spermatogenesis. This event, termed the A to A1 transition, is essential to the production of spermatozoa as spermatogenesis cannot proceed if it does not occur. Investigations utilizing genetically engineered mouse models containing cell-specific deletions have highlighted the critical role played by *at*RA synthesizing enzymes, notably in Sertoli cells. Understanding the triggers that underlie the A to A1 transition are essential, as any discoveries into this area of reproductive biology may result in new treatments for male infertility, or in the development of a novel male contraceptive by blocking spermatogenesis at an early step. WIN 18,446 was one such potential avenue for a male contraceptive, but following the discovery of some harmful side-effects that disqualified its use as a pharmaceutical, this compound has now been reappropriated into an essential tool for male reproductive biologists. Studies utilizing WIN 18,446-induced spermatogenic synchrony in animal models can provide temporal specificity that cannot be reached in an unsynchronized environment. They will expand our knowledge of the cycle of the seminiferous epithelium and of germ cell-specific functions throughout spermatogenesis, hopefully leading to new discoveries that may positively shape the landscape of male reproductive research and medicine for generations to come.

References

Adams, M. K., Belyaeva, O. V., Wu, L., & Kedishvili, N. Y. (2014). The retinaldehyde reductase activity of DHRS3 is reciprocally activated by retinol dehydrogenase 10 to control retinoid homeostasis *. *Journal of Biological Chemistry, 289*, 14868–14880. https://doi.org/10.1074/jbc.M114.552257.

Agarwal, A., Mulgund, A., Hamada, A., & Chyatte, M. R. (2015). A unique view on male infertility around the globe. *Reproductive Biology and Endocrinology: RB&E, 13*, 37. https://doi.org/10.1186/s12958-015-0032-1.

Agrimson, K. S., Onken, J., Mitchell, D., Topping, T. B., Chiarini-Garcia, H., Hogarth, C. A., & Griswold, M. D. (2016). Characterizing the spermatogonial response to retinoic acid during the onset of spermatogenesis and following synchronization in the neonatal mouse testis. *Biology of Reproduction, 95*, 81. https://doi.org/10.1095/biolreprod.116.141770.

Alberts, B., Johnson, A., Lewis, J., Raff, M., Roberts, K., & Walter, P. (2002). Primordial germ cells and sex determination in mammals. In *Molecular biology of the cell*. Garland Science.

Allenby, G., Bocquel, M. T., Saunders, M., Kazmer, S., Speck, J., Rosenberger, M., ... Chambon, P. (1993). Retinoic acid receptors and retinoid X receptors: Interactions with endogenous retinoic acids. *Proceedings of the National Academy of Sciences of the United States of America, 90*, 30–34.

Amengual, J., Zhang, N., Kemerer, M., Maeda, T., Palczewski, K., & Von Lintig, J. (2014). STRA6 is critical for cellular vitamin A uptake and homeostasis. *Human Molecular Genetics, 23*, 5402–5417. https://doi.org/10.1093/hmg/ddu258.

Amory, J. K., Muller, C. H., Shimshoni, J. A., Isoherranen, N., Paik, J., Moreb, J. S., ... Griswold, M. D. (2011). Suppression of spermatogenesis by bisdichloroacetyldiamines is mediated by inhibition of testicular retinoic acid biosynthesis. *Journal of Andrology, 32*, 111–119. https://doi.org/10.2164/jandrol.110.010751.

Anderson, E. L., Baltus, A. E., Roepers-Gajadien, H. L., Hassold, T. J., de Rooij, D. G., van Pelt, A. M. M., & Page, D. C. (2008). Stra8 and its inducer, retinoic acid, regulate meiotic initiation in both spermatogenesis and oogenesis in mice. *Proceedings of the National Academy of Sciences of the United States of America, 105*, 14976–14980. https://doi.org/10.1073/pnas.0807297105.

Arnold, S. L., Kent, T., Hogarth, C. A., Schlatt, S., Prasad, B., Haenisch, M., ... Isoherranen, N. (2015). Importance of ALDH1A enzymes in determining human testicular retinoic acid concentrations. *Journal of Lipid Research, 56*, 342–357. https://doi.org/10.1194/jlr.M054718.

Bastien, J., & Rochette-Egly, C. (2004). Nuclear retinoid receptors and the transcription of retinoid-target genes. *Gene, 328*, 1–16. https://doi.org/10.1016/j.gene.2003.12.005.

Bellvé, A. R., Cavicchia, J. C., Millette, C. F., O'Brien, D. A., Bhatnagar, Y. M., & Dym, M. (1977). Spermatogenic cells of the prepuberal mouse. Isolation and morphological characterization. *The Journal of Cell Biology, 74*, 68–85. https://doi.org/10.1083/jcb.74.1.68.

Belyaeva, O. V., Wu, L., Shmarakov, I., Nelson, P. S., & Kedishvili, N. Y. (2018). Retinol dehydrogenase 11 is essential for the maintenance of retinol homeostasis in liver and testis in mice. *Journal of Biological Chemistry, 293*, 6996–7007. https://doi.org/10.1074/jbc.RA117.001646.

Busada, J. T., Chappell, V. A., Niedenberger, B. A., Kaye, E. P., Keiper, B. D., Hogarth, C. A., & Geyer, C. B. (2015). Retinoic acid regulates Kit translation during spermatogonial differentiation in the mouse. *Developmental Biology, 397*, 140–149. https://doi.org/10.1016/j.ydbio.2014.10.020.

Chambon, P. (1996). A decade of molecular biology of retinoic acid receptors. *The FASEB Journal, 10*, 940–954.

Chen, Y., Zheng, Y., Gao, Y., Lin, Z., Yang, S., Wang, T., ... Tong, M.-H. (2018). Single-cell RNA-seq uncovers dynamic processes and critical regulators in mouse spermatogenesis. *Cell Research, 28*, 879–896. https://doi.org/10.1038/s41422-018-0074-y.

Cheng, C. Y., & Mruk, D. D. (2012). The blood-testis barrier and its implications for male contraception. *Pharmacological Reviews, 64*, 16–64. https://doi.org/10.1124/pr.110.002790.

Chiarini-Garcia, H., & Russell, L. D. (2001). High-resolution light microscopic characterization of mouse spermatogonia. *Biology of Reproduction, 65*, 1170–1178. https://doi.org/10.1095/biolreprod65.4.1170.

Coulston, F., Beyler, A. L., & Drobeck, H. P. (1960). The biologic actions of a new series of bis(dichloroacetyl) diamines. *Toxicology and Applied Pharmacology, 2*, 715–731. https://doi.org/10.1016/0041-008x(60)90088-0.

Culty, M. (2009). Gonocytes, the forgotten cells of the germ cell lineage. *Birth Defects Research Part C: Embryo Today: Reviews, 87*, 1–26. https://doi.org/10.1002/bdrc.20142.

Duester, G. (2008). Retinoic acid synthesis and signaling during early organogenesis. *Cell, 134*, 921–931. https://doi.org/10.1016/j.cell.2008.09.002.

Dupé, V., Matt, N., Garnier, J.-M., Chambon, P., Mark, M., & Ghyselinck, N. B. (2003). A newborn lethal defect due to inactivation of retinaldehyde dehydrogenase type 3 is prevented by maternal retinoic acid treatment. *Proceedings of the National Academy of Sciences of the United States of America, 100*, 14036–14041. https://doi.org/10.1073/pnas.2336223100.

Endo, T., Romer, K. A., Anderson, E. L., Baltus, A. E., Rooij, D. G., de, & Page, D. C. (2015). Periodic retinoic acid–STRA8 signaling intersects with periodic germ-cell competencies to regulate spermatogenesis. *PNAS, 112*, E2347–E2356. https://doi.org/10.1073/pnas.1505683112.

Evans, E., Hogarth, C., Mitchell, D., & Griswold, M. (2014). Riding the spermatogenic wave: Profiling gene expression within neonatal germ and sertoli cells during a synchronized initial wave of spermatogenesis in mice. *Biology of Reproduction, 90*, 108. https://doi.org/10.1095/biolreprod.114.118034.

Fan, X., Molotkov, A., Manabe, S.-I., Donmoyer, C. M., Deltour, L., Foglio, M. H., ... Duester, G. (2003). Targeted disruption of Aldh1a1 (Raldh1) provides evidence for a complex mechanism of retinoic acid synthesis in the developing retina. *Molecular and Cellular Biology, 23*, 4637–4648. https://doi.org/10.1128/MCB.23.13.4637-4648.2003.

Farjo, K. M., Moiseyev, G., Nikolaeva, O., Sandell, L. L., Trainor, P. A., & Ma, J. (2011). RDH10 is the primary enzyme responsible for the first step of embryonic Vitamin A metabolism and retinoic acid synthesis. *Developmental Biology, 357*, 347–355. https://doi.org/10.1016/j.ydbio.2011.07.011.

Firlit, C. F., & Davis, J. R. (1965). Morphogenesis of the residual body of the mouse testis. *Journal of Cell Science s3-106*, 93–98. https://doi.org/10.1242/jcs.s3-106.73.93.

França, L. R., Hess, R. A., Dufour, J. M., Hofmann, M. C., & Griswold, M. D. (2016). The Sertoli cell: One hundred fifty years of beauty and plasticity. *Andrology, 4*, 189–212. https://doi.org/10.1111/andr.12165.

Gaemers, I., Sonneveld, E., van Pelt, A., Schrans, B., Themmen, A., van der Saag, P., & Rooij, D. (1998). The effect of 9-cis-retinoic acid on proliferation and differentiation of A spermatogonia and retinoid receptor gene expression in the vitamin A- deficient mouse testis. *Endocrinology, 139*, 4269–4276. https://doi.org/10.1210/en.139.10.4269.

Gewiss, R., Topping, T., & Griswold, M. D. (2020). Cycles, waves, and pulses: Retinoic acid and the organization of spermatogenesis. *Andrology, 8*, 892–897. https://doi.org/10.1111/andr.12722.

Gewiss, R. L., Law, N. C., Helsel, A. R., Shelden, E. A., & Griswold, M. D. (2021). Two distinct Sertoli cell states are regulated via germ cell crosstalk. *Biology of Reproduction, 105*, 1591–1602. https://doi.org/10.1093/biolre/ioab160.

Gewiss, R. L., Shelden, E. A., & Griswold, M. D. (2021). STRA8 induces transcriptional changes in germ cells during spermatogonial development. *Molecular Reproduction and Development, 88*, 128–140. https://doi.org/10.1002/mrd.23448.

Ghyselinck, N. B., & Duester, G. (2019). Retinoic acid signaling pathways. *Development (Cambridge, England), 146*, dev167502. https://doi.org/10.1242/dev.167502.

Green, C. D., Ma, Q., Manske, G. L., Shami, A. N., Zheng, X., Marini, S., ... Hammoud, S. S. (2018). A comprehensive roadmap of murine spermatogenesis defined by single-cell RNA-seq. *Developmental Cell, 46*, 651–667.e10. https://doi.org/10.1016/j.devcel.2018.07.025.

Griswold, M. D. (2016). Spermatogenesis: The commitment to meiosis. *Physiological Reviews, 96*, 1–17. https://doi.org/10.1152/physrev.00013.2015.

Griswold, M. D. (1988). Protein secretions of Sertoli cells. *International Review of Cytology, 110*, 133–156. https://doi.org/10.1016/s0074-7696(08)61849-5.

Hamada, A., Esteves, S. C., Nizza, M., & Agarwal, A. (2012). Unexplained Male infertility: Diagnosis and Management. *International Braz j Urol: Official Journal of the Brazilian Society of Urology, 38*, 576–594. https://doi.org/10.1590/S1677-55382012000500002.

Heller, C. G., Moore, D. J., & Paulsen, C. A. (1961). Suppression of spermatogenesis and chronic toxicity in men by a new series of bis(dichloroacetyl) diamines. *Toxicology and Applied Pharmacology, 3*, 1–11. https://doi.org/10.1016/0041-008x(61)90002-3.

Hess, R. A., & de Franca, L. R. (2008). Spermatogenesis and cycle of the seminiferous epithelium. In C. Y. Cheng (Ed.). *Molecular mechanisms in spermatogenesis, advances in experimental medicine and biology* (pp. 1–15). New York, NY: Springer. https://doi.org/10.1007/978-0-387-09597-4_1.

Hogarth, C. A., Arnold, S., Kent, T., Mitchell, D., Isoherranen, N., & Griswold, M. D. (2015a). Processive pulses of retinoic acid propel asynchronous and continuous murine sperm production1. *Biology of Reproduction, 92*(37), 1–11. https://doi.org/10.1095/biolreprod.114.126326.

Hogarth, C. A., Evanoff, R., Mitchell, D., Kent, T., Small, C., Amory, J. K., & Griswold, M. D. (2013). Turning a spermatogenic wave into a tsunami: Synchronizing murine spermatogenesis using WIN 18,4461. *Biology of Reproduction, 88*(40), 1–49. https://doi.org/10.1095/biolreprod.112.105346.

Hogarth, C. A., Evanoff, R., Snyder, E., Kent, T., Mitchell, D., Small, C., ... Griswold, M. D. (2011). Suppression of Stra8 expression in the mouse gonad by WIN 18,446. *Biology of Reproduction, 84*, 957–965. https://doi.org/10.1095/biolreprod.110.088575.

Hogarth, C. A., Evans, E., Onken, J., Kent, T., Mitchell, D., Petkovich, M., & Griswold, M. D. (2015b). CYP26 enzymes are necessary within the postnatal seminiferous epithelium for normal murine spermatogenesis. *Biology of Reproduction, 93*, 19. https://doi.org/10.1095/biolreprod.115.129718.

Hu, Y.-C., Nicholls, P. K., Soh, Y. Q. S., Daniele, J. R., Junker, J. P., van Oudenaarden, A., & Page, D. C. (2015). Licensing of primordial germ cells for gametogenesis depends on genital ridge signaling. *PLoS Genetics, 11*, e1005019. https://doi.org/10.1371/journal.pgen.1005019.

Isoherranen, N., & Zhong, G. (2019). Biochemical and physiological importance of the CYP26 retinoic acid hydroxylases. *Pharmacology & Therapeutics, 204*, 107400. https://doi.org/10.1016/j.pharmthera.2019.107400.

Johnson, T. A., Niedenberger, B. A., Kirsanov, O., Harrington, E. V., Malachowski, T., & Geyer, C. B. (2023). Differential responsiveness of spermatogonia to retinoic acid dictates precocious differentiation but not meiotic entry during steady-state spermatogenesis. *Biology of Reproduction, 108*, 822–836. https://doi.org/10.1093/biolre/ioad010.

Kaur, G., Thompson, L. A., & Dufour, J. M. (2014). Sertoli cells- Immunological sentinels of spermatogenesis. *Seminars in Cell & Developmental Biology, 36*–44. https://doi.org/10.1016/j.semcdb.2014.02.011.

Kent, T., Arnold, S. L., Fasnacht, R., Rowsey, R., Mitchell, D., Hogarth, C. A., ... Griswold, M. D. (2016). ALDH Enzyme Expresion Is Independent of the Spermatogenic Cycle, and Their Inhibition Causes Misregulation of Murine Spermatogenic Processes. *Biol Reprod, 94*(1), 12. https://doi.org/10.1095/biolreprod.115.131458.

Kirsanov, O., Johnson, T. A., Niedenberger, B. A., Malachowski, T. N., Hale, B. J., Chen, Q., ... Geyer, C. B. (2023). Retinoic acid is dispensable for meiotic initiation but required for spermiogenesis in the mammalian testis. *Development (Cambridge, England), 150*, dev201638. https://doi.org/10.1242/dev.201638.

Kothandaraman, N., Agarwal, A., Abu-Elmagd, M., & Al-Qahtani, M. H. (2016). Pathogenic landscape of idiopathic male infertility: New insight towards its regulatory networks. *npj Genomic Medicine, 1*, 16023. https://doi.org/10.1038/npjgenmed.2016.23.

Laudet, V., & Gronemeyer, H. (2002). RAR. In V. Laudet, & H. Gronemeyer (Eds.). *The nuclear receptor FactsBook, Factsbook* (pp. 113–140). London: Academic Press. https://doi.org/10.1016/B978-012437735-6/50014-X.

Loveland, K. L., & Schlatt, S. (1997). Stem cell factor and c-kit in the mammalian testis: Lessons originating from Mother Nature's gene knockouts. *Journal of Endocrinology, 153*, 337–344. https://doi.org/10.1677/joe.0.1530337.

Luaces, J. P., Toro-Urrego, N., Otero-Losada, M., & Capani, F. (2023). What do we know about blood-testis barrier? Current understanding of its structure and physiology. *Frontiers in Cell and Developmental Biology, 11*.

Manova, K., Huang, E. J., Angeles, M., De Leon, V., Sanchez, S., Pronovost, S. M., ... Bachvarova, R. F. (1993). The expression pattern of the c-kit ligand in gonads of mice supports a role for the c-kit receptor in oocyte growth and in proliferation of spermatogonia. *Developmental Biology, 157*, 85–99. https://doi.org/10.1006/dbio.1993.1114.

Mark, M., Ghyselinck, N. B., & Chambon, P. (2006). Function of retinoid nuclear receptors: Lessons from genetic and pharmacological dissections of the retinoic acid signaling pathway during mouse embryogenesis. *Annual Review of Pharmacology and Toxicology, 46*, 451–480. https://doi.org/10.1146/annurev.pharmtox.46.120604.141156.

Mark, M., Jacobs, H., Oulad-Abdelghani, M., Dennefeld, C., Féret, B., Vernet, N., ... Ghyselinck, N. B. (2008). STRA8-deficient spermatocytes initiate, but fail to complete, meiosis and undergo premature chromosome condensation. *Journal of Cell Science, 121*, 3233–3242. https://doi.org/10.1242/jcs.035071.

Mark, M., Teletin, M., Vernet, N., & Ghyselinck, N. B. (2015). Role of retinoic acid receptor (RAR) signaling in post-natal male germ cell differentiation. *Biochimica et Biophysica Acta (BBA) – Gene Regulatory Mechanisms, Nuclear Receptors in Animal Development, 1849*, 84–93. https://doi.org/10.1016/j.bbagrm.2014.05.019.

Matt, N., Dupé, V., Garnier, J.-M., Dennefeld, C., Chambon, P., Mark, M., & Ghyselinck, N. B. (2005). Retinoic acid-dependent eye morphogenesis is orchestrated by neural crest cells. *Development (Cambridge, England), 132*, 4789–4800. https://doi.org/10.1242/dev.02031.

Meistrich, M. L., & Hess, R. A. (2013). Assessment of spermatogenesis through staging of seminiferous tubules. In D. T. Carrell, & K. I. Aston (Eds.). *Spermatogenesis: Methods and protocols, methods in molecular biology* (pp. 299–307). Totowa, NJ: Humana Press. https://doi.org/10.1007/978-1-62703-038-0_27.

Mouchon, A., Delmotte, M.-H., Formstecher, P., & Lefebvre, P. (1999). Allosteric regulation of the discriminative responsiveness of retinoic acid receptor to natural and synthetic ligands by retinoid X receptor and DNA. *Molecular and Cellular Biology, 19*, 3073–3085.

Mruk, D. D., & Cheng, C. Y. (2015). The mammalian blood-testis barrier: Its biology and regulation. *Endocrine Reviews, 36*, 564–591. https://doi.org/10.1210/er.2014-1101.

Mruk, D. D., & Cheng, C. Y. (2004). Sertoli-sertoli and sertoli-germ cell interactions and their significance in germ cell movement in the seminiferous epithelium during spermatogenesis. *Endocrine Reviews, 25*, 747–806. https://doi.org/10.1210/er.2003-0022.

Napoli, J. L. (1996). Retinoic acid biosynthesis and metabolism. *The FASEB Journal, 10*, 993–1001. https://doi.org/10.1096/fasebj.10.9.8801182.

Newcomer, M. E., Ong, D. E. (2013). Retinol binding protein and its interaction with transthyretin. In *Madame Curie Bioscience Database* [Internet]. Landes Bioscience.

Niederreither, K., Subbarayan, V., Dollé, P., & Chambon, P. (1999). Embryonic retinoic acid synthesis is essential for early mouse post-implantation development. *Nature Genetics, 21*, 444–448. https://doi.org/10.1038/7788.

Nikolic, A., Volarevic, V., Armstrong, L., Lako, M., & Stojkovic, M. (2016). Primordial germ cells: Current knowledge and perspectives. *Stem Cells International 2016*, 1741072. https://doi.org/10.1155/2016/1741072.

O'Byrne, S. M., & Blaner, W. S. (2013). Retinol and retinyl esters: Biochemistry and physiology: Thematic Review Series: Fat-Soluble Vitamins: Vitamin A. *Journal of Lipid Research, 54*, 1731–1743. https://doi.org/10.1194/jlr.R037648.

O'Connor, C., Varshosaz, P., & Moise, A. R. (2022). Mechanisms of feedback regulation of vitamin A metabolism. *Nutrients, 14*, 1312. https://doi.org/10.3390/nu14061312.

Paik, J., Haenisch, M., Muller, C. H., Goldstein, A. S., Arnold, S., Isoherranen, N., ... Amory, J. K. (2014). Inhibition of retinoic acid biosynthesis by the bisdichloroacetyldiamine WIN 18,446 markedly suppresses spermatogenesis and alters retinoid metabolism in mice. *The Journal of Biological Chemistry, 289*, 15104–15117. https://doi.org/10.1074/jbc.M113.540211.

Petersen, C., & Söder, O. (2006). The sertoli cell – A hormonal target and 'super' nurse for germ cells that determines testicular size. *Hormone Research, 66*, 153–161. https://doi.org/10.1159/000094142.

Raverdeau, M., Gely-Pernot, A., Féret, B., Dennefeld, C., Benoit, G., Davidson, I., ... Ghyselinck, N. B. (2012). Retinoic acid induces Sertoli cell paracrine signals for spermatogonia differentiation but cell autonomously drives spermatocyte meiosis. *Proceedings of the National Academy of Sciences, 109*, 16582–16587. https://doi.org/10.1073/pnas.1214936109.

Rhinn, M., Schuhbaur, B., Niederreither, K., & Dollé, P. (2011). Involvement of retinol dehydrogenase 10 in embryonic patterning and rescue of its loss of function by maternal retinaldehyde treatment. *Proceedings of the National Academy of Sciences of the United States of America, 108*, 16687–16692. https://doi.org/10.1073/pnas.1103877108.

Richburg, J. H., Redenbach, D. M., & Boekelheide, K. (1994). Seminiferous tubule fluid secretion is a Sertoli cell microtubule-dependent process inhibited by 2,5-hexanedione exposure. *Toxicology and Applied Pharmacology, 128*, 302–309. https://doi.org/10.1006/taap.1994.1210.

Rochette-Egly, C., & Germain, P. (2009). Dynamic and combinatorial control of gene expression by nuclear retinoic acid receptors (RARs). *Nuclear Receptor Signaling, 7*, e005. https://doi.org/10.1621/nrs.07005.

Russell, L. D., Ettlin, R. A., Hikim, A. P. S., & Clegg, E. D. (1993). Histological and histopathological evaluation of the testis. *International Journal of Andrology, 16*, 83. https://doi.org/10.1111/j.1365-2605.1993.tb01156.x.

Sandell, L. L., Lynn, M. L., Inman, K. E., McDowell, W., & Trainor, P. A. (2012). RDH10 oxidation of vitamin A is a critical control step in synthesis of retinoic acid during mouse embryogenesis. *PLoS One, 7*, e30698. https://doi.org/10.1371/journal.pone.0030698.

Sandell, L. L., Sanderson, B. W., Moiseyev, G., Johnson, T., Mushegian, A., Young, K., ... Trainor, P. A. (2007). RDH10 is essential for synthesis of embryonic retinoic acid and is required for limb, craniofacial, and organ development. *Genes & Development, 21*, 1113–1124. https://doi.org/10.1101/gad.1533407.

Schleif, C., Gewiss, R., & Griswold, M. (2023). Chromatin remodeling via retinoic acid action during murine spermatogonial development. *Life (Basel), 13*, 690. https://doi.org/10.3390/life13030690.

Schleif, M. C., Havel, S. L., & Griswold, M. D. (2022). Function of retinoic acid in development of male and female gametes. *Nutrients, 14*, 1293. https://doi.org/10.3390/nu14061293.

Schrans-Stassen, B. H. G. J., Van De Kant, H. J. G., De Rooij, D. G., & Van Pelt, A. M. M. (1999). Differential expression of c-kit in mouse undifferentiated and differentiating type A spermatogonia. *Endocrinology, 140*, 5894–5900. https://doi.org/10.1210/endo.140.12.7172.

Shingleton, J. L., Skinner, M. K., & Ong, D. E. (1989). Retinol esterification in Sertoli cells by lecithin-retinol acyltransferase. *Biochemistry, 28*, 9647–9653. https://doi.org/10.1021/bi00451a016.

Snyder, E. M., Small, C., & Griswold, M. D. (2010). Retinoic acid availability drives the asynchronous initiation of spermatogonial differentiation in the mouse. *Biology of Reproduction, 83*, 783–790. https://doi.org/10.1095/biolreprod.110.085811.

Stokes, M.; Abdijadid, S. (2023). Disulfiram. In: StatPearls. StatPearls Publishing, Treasure Island (FL).

Teletin, M., Vernet, N., Yu, J., Klopfenstein, M., Jones, J. W., Féret, B., ... Mark, M. (2019). Two functionally redundant sources of retinoic acid secure spermatogonia differentiation in the seminiferous epithelium. *Development (Cambridge, England), 146*, dev170225. https://doi.org/10.1242/dev.170225.

Thatcher, J. E., & Isoherranen, N. (2009). The role of CYP26 enzymes in retinoic acid clearance. *Expert Opinion on Drug Metabolism & Toxicology, 5*, 875–886. https://doi.org/10.1517/17425250903032681.

Tong, M.-H., Yang, Q.-E., Davis, J. C., & Griswold, M. D. (2013). Retinol dehydrogenase 10 is indispensible for spermatogenesis in juvenile males. *Proceedings of the National Academy of Sciences, 110*, 543–548. https://doi.org/10.1073/pnas.1214883110.

Topping, T., & Griswold, M. D. (2022). Global deletion of ALDH1A1 and ALDH1A2 genes does not affect viability but blocks spermatogenesis. *Frontiers in Endocrinology, 13*.

Vernet, N., Dennefeld, C., Rochette-Egly, C., Oulad-Abdelghani, M., Chambon, P., Ghyselinck, N. B., & Mark, M. (2006). Retinoic acid metabolism and signaling pathways in the adult and developing mouse testis. *Endocrinology, 147*, 96–110. https://doi.org/10.1210/en.2005-0953.

Yoshida, S., Sukeno, M., Nakagawa, T., Ohbo, K., Nagamatsu, G., Suda, T., & Nabeshima, Y. (2006). The first round of mouse spermatogenesis is a distinctive program that lacks the self-renewing spermatogonia stage. *Development (Cambridge, England), 133*, 1495–1505. https://doi.org/10.1242/dev.02316.

Yoshinaga, K., Nishikawa, S., Ogawa, M., Hayashi, S.-I., Kunisada, T., Fujimoto, T., & Nishikawa, S.-I. (1991). Role of c-kit in mouse spermatogenesis: Identification of spermatogonia as a specific site of c-kit expression and function. *Development (Cambridge, England), 113*, 689–699. https://doi.org/10.1242/dev.113.2.689.

Zhou, Q., Nie, R., Li, Y., Friel, P., Mitchell, D., Hess, R. A., ... Griswold, M. D. (2008). Expression of stimulated by retinoic acid gene 8 (Stra8) in spermatogenic cells induced by retinoic acid: An in vivo study in vitamin a-sufficient postnatal murine testes. *Biology of Reproduction, 79*, 35–42. https://doi.org/10.1095/biolreprod.107.066795.

CHAPTER SEVEN

The interplay between retinoic acid binding proteins and retinoic acid degrading enzymes in modulating retinoic acid concentrations☆

Nina Isoherranen* and Yue Winnie Wen
Department of Pharmaceutics, School of Pharmacy, University of Washington
*Corresponding author. e-mail address: ni2@uw.edu

Contents

1. Introduction	168
2. Tissue and cell partitioning and protein binding of *at*RA	172
2.1 Binding of *at*RA in plasma and passive distribution to tissues	172
2.2 *at*RA binding to CRABPs	176
3. Pathways of retinoic acid clearance and patterns of enzyme expression	181
3.1 Enzymes that metabolize *at*RA	181
3.2 Role of CYP26 enzymes in *at*RA metabolism	183
4. Interplay of CYP26 enzymes and CRABPs in regulating tissue *at*RA concentrations	188
5. Conclusions	192
References	193

Abstract

The active metabolite of vitamin A, all-*trans*-retinoic acid (*at*RA), is critical for maintenance of many cellular processes. Although the enzymes that can synthesize and clear *at*RA in mammals have been identified, their tissue and cell-type specific roles are still not fully established. Based on the plasma protein binding, tissue distribution and lipophilicity of *at*RA, *at*RA partitions extensively to lipid membranes and other neutral lipids in cells. As a consequence, free *at*RA concentrations in cells are expected to be exceedingly low. As such mechanisms must exist that allow sufficiently high *at*RA concentrations to occur for binding to retinoic acid receptor (RARs) and for RAR mediated signaling. Kinetic simulations suggest that cellular retinoic acid binding proteins (CRABPs) provide a cytosolic reservoir for *at*RA to allow high enough

---

☆ The work presented was supported in part by funding from the National Institutes of Health grant R01GM111772. The authors declare no conflicts of interest with this work.

Current Topics in Developmental Biology, Volume 161
ISSN 0070-2153, https://doi.org/10.1016/bs.ctdb.2024.09.001
Copyright © 2025 Elsevier Inc. All rights are reserved, including those for text and data mining, AI training, and similar technologies.

cytosolic concentrations that enable RAR signaling. Yet, the different CRABP family members CRABP1 and CRABP2 may serve different functions in this context. CRABP1 may reside in the cytosol as a member of a cytosolic signalosome and CRABP2 may bind atRA in the cytosol and localize to the nucleus. Both CRABPs appear to interact with the atRA-degrading cytochrome P450 (CYP) family 26 enzymes in the endoplasmic reticulum. These interactions, together with the expression levels of the CRABPs and CYP26s, likely modulate cellular atRA concentration gradients and tissue atRA concentrations in a tightly coordinated manner. This review provides a summary of the current knowledge of atRA distribution, metabolism and protein binding and how these characteristics may alter tissue atRA concentrations.

1. Introduction

Vitamin A is an essential fat-soluble vitamin (O'Byrne & Blaner, 2013). Retinol (Fig. 1) is the main circulating form of vitamin A, but retinol is considered pharmacologically and biochemically inactive. Retinol requires metabolism to retinaldehyde to support vision, and to retinoic acid (RA) (Fig. 1) to regulate gene expression, cell cycle and cell differentiation. RA exists as different geometric isomers (all-*trans*, 13-*cis*, 9-*cis*, 11-*cis* and 9, 13-*dicis*), of which the all-*trans*- isomer has been assumed the predominant active isomer in mammals (Blomhoff & Blomhoff, 2006; O'Byrne & Blaner, 2013) and is the focus of this review.

All-*trans*-RA (atRA) is a ligand for the nuclear retinoic acid receptors (RARs) (Petkovich & Chambon, 2022). The binding affinity of atRA to the human RARα, RARβ and RARγ has been reported as an IC$_{50}$ of 4–5 nM (Allenby et al., 1994) or as a K$_d$ of 0.3–0.8 nM (Allegretto et al., 1993). In cell-based reporter assays, atRA activates RARs with EC$_{50}$ values of 4–19 nM (Topletz et al., 2015). As RARs are ligand activated, RAR mediated induction of gene transcription requires nuclear atRA concentrations to be above the binding affinities measured (0.3–5 nM). Other mechanisms of action besides RAR binding may also exist for atRA (Blomhoff & Blomhoff, 2006; Nhieu, Lin, & Wei, 2022). Nevertheless, the variety of effects are atRA concentration dependent. As such, spatiotemporal regulation of atRA concentrations in a cell is critical for appropriate atRA signaling, and for maintaining the processes regulated by atRA (Duester, 2008; Hogarth et al., 2014; Maden, 2007; Raverdeau et al., 2012). Despite years of work identifying and characterizing the enzymes responsible for vitamin A metabolic flux (Kedishvili, 2013; Napoli, 2022), the mechanisms that drive temporal and spatial variations of atRA concentrations needed for atRA signaling are still not

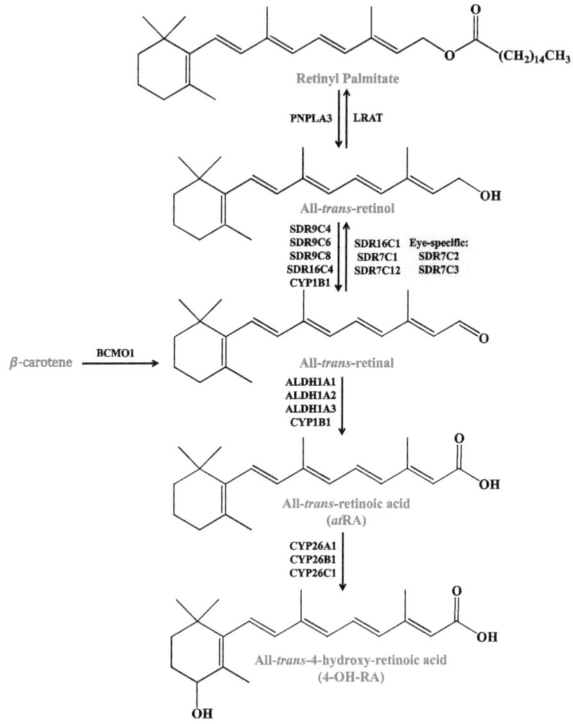

Modulating RA concentrations 169

Fig. 1 Chemical structures of vitamin A and its major metabolites and the overall enzymes contributing to vitamin A metabolic pathway. The enzymes indicated in the individual metabolic steps are labeled next to the pathway. BCMO1: *β*-carotene monooxygenase type 1; PNPLA3: Patatin-like phospholipase domain-containing 3; LRAT: Lecithin retinol acyltransferase; CYP: cytochrome P450; ALDH aldehyde dehydrogenase. For detailed description of the short chain dehydrogenase reductase (SDR) enzymes see (Kedishvili, 2013). Some SDR enzymes are also commonly known as retinol dehydrogenase (RDH) or as dehydrogenase/reductase (DHRS) and are as follows: SDR9C4 is also known as DHRS9 or RDH15; SDR9C8 as RDH16; SDR16C4 as RDH10; SDR16C1 as DHRS3; SDR7C1 as RDH11; SDR7C12 as RDH14; SDR7C2 as RDH12; and SDR7C3 as RDH13. In addition, SDR9C6 is also known as Hydroxysteroid 17-β Dehydrogenase 6 (HSD17β6). *Chemical structures were drawn in ChemDraw 21.0.0.*

fully understood. Fig. 1 summarizes the key metabolic steps in retinoid flux. The quantitative importance of the individual enzymes may, however, vary from tissue to tissue and with time within a tissue.

Mammals cannot synthesize vitamin A and hence adequate vitamin A intake from the diet is necessary (Lerner, 2024). Despite variety of dietary sources of vitamin A, Vitamin A deficiency has been common throughout human history (Wolf, 1996) and remains a major global health problem. The mechanisms that lead to night blindness as a consequence of vitamin A deficiency have been extensively studied. The night blindness is caused by insufficient retinol and retinaldehyde in the eye and depletion of rhodopsin (Daruwalla, Choi, Palczewski, & Kiser, 2018). It can be hypothesized that vitamin A deficiency not only depletes tissue stores of retinyl esters but also limits the availability of retinol for *at*RA synthesis. Limited availability or decreased concentrations of retinol would be expected to result in decreased tissue *at*RA concentrations leading to weakened immune response along with teratogenicity and male infertility that are associated with vitamin A deficiency. However, in experimental models of vitamin A deficiency (Honarbakhsh et al., 2021) the decrease in *at*RA concentrations in tissues was relatively modest even when retinol and retinyl ester concentrations were significantly decreased. This raises some profound questions on how tissue *at*RA concentrations are regulated and maintained and how dietary vitamin A intake or supplementation relates to tissue *at*RA concentrations and *at*RA signaling.

A majority of the vitamin A in the body is stored in the liver stellate cells (O'Byrne & Blaner, 2013). The concentrations of retinyl esters in the liver are much higher than in other organs studied. For example in mice, liver retinyl ester concentrations range from 300–1300 nmol/g while concentrations in the adipose, skin, testes, spleen, kidney and intestines are around 0.1–13 nmol/g (Obrochta, Kane, & Napoli, 2014; Snyder et al., 2020; Zhong, Hogarth et al., 2019). The livers of artic top predators contain up to 40-fold higher liver concentrations of retinyl esters than their continental counterparts or other arctic animals (Penniston & Tanumihardjo, 2006; Senoo et al., 2012). Yet, these animals do not show signs of hypervitaminosis A. Retinyl ester concentrations in human livers measured byliquid chromatography-mass spectrometry (LC-MS/MS) varied by 400-fold between donors (12.6–4282 nmol/g) (Zhong, Kirkwood et al., 2019), with the highest liver concentrations being similar to those in arctic bearded seal (4700 nmol/g) or glaucous gull (6840 nmol/g) (Senoo et al., 2012). In mice fed with standard mouse chow, liver retinyl ester concentrations measured by LC-MS/MS were similar (300–1200 nmol/g (Snyder et al., 2020; Zhong, Hogarth et al., 2019)) to those measured in the various nonpredatory arctic and continental animals (400–1500 nmol/g). This may be surprising as the vitamin A content in the standard mouse chow has been criticized for being too high (Yoo, Cockrum, & Napoli, 2023), suggesting the mice should accumulate retinyl esters in the liver similar to arctic top predators or humans with high dietary intake. Despite the high intake of vitamin A from standard chow, the liver vitamin A stores in mice are low compared to humans. The liver *at*RA concentrations (8.1–26.7 pmol/g) in the mice were also overall much lower than in the human liver. Human liver *at*RA concentrations were surprisingly variable and high (14–580 pmol/g) (Zhong, Hogarth et al., 2019; Zhong, Kirkwood et al., 2019). Taken together this suggests that in the liver, concentrations and availability of retinyl esters alter *at*RA concentrations. Same might not be true in other organs as human testis *at*RA concentrations are not as high and human and mouse *at*RA testis concentrations are strikingly similar (2–10 pmol/g) (Arnold, Griswold et al., 2015; Arnold, Schlatt et al., 2015).

The accumulation of vitamin A in the liver makes the livers of top predators such as polar bears dangerously high in vitamin A (or perhaps *at*RA) for human consumption. European arctic explorers who consumed polar bear or sledge dog livers appeared to experience many of the classic symptoms of acute vitamin A (or *at*RA) toxicity (Horowitz, 2022; Lips, 2003). The symptoms observed included peeling of the skin and nausea, vomiting, headaches, seizures and confusion corresponding to the idiopathic intracranial hypertension related to acute vitamin A toxicity (Horowitz, 2022). The mechanisms and signaling pathways that cause these symptoms are not all defined. It can be speculated that the excess intake of vitamin A results in excessive *at*RA synthesis and RAR activation in the target tissues, but the specific cause of death even in animal models has not been delineated. It is surprising that excess retinyl esters can cause such severe adverse effects, as retinyl esters and retinol are not active but need to be metabolized to *at*RA. How the concentrations of *at*RA become dysregulated in acute vitamin A poisoning remains to be determined.

The mechanisms of chronic hypervitaminosis A are also poorly defined but the phenotype appears distinct. Dietary assessment of vitamin A intake suggests that up to 75 % of the population in high income countries is getting more vitamin A in the diet than the recommended daily intake leading to subtoxicity (Penniston & Tanumihardjo, 2006). In clinical studies and in preclinical models chronic high intake of vitamin A has been related to increased risk of fractures, decreased bone mineral density and development of osteoporosis (Lerner, 2024; Lips, 2003; Penniston & Tanumihardjo, 2006). The molecular mechanisms leading to the bone loss due to chronic high consumption of retinyl esters are not known. It is the active metabolite *at*RA that is needed for signaling and its synthesis is carefully regulated (Gudas, 2022). How high dietary intake of vitamin A bypasses the regulatory mechanisms of *at*RA synthesis is puzzling. One mechanism that may explain this is the ability of liver alcohol dehydrogenase 1 (ADH1) to oxidize retinol to retinaldehyde when retinol is in high levels, followed by oxidation of retinaldehyde to *at*RA by aldehyde dehydrogenase 1A1 (Molotkov & Duester, 2003) or aldehyde oxidase (Zhong et al., 2021) both expressed highly in the liver.

It is noteworthy that humans are relatively tolerant to increased concentrations of *at*RA. For example, following therapeutic administration of RA in the form of 13-*cis*-RA (isotretinoin) or as *at*RA (tretinoin), circulating *at*RA concentrations are increased by about 10-fold (Amory et al., 2017; Jing et al., 2017; Stevison et al., 2019). The main side effects resulting from isotretinoin treatment are dry skin and chapped lips that may be considered the lowest intensity epithelial "toxicity" of vitamin A, and increase in triglycerides likely related to RA signaling in the liver (Amory et al., 2017; Stevison et al., 2019). This is in striking contrast to the observed phenotype of acute vitamin A toxicity and raises questions on whether the pathways of retinoid signaling and metabolism are fully characterized. The following sections provide a review of how physicochemical properties, protein binding and metabolism of *at*RA regulate tissue *at*RA concentrations, and explore how cellular mechanisms including binding proteins and metabolic enzymes regulate *at*RA concentrations available for signaling.

2. Tissue and cell partitioning and protein binding of *at*RA

2.1 Binding of *at*RA in plasma and passive distribution to tissues

The extent of distribution of small molecules to different tissues is governed by the partitioning of the compound to lipids and membranes and by protein binding in the tissue and plasma (Fig. 2A). Distribution of small molecules to different tissues and organs follows thermodynamic equilibrium principles and the compounds passively diffuse across membranes to reach equilibrium (Holt, Nagar, & Korzekwa, 2019; Rodgers & Rowland, 2006, 2007; Rodgers, Leahy, & Rowland, 2005). At equilibrium, the unbound (free) concentrations are equal at both sides of a bio-membrane. This equilibrium can be altered by the presence of active uptake or efflux transporters or by cellular metabolism resulting in a gradient of the unbound concentrations between the cytoplasm and plasma. When membrane permeability is not rate limiting (highly permeable compounds), the rate of distribution is defined by the perfusion of the tissue (blood flow) and the partitioning of the

Modulating RA concentrations 173

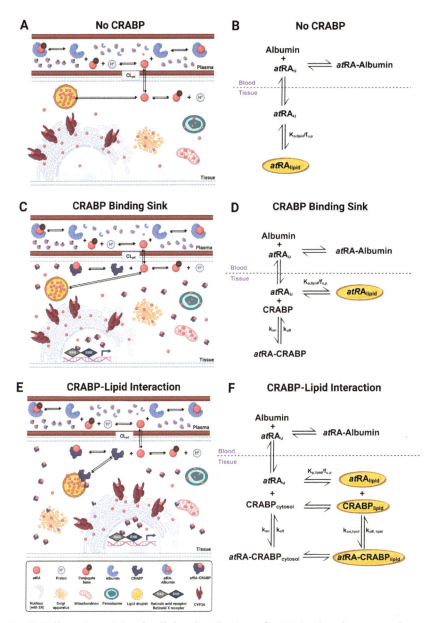

Fig. 2 **Different models of cellular distribution of *at*RA in the absence and presence of CRABP expression.** Panel (A) shows the distribution of *at*RA between plasma and a tissue without CRABP expression. The binding of *at*RA to albumin in plasma and partitioning of *at*RA to cellular lipids in different cellular organelles is illustrated. Panel (B) shows the kinetic scheme corresponding to the conceptual model in A. Panel (C) shows the predicted impact of CRABP expression on *at*RA

(Continued)

compounds to the tissue (K_p, tissue to plasma equilibrium concentration ratio). These principles of distribution kinetics can be effectively applied to atRA distribution.

atRA is a lipophilic (logP 5.6) organic acid with a pK_a of 4.7–5 that is mostly ionized at physiological pH. In Caco-2 cells atRA has an apparent permeability P_{app} of 3.6×10^{-6} cm/s (Jing et al., 2017) and hence the rate of distribution of atRA is expected to be perfusion rate limited. The overall partitioning (K_p) of atRA to tissues is relatively low following dosing of exogenous atRA (Table 1). In mice, the measured K_p values range from 1 to 4 after dosing of atRA (Jing et al., 2017; Kurlandsky, Gamble, Ramakrishnan, & Blaner, 1995; Wang, Campbell, Furner, & Hill, 1980). This extent of tissue distribution is typical for organic acids that bind extensively to plasma proteins (Rodgers & Rowland, 2006). In human plasma, atRA is very highly protein bound with an unbound fraction (f_u) of 0.0002 (Stevison et al., 2019). The binding is mainly to albumin (K_a of 3.33×10^5 M^{-1}) (Belatik, Hotchandani, Bariyanga, & Tajmir-Riahi, 2012; N'soukpoé-Kossi, Sedaghat-Herati, Ragi, Hotchandani, & Tajmir-Riahi, 2007). The plasma protein binding of atRA is unlikely to be saturated even in acute vitamin A toxicity as plasma albumin concentration (~700 µM) greatly exceeds atRA concentrations found in circulation.

When the lipophilicity and plasma protein binding of atRA are considered, the classic mathematical methods (Rodgers & Rowland, 2006, 2007) predict K_p values lower than those measured after dosing of deuterium labeled atRA in mice for liver and kidney and similar to those measured in blood (Table 1). The predicted and observed values suggest that the distribution of atRA from plasma to tissues is restricted by binding to plasma albumin, and that atRA partitions extensively to cellular lipid pools (Fig. 2A). Surprisingly, the lipid partitioning of atRA in cells has not been extensively considered. Based on the average concentrations of atRA measured in adult human serum (2–10 nM) (Arnold, Amory, Walsh, & Isoherranen, 2012; Czuba et al., 2022; Jeong et al., 2023) and the f_u of 0.0002, the free concentrations of atRA in plasma are 0.4–1 pM, nearly 1000-fold lower than the

Fig. 2—Cont'd distribution when a model in which CRABP cannot interact with the lipid membranes and organelles is assumed. In this model CRABP acts as a cytosolic binding sink. Panel (D) shows the kinetic scheme corresponding to the conceptual model in C. Panel (E) shows the scenario in which CRABP interacts with the lipid membrane (CRABP-lipid interaction) and atRA partitioned into the lipid can bind to CRABP at the lipid membrane. Panel (F) shows the kinetic scheme corresponding to the conceptual model in E. atRA: all-*trans*-retinoic acid; CRABP: cellular retinoic acid binding protein. *Figure was created with Biorender.com.*

Table 1 Predicted and observed tissue to plasma concentration ratios (K_ps) of atRA in different mouse tissues.

Tissue	Predicted K_p	Observed K_p
Blood	3.89	3.9[a]
Liver	0.09	3.8[a] 3.5[b] 4.4[c]
Kidney	0.14	2.1[a]
Pancreas	0.07	1.1[a]
Skin	0.28	0.7[b] 1.3[c]
Spleen	0.10	0.9[b] 1.2[c]

The predicted K_p values were calculated using methods described in Rodgers and Rowland (2006) and assuming plasma f_u of 0.0002 and blood to plasma ratio of 2.3.
[a] Reported following exogenous atRA administration to mice in (Jing et al., 2017).
[b] Calculated from tissue to plasma concentration ratio in the control male and female mice in (Zhong, hogarth et al., 2019).
[c] Calculated from tissue to plasma concentration ratio in the Cyp26a1$^{-/-}$Cyp26b1$^{-/-}$ male and female mice (Snyder et al., 2020).

K_d values (0.3–0.8 nM (Allegretto et al., 1993)) for RAR activation. This explains how autocrine signaling of atRA is established in post-natal animals and humans. Passive diffusion of atRA from plasma can only establish cellular free concentrations equivalent to plasma which are insufficient to drive retinoid signaling within a cell. Hence, each cell or tissue must synthesize atRA within the tissue to reach atRA concentrations that activate RARs, or active uptake transporters for atRA must exist to import atRA from outside the cell. Notably, the K_p values for atRA target organs such as the pancreas, skin and spleen are underpredicted (Table 1) suggesting a local regulation of tissue atRA concentrations by synthesis, metabolism and/or other factors not considered in the prediction methods. During development and during early organogenesis when blood flows and circulatory system have not fully developed, atRA concentration gradients and signaling are, however, likely established in a paracrine context. Similar to diffusion from blood in adult animals, thermodynamics will drive passive diffusion of atRA from cells that synthesize atRA to adjacent cells resulting in paracrine atRA signaling. The role of these retinoid gradients that occur during development is well established (Maden, 2007; Sirbu, Gresh, Barra, & Duester, 2005).

In adult humans, one can consider the measured human liver atRA total concentration of ~40 nM (Zhong, Kirkwood et al., 2019), and a human plasma free concentration of 1 pM in light of equilibrium partitioning. Assuming distribution equilibrium between plasma and liver (free liver concentration is also 1 pM) yields a tissue free fraction ($f_{u,t}$) of atRA in the liver of 0.000025. The overall concentration of atRA in the cellular lipids would then be 39.999 nM if the lipid was uniformly distributed in the cell. Therefore, the overall tissue atRA concentrations reflect almost entirely atRA partitioned to lipids. Total tissue atRA concentrations are hence likely inactive for atRA signaling, as the lipid partitioned atRA is not expected to interact with RARs. For atRA signaling to occur, additional processes are needed to sequester atRA from cellular lipids to the cytosol to allow RAR activation.

2.2 atRA binding to CRABPs

Free atRA binds to the cellular retinoic acid binding proteins (CRABPs). Together with the fatty acid binding proteins (FABPs) CRABPs form the family of intracellular lipid binding proteins (Yabut & Isoherranen, 2023). The structural features of atRA binding to the CRABPs have been extensively studied. Many crystal structures of both CRABP1 and CRABP2 with atRA bound exist, and the amino acid residues in CRABP1 and CRABP2 that interact with atRA are well defined (Kleywegt et al., n.d.; Vaezeslami, Mathes, Vasileiou, Borhan, & Geiger, 2006). The structural flexibility and changes in CRABP structure upon ligand binding have also been extensively studied using nuclear magnetic resonance (NMR) (Norris et al., 1995). These structural features were recently reviewed (Yabut & Isoherranen, 2023) and the reader is referred to the prior review and references therein for the characteristics of atRA binding to CRABPs.

All vertebrates express two CRABPs, CRABP1 and CRABP2 that share about 70% sequence identity (Lampron et al., 1995; Li & Norris, 1996). Each CRABP is highly conserved across species (Li & Norris, 1996). CRABP1 sequence is identical in humans and pigs and the human CRABP1 differs by a single amino acid (A86P) from rat, mouse and bovine CRABP1 (Nhieu et al., 2022). CRABP2 is also nearly identical in humans and rodents and has 74% sequence identity between zebrafish and humans (Sharma, Denovan-Wright, Boudreau, & Wright, 2003). The high conservation of sequences suggests that both proteins serve a critical function and are essential for life (Gorry et al., 1994; Lampron et al., 1995; Li & Norris, 1996; Nhieu et al., 2022). However, when either CRABP1 or CRABP2 or both were knocked out, the mice were phenotypically indistinguishable from wild-type mice (Gorry et al., 1994; Lampron et al., 1995). The only abnormalities found in the CRABP2$^{-/-}$ and CRABP1$^{-/-}$CRABP2$^{-/-}$ mice were limb abnormalities such as limb outgrowths (Lampron et al., 1995). However, at six weeks of age the CRABP1$^{-/-}$/CRABP2$^{-/-}$ mice had significantly lower viability than controls (9 % vs 2 % death, respectively) (Lampron et al., 1995). Yet, the knock-out mice were not more sensitive to *at*RA dosing and toxicity than control mice. The authors concluded that the lack of distinct phenotype suggests CRABPs are dispensable during mouse development and adult life and are not critical for *at*RA signaling pathway. They also noted that based on the results, the proposed functions of CRABPs as "buffers" to control free *at*RA concentration via sequestering *at*RA to the cytosol or modulating its metabolism, cannot explain the evolutionary conservation of CRABPs (Lampron et al., 1995). Later studies have, however, found a phenotype of the adult CRABP1$^{-/-}$ mice and shown variety of physiological and disease effects of CRABP1$^{-/-}$ (Nhieu et al., 2022). Novel functions of CRABP1 as part of a signalosome have been proposed (Nhieu et al., 2022) but the high conservation of CRABP1 and CRABP2 remains an enigma.

Many biochemical and cell studies have supported the role of CRABPs in regulating *at*RA signaling and metabolism (Napoli, 2012, 2017; Napoli & Yoo, 2020). Tight binding of *at*RA to both CRABP1 and CRABP2 has been unequivocally shown (Dong, Ruuska, Levinthal, & Noy, 1999; Li & Norris, 1996; Napoli & Yoo, 2020; Noy, 2000), although numerous technical challenges have prevented estimation of the exact binding affinity of *at*RA to CRABPs (Yabut & Isoherranen, 2022, 2023). For example, the binding affinity of *at*RA to CRABPs is usually detected via quenching of tryptophan fluorescence by *at*RA. Most fluorescence spectrophotometers are not sensitive enough to allow spectral titrations with protein concentrations in low pM range that would be needed to meet equilibrium assumptions. Hence the CRABP concentrations in the spectral titration experiments are typically much higher than the binding affinity of *at*RA to CRABP1 and CRABP2. This can, to some degree, be addressed by careful kinetic analysis but the parameter estimates still lack confidence (Yabut & Isoherranen, 2022, 2023). Stopped flow experiments have been used to calculate K_d values of 60 pM for CRABP1 and 130 pM for CRABP2 (Dong et al., 1999). However, the k_{on} published should be considered the k_{obs} based on the published methodology. Thus, some uncertainty remains on the exact kinetics of *at*RA binding to CRABPs. The uncertainty of the binding kinetics is increased if CRABPs can interact with the lipid membrane, a possibility not included in the kinetic assessments reported (Dong, Ruuska, Levinthal, & Noy, 1999; Yabut & Isoherranen, 2022). Nevertheless, the stopped flow experiments together with many fluorescence titration experiments provide clear data that *at*RA binding affinity to CRABPs is high.

CRABP expression levels in adult rats were reported in the reproductive organs (testes, epididymis, seminal vesicle, ovary), and in the kidney, brain, lung, adrenal gland and skin (Chytil & Ong, 1987; Ong & Chytil, 1975). CRABPs have also been detected in neonatal rats suggesting that CRABPs are expressed in distinct tissues and organs during embryonic development and fetal growth (Bailey & Siu, 1988, 1990). In adult rats, the reported expression levels were typically in the low pmol/g range (2–85 pmol/g) resulting in CRABP expression levels in the target tissues of 2–90 nM (Bailey & Siu, 1990; Ong, Crow, & Chytil, 1982). The measured expression levels do vary considerably, and little is known about dynamics of these expression levels.

Human tissue RNAseq analysis indicates that CRABP1 is robustly expressed in the brain, kidney, placenta, skin, spleen and thyroid with low detection in the GI tract and adrenal (Fagerberg et al., 2014). In contrast, high mRNA expression of CRABP2 was found in human endometrium, esophagus, skin and bladder with lower expression in adrenal, fat, gallbladder, placenta, prostate, salivary gland and thyroid (Fagerberg et al., 2014). Both CRABP1 and CRABP2 mRNAs were detected in some organs such as skin, adrenal gland, and thyroid. Other organs such as brain and kidney appear to express only CRABP1 and epithelial tissues such as endometrium, esophagus and bladder express only CRABP2 mRNA. Collectively CRABP1 and CRABP2 are expressed in the organs that are sensitive to vitamin A toxicity and in which *at*RA signaling is critical such as the skin and brain. This points to a role of CRABPs in retinoid signaling.

The expression of CRABPs may alter *at*RA partitioning and distribution in the cell. Three different models of tissue distribution of *at*RA are put forward here (Fig. 2). In the first scenario (Fig. 2A and B) *at*RA distribution is considered in a tissue that does not express any CRABPs or other proteins that would bind *at*RA. In the second scenario (Fig. 2C and D) CRABPs are expressed in the cytosol and bind free *at*RA available in the cytosol. This model is called the "binding sink" model and CRABP is considered in this model to act as the "buffer" that binds free *at*RA as has been proposed (Budhu & Noy, 2002; Lampron et al., 1995). In the third model CRABP interacts with the lipid membranes (Fig. 2E and F). This model reflects the observations made using fluorescence microscopy that CRABP2 is associated with the endoplasmic reticulum (Majumdar, Petrescu, Xiong, & Noy, 2011). This model assumes that CRABP can directly deliver atRA to or bind atRA from the lipid membrane via collisional interactions. The distribution of atRA in a model tissue (skin) that expresses CRABP1 and in which CRABP1 acts according to the "binding sink" model was simulated to explore conceptually the impact that CRABPs may have on cellular, cytosolic and lipid concentrations of atRA.

The total tissue, lipid partitioned, CRABP1 bound, free and cytosolic concentrations of atRA were simulated at equilibrium (Fig. 3). The predicted

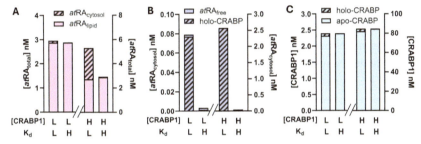

Fig. 3 Simulation of the effect of CRABP expression on atRA concentrations in different cellular compartments. The concentrations of free unbound atRA (atRA$_{free}$), atRA partitioned to lipid (atRA$_{lipid}$), unbound CRABP (apo-CRABP), atRA bound to CRABP (holo-CRABP), and total cytosolic atRA (atRA$_{cytosol}$) calculated as the sum of atRA$_{free}$ and holo-CRABP, were simulated at equilibrium assuming the binding sink kinetic model shown in Fig. 2D. The expression level of CRABP in the simulations was set either as low (L), 2.4 nM, based on reported CRABP1 expression in rat skin (Ong et al., 1982) or as high (H), 84.7 nM, based on separate report of the CRABP1 expression in rat skin (Bailey & Siu, 1990). The binding constant of atRA to CRABP1 was set either as low K$_d$ (L) of 60 pM based on stopped flow measurements reported (Dong et al., 1999) or as high K$_d$ (H) of 4.1 nM based on binding experiments and fluorescence titrations reported (Yabut & Isoherranen, 2022). The atRA concentrations in the model were simulated to steady state. Panel (A) shows the overall simulated tissue concentrations of atRA distributed between cytosolic atRA (atRA$_{free}$ plus holo-CRABP) and lipid partitioned atRA when either high or low CRABP expression were considered and the different binding constants of atRA to CRABP were incorporated. Panel (B) shows the simulated concentrations of atRA in cytosol distributed between holo-CRABP and free atRA when either high or low CRABP expression was considered and the different binding constants of atRA to CRABP were incorporated. Panel (C) shows the concentrations of apo- and holo-CRABP under the simulated conditions in which atRA partitions to the lipid and only free atRA in cytosol can bind to CRABP. Simulations in panels (A–C) were performed in MATLAB and Simulink (R2023b) and plotted using GraphPad Prism (V10.1.2).

K_p value of 0.28 for *at*RA between skin and plasma was assumed in the model and considered to reflect the lipid partitioning of *at*RA within the cell. As described above for the liver, based on the physicochemical characteristics of *at*RA, these simulations predict that majority of *at*RA in the tissue will be within the cellular lipids (Fig. 3A). In the absence of CRABP1 expression, only 0.07 % of the total tissue *at*RA is free in the cytosol. Expression of CRABP1 at 2.4 nM (assuming a K_d of 60 pM) is predicted to lead to a 38-fold increase in cytosolic *at*RA. This is due to the binding sink effect of CRABP1. When the CRABP1 is expressed, cytosolic *at*RA is almost entirely comprised of the CRABP1 bound *at*RA (holo-CRABP) with a small fraction free in the cytosol (Fig. 3B). With increasing CRABP1 expression (from 2.4 nM to 84.7 nM) the cytosolic *at*RA concentration increases accordingly (~33-fold) due to the increase in holo-CRABP1 concentration (Fig. 3B). The binding affinity of *at*RA to CRABP1 also has a significant impact on cytosolic *at*RA concentrations. Lower binding affinity (high K_d) results in much lower cytosolic *at*RA (and holo-CRABP1) concentrations (Fig. 3A and B). This is mainly due to the very low free concentrations of *at*RA in the cytosol that are below the K_d value for *at*RA binding to CRABPs and that act as the driving force for CRABP binding.

Binding of *at*RA to CRABP1 results in an increase in the fraction of total tissue *at*RA that resides in the cytosol. A considerable fraction (~49 %) of tissue *at*RA is predicted to be cytosolic only when a high CRABP1 expression level of 84.7 nM and a high binding affinity (K_d of 60 pM) of *at*RA to CRABP1 are considered (Fig. 3A). Although 51 % of the *at*RA is still predicted to reside in the lipids, this reflects a > 1000-fold increase in the cytosolic *at*RA concentration when compared to the situation where no CRABP1 is present. This is even though the majority of CRABP1 in these simulations is predicted to be in the apo-, ligand free, form (Fig. 3C). A lower *at*RA binding affinity (K_d = 4.1 nM) is predicted to result in nearly all of the cellular CRABP1 to be present in the apo-CRABP1 form. While these simulations are conceptual, and use theoretical values for lipid partitioning, they provide clear insight into how CRABPs may regulate cytosolic *at*RA concentrations available for signaling. An important finding from these simulations is that the presence of CRABPs allows cytosolic *at*RA concentrations to reach the concentrations that are required for RAR activation. Hence CRABPs provide the necessary sequestration of *at*RA in the tissue to enable retinoid signaling. In the context of vitamin A deficiency, CRABPs likely serve a role in maintaining cytosolic *at*RA concentrations as long as possible when tissue retinol stores are depleted.

Some FABPs have been shown to interact with cellular membranes and transfer fatty acids from the FABP to the membranes via direct collisional interactions between the FABP and the membrane (de Gerónimo et al., 2014; Hsu & Storch, 1996; Storch & Corsico, 2008). It is not known whether CRABPs interact with lipid membranes and extract *at*RA from the membrane. The model of CRABPs interacting with the membrane (Fig. 2C) is intriguing considering the observations that CRABP2 translocates from the ER membrane to the nucleus (Budhu & Noy, 2002; Dong et al., 1999; Majumdar et al., 2011) upon addition of *at*RA. In the absence of any kinetic data on whether holo-CRABP interacts with the membranes and what the binding kinetics between *at*RA in the lipid and CRABP interacting with the lipid would be, simulations of the impact of the CRABP-lipid interactions on cellular *at*RA distribution cannot be conducted. Further studies of this mechanism and the kinetics are warranted.

Binding of *at*RA to FABP5 has also been proposed (Schug, Berry, Shaw, Travis, & Noy, 2007). The direct *at*RA binding to FABP5 could, however, not be reproduced when tested using a different fluorescent probe than the original studies (Yabut & Isoherranen, 2022). Examination of direct ligand binding to FABPs is challenging because FABPs, unlike the CRABPs, do not contain fluorescent tryptophans and hence direct fluorescence quenching assays cannot be done with FABPs. The binding of *at*RA to FABP5 was shown using ANS as a fluorescent probe. The fluorescence spectrum of ANS overlaps with that of *at*RA making analysis of binding challenging. Yet, FABPs are promiscuous binding proteins and numerous free fatty acids bind to the FABPs. As such, low affinity binding of *at*RA to different FABPs is highly likely. Yet the biological significance of FABP binding is unclear. Whether *at*RA can outcompete other fatty acids that bind to FABPs in vivo also remains to be evaluated. The observed effects of FABP5 expression on *at*RA signaling (Schug et al., 2007) may be due to direct binding of *at*RA to FABP5 or other less direct mechanisms relating to altered lipid homeostasis in the presence of FABP5.

3. Pathways of retinoic acid clearance and patterns of enzyme expression

3.1 Enzymes that metabolize *at*RA

In humans *at*RA is efficiently cleared via several parallel processes. In humans and in cells, *at*RA readily isomerizes to 13-*cis*-RA (Lansink, Van Bennekum,

Blaner, & Kooistra, 1997; Muindi et al., 2008; Tsukada et al., 2000), a reaction possibly catalyzed by glutathione-S-transferases (Chen & Juchau, 1997, 1998) or other sulfhydryl groups (Urbach & Rando, 1994). This isomerization may contribute to *at*RA clearance in humans but is not detected for endogenous *at*RA in mice (Zhong. Kirkwood et al., 2019). Following administration of exogenous *at*RA to mice, 13-*cis*-RA is detected (Jing et al., 2017; Zhong, Hogarth et al., 2019). Clearly, in humans 13-*cis*-RA also isomerizes to *at*RA and there appears to be an equilibrium between the all-*trans* and 13-*cis* isomers (Muindi et al., 2008; Stevison et al., 2019). The fraction of *at*RA that is eliminated via the isomerization to 13-*cis*-RA is likely minor as the clearance of 13-*cis*-RA is much lower than that of *at*RA (Muindi et al., 2008).

Glucuronidation may provide an important elimination pathway for *at*RA. *at*RA is glucuronidated by the UGT enzymes forming an acyl glucuronide (Samokyszyn et al., 2000; Sass, Forster, Bock, & Nau, 1994). In rats these acyl glucuronides were detected in bile (Zile, Inhorn, & DeLuca, 1982) and biliary excretion of 13-*cis*-RA acyl glucuronide has been detected in humans (Vane, Bugge, Rodriguez, Rosenberger, & Doran, 1990). Acyl glucuronides in general undergo acyl migration that may result in formation of protein adducts (Hammond et al., 2014; Bradow, Kan, & Fenselau, 1989). This instability of acyl glucuronides can make their quantification challenging. Of the human liver UGT enzymes UGT1A3 and UGT2B7 have been shown to form the acyl glucuronide of *at*RA (Samokyszyn et al., 2000), but other UGT enzymes in the liver may also form this glucuronide. UGT enzymes are broadly expressed in different organs including many epithelial tissues and hence may contribute to *at*RA elimination in target tissues.

Oxidative metabolism is well known as a clearance pathway of *at*RA and 13-*cis*-RA. In the adult human liver many cytochrome P450 enzymes (CYPs) oxidize *at*RA (and 13-*cis*-RA) to the primary 4-OH-RA metabolite (Fig. 1) with CYP3A4 and CYP2C8 having the highest efficiency (Marill et al., 2002; Thatcher, Zelter, & Isoherranen, 2010). The 4-hydroxylation activity is perhaps not surprising due to the conjugation of the C4 in the β-ionone ring of *at*RA. Hence, *at*RA hydroxylation activity alone should not be considered as evidence for a specific enzyme having a role as important *at*RA hydroxylase in vivo. Unlike the UGT enzymes that have broad extrahepatic expression, the expression of CYP3A4 and CYP2C8 is limited to the liver and the small intestine. It is possible that in these two organs CYP3A4 and CYP2C8 contribute to *at*RA hydroxylation, but it is highly unlikely that these enzymes affect *at*RA concentrations in target tissues where retinoid signaling is critical. In the human fetal liver, CYP3A7 appears to be the main fetal liver *at*RA hydroxylase (Chen, Fantel, & Juchau, 2000; Topletz, Zhong, & Isoherranen, 2019) and *at*RA hydroxylation correlates with CYP3A7 activity in fetal livers (Topletz et al., 2019). The CYP3A7 activity was not predicted to be sufficient, however, to prevent maternal *at*RA reaching the fetus as the predicted fetal liver extraction ratio was about 0.03. There is essentially no information of CYP expression in the human fetal liver during the period of organogenesis and during first trimester and hence presence of CYP3A7 and its importance in *at*RA metabolism during human organogenesis is unknown.

CYP2C8, CYP3A4 and CYP3A7 are not conserved across species and correspond to large subfamilies of Cyp2c and Cyp3a enzymes in rodents. The lack of conservation across species, in addition to the expression of CYP3A4 and CYP2C8 restricted to the liver and intestine, suggest that these enzymes are not critical for clearing endogenous *at*RA. In contrast, the enzymes in the CYP26 family are highly conserved across species and are expressed in many extrahepatic tissues including organs and tissues that are central to *at*RA signaling (MacLean et al., 2001; Taimi et al., 2004; White et al., 2000). The CYP26 family includes three isoforms in all chordates, CYP26A1, CYP26B1 and CYP26C1 (Isoherranen & Zhong, 2019). During fetal development, both Cyp26a1 and Cyp26b1 are required, and knockout mice of Cyp26a1 or Cyp26b1 are not viable (Abu-Abed et al., 2001; Dranse, Sampaio, Petkovich, & Underhill, 2011; Maclean, Dollé, & Petkovich, 2009; Pennimpede et al., 2010; Yashiro et al., 2004). The biochemical function, expression patterns, genetics and regulation of the CYP26 enzymes was recently reviewed and the reader is referred to the prior review for details of these enzymes (Isoherranen & Zhong, 2019).

3.2 Role of CYP26 enzymes in *at*RA metabolism

The key characteristics of CYP26A1 as *at*RA hydroxylase are its high inducibility by *at*RA in cell culture and in vivo via a retinoic acid response element (RARE) (Loudig et al., 2000; White et al. 1996, 1997), the high affinity of *at*RA to CYP26A1 ($K_m < 10$ nM) (Lutz et al., 2009), and the fact that knockout of Cyp26a1 in mice is not compatible with normal development (Abu-Abed et al., 2001; Pennimpede et al., 2010). In the $Cyp26a1^{-/-}$ mice, the observed malformations closely correlate to dysregulated *at*RA gradients during development resulting in typical retinoid teratogenicity. During postnatal life, one may propose a model in which CYP26A1 expression is rapidly induced by increasing *at*RA concentrations (exogenous or endogenous) in an RAR mediated manner to "normalize" *at*RA concentrations. Other regulatory factors such as glucocorticoids or energy balance may also alter CYP26A1 expression to drive changes in *at*RA concentrations (Yoo et al. 2022, 2023).

CYP26B1 also metabolizes *at*RA efficiently and *at*RA has high affinity to CYP26B1 (Topletz et al., 2012). However, regulation of *CYP26B1* by *at*RA is equivocal. No RARE has been reported in *CYP26B1* promoter. Although *Cyp26b1* mRNA expression is increased dose dependently in response to *at*RA treatment (Holloway et al., 2022), inducibility of *CYP26B1* mRNA by *at*RA is not as profound as *CYP26A1* (Tay, Dickmann, Dixit, & Isoherranen, 2010). The critical role of Cyp26b1 in regulating *at*RA concentrations is supported by the fact that *Cyp26b1* knockout mice are not viable showing limb malformations that are typical for excess *at*RA (Yashiro et al., 2004). Notably, the patterns of developmental defects in the $Cyp26a1^{-/-}$ and $Cyp26b1^{-/-}$ mice are different from each other and the two enzymes are expressed in different tissues at different times during embryogenesis and fetal development. It is interesting that both Cyp26a1 and Cyp26b1 are required during fetal development at distinct spatiotemporal pattern even though both have been shown to hydroxylate *at*RA to the 4-OH-RA. It has been proposed that this may be due to Cyp26b1 metabolizing a different substrate than *at*RA (Kumar et al., 2011), but that substrate is still unidentified. The current consensus is that Cyp26a1 and Cyp26b1 are both required due to specific tissue localization and activity of each CYP in *at*RA clearance.

The role of CYP26C1 as *at*RA hydroxylase is not clear. CYP26C1 metabolizes *at*RA and the other RA isomers 9-*cis*-RA and 13-*cis*-RA, (Taimi et al., 2004; Zhong, Ortiz, Zelter, Nath, & Isoherranen, 2018). Knockout of *Cyp26c1* is largely inconsequential (Uehara et al., 2007). The expression of Cyp26c1 in animal models and humans is limited. During mouse development the expression of Cyp26c1 mimics that of Cyp26a1 (Uehara et al., 2007).

Conditional (inducible) *Cyp26* knockout models have provided valuable insights to the role of Cyp26 enzymes in post-natal life. Based on the phenotype of the global $Cyp26a1^{-/-}$ mice, an inducible, global *Cyp26a1* knockout was expected to result in significant retinoid toxicity. However, the mice were healthy and lacked adverse phenotype except a minor effect on hematopoiesis (Zhong, Hogarth et al., 2019). The conditional knockout was confirmed by reverse transcription PCR analysis and the dose of tamoxifen was optimized to ensure robust induction of the knockout. The clearance of exogenous atRA was significantly decreased in these mice confirming the role of CYP26A1 in RA clearance (Zhong, Hogarth et al., 2019). The study has been criticized to not have been sufficiently sensitive as the mice were fed standard mouse chow with relatively high vitamin A content (Yoo et al., 2023). This criticism is somewhat counterintuitive as high vitamin A content in the diet would be expected to challenge the mice for retinoid toxicity and exacerbate a phenotype if one was present. The initial study of the conditional knockouts (Zhong, Hogarth et al., 2019) did not explore nuanced effects on metabolic function such as blood glucose levels or retinoid concentrations under fasting and refeeding but was focused on gross phenotypic effects. If glucose and energy homeostasis was severely altered in these mice, it may be expected that liver weights, body weights and growth of the mice would have been affected. However, this was not observed. Careful studies in fasted and refed mice including conditional $Cyp26a1^{-/-}$ mice have subsequently shown a role of Cyp26a1 in regulating hepatic atRA concentrations and glucose homeostasis (Yoo et al., 2023) possibly pointing to an expanded role of Cyp26a1 in regulating energy balance.

In contrast to the *Cyp26a1* conditional knockouts, the *Cyp26b1* conditional knockout mice showed a distinct phenotype that was observed few days after induction of the knockout and worsened with time (Snyder et al., 2020). These mice were fed the same diet as the conditional $Cyp26a1^{-/-}$ mice in the same vivarium around the same time. The conditional $Cyp26b1^{-/-}$ mice did not lose weight significantly nor did the knockout of Cyp26b1 result in decreased survival rate of the mice (Snyder et al., 2020). The most striking phenotype of these mice was observed in the skin. The $Cyp26b1^{-/-}$ mice had higher skin atRA concentrations than wild type mice, clear dermatitis, blepharitis, skin ulceration and inflammation (Snyder et al., 2020). The skin lesions were phenotypically similar to retinoid and vitamin A toxicity (Biesalski, 1989; Kretzschmar & Leuschner, 1975; Kurtz, Emmerling, & Donofrio, 1984) supporting the interpretation that skin toxicity after hypervitaminosis A is due to high atRA concentrations in the skin. The $Cyp26b1^{-/-}$ mice also showed inflammation of the epithelium of the GI tract and significant enlargement of the spleen. These findings support the critical role of Cyp26b1 in modulating endogenous atRA concentrations and signaling in tissues that are classic retinoid responsive organs.

When knockout of both *Cyp26a1* and *Cyp26b1* was induced in postnatal mice, a severe phenotype was observed (Snyder et al., 2020). These mice did not gain weight and their survival rate was significantly decreased regardless of whether the knockout was induced in juvenile or adult mice. The mice had severe dermatitis and splenomegaly, adipose tissue inflammation and inflammation and hyper-keratinization of epithelia. The movements of the mice appeared somewhat slow and impaired (unpublished observations), but it was impossible to determine whether this was due to the otitis media, severe inflammation or a CNS effect. One may speculate that the mice suffered from intracranial hypertension seen in acute vitamin A toxicity. Notably these mice were also hypersensitive to exogenous *at*RA. Normal doses of *at*RA were unexpectedly toxic to these mice (Snyder et al., 2020). These findings are consistent with acute vitamin A toxicity but despite extensive histopathology, no clear cause of death could be identified. Interestingly, while liver *at*RA concentrations were increased, the livers of these mice were completely healthy with no histopathological findings. Similarly, no histopathological abnormalities were observed in most major organs such as the kidney, heart, brain, lungs and the pancreas after two months from induction of the knockout (Snyder et al., 2020). No histopathological abnormalities were observed in these organs even in the mice that died or had to be euthanized due to the retinoid toxicity. This points to the highly cell and tissue specific effects of *at*RA in adult animals and to the local regulation of *at*RA concentrations and signaling. It is interesting to note that as the CYP26 enzymes were knocked out, the K_p values of *at*RA for the target tissues increased (Table 1) supporting the role of CYP26 enzymes regulating the tissue *at*RA concentrations.

During postnatal life CYP26 enzymes appear to share the tissue expression patterns of the CRABPs. CYP26A1 mRNA was found in adult mice in the liver, testes, ovaries, thymus, subcutaneous fat pad, and placenta (skin was not tested) (Yue et al., 2014). In humans, of the organs tested in single donors, CYP26A1 protein was detected at relatively high levels in the liver, lung, pancreas, skin and uterus and at lower levels in the adipose, intestine, kidney and spleen (Topletz et al., 2012). In comparison, CYP26B1 mRNA was detected in most organs evaluated in mice, with the highest expression in the cerebellum, heart, lung and testes (skin not evaluated) (Yue et al., 2014). Whether this broad mRNA expression translates to protein is unclear. CYP26B1 protein was not detected in human liver, but mRNA was detected in the mouse and human liver.

CYP26B1 protein was also detected at high levels in the lung, pancreas, skin and uterus and at lower levels in the adipose, intestine, kidney and spleen (Topletz et al., 2012). To fully establish the role of the CYP26 enzymes in *at*RA clearance, their expression patterns at the protein level and cell type specific expression need to be established in tissues from healthy donors and/or in animal models.

The expression pattern of the CYP26 enzymes in the liver is an excellent example of the challenges in understanding the cell type and tissue specific roles of the CYP26 enzymes. Studies in whole human liver preparations originally suggested that CYP26A1 is the human liver *at*RA hydroxylase while CYP26B1 is not expressed in the human liver (Tay, Dickmann, Dixit, & Isoherranen, 2010). Robust CYP26A1 protein expression was detected in majority of human liver donors although the expression levels were variable (Thatcher, Zelter, & Isoherranen, 2010). The protein expression corresponded to high mRNA expression (Tay, Dickmann, Dixit, & Isoherranen, 2010). *CYP26A1* is also clearly inducible by *at*RA in liver-based cell lines such as HepG2 cells and in human hepatocytes and in vivo in mouse liver (Stevison et al., 2019; Stevison, Hogarth, Tripathy, Kent, & Isoherranen, 2017; Tay, Dickmann, Dixit, & Isoherranen, 2010). In contrast, CYP26B1 protein was not detected in the human livers (Topletz et al., 2012), and the mRNA, while detectable, was very low (Tay, Dickmann, Dixit, & Isoherranen, 2010). However, recent data suggest that these findings may be specific to hepatocytes (Czuba & Isoherranen, 2024; Czuba et al., 2021). Hepatocytes constitute the majority of the liver and in any whole liver preparation protein from the hepatocytes will dominate the overall assessment of protein expression. Yet, liver stellate cells are more central to liver vitamin A and retinoid homeostasis. *CYP26B1* mRNA was detected in quiescent LX-2 cells that are a model of stellate cells while *CYP26A1* was not present (Czuba & Isoherranen, 2024). When the LX-2 cells were activated, *CYP26A1* mRNA was detected together with *CYP26B1*. When expression of *CYP26A1* mRNA in different hepatic cell types isolated from liver was evaluated, *CYP26A1* mRNA was detected in hepatocytes and in some donors in stellate cells and Kupffer cells (Czuba et al., 2021). Unfortunately, *CYP26B1* mRNA expression was not evaluated. Taken together, these findings suggest that CYP26A1 may be the hepatocyte specific CYP26 while CYP26B1 is in the stellate cells in the liver. Clearly, measurements of whole organ mRNA expression of enzymes contributing to retinoid homeostasis can be misleading. Similar cell type specific expression has been shown in the testis

(Hogarth & Griswold, 2013) where CYP26A1 and CYP26B1 have cell type specific roles (Hogarth et al., 2015). This consideration for cell type specific expression in various tissues is important when the role of CYP26 enzymes in regulating retinoid signaling in various tissues is considered.

4. Interplay of CYP26 enzymes and CRABPs in regulating tissue *at*RA concentrations

CYP26A1 and CYP26B1 metabolize *at*RA to the 4-OH-*at*RA and other primary hydroxylation products (Lutz et al., 2009; MacLean et al., 2001; Thatcher et al., 2011; Topletz et al., 2012; White et al. 1996, 1997, 2000). The functional similarity was unexpected as the CYP26 enzymes only share 40–45% sequence similarity between the isoforms (Isoherranen & Zhong, 2019). Even small differences in the overall amino acid sequence between CYPs often change their substrate specificity. The relatively low sequence identity between CYP26A1 and CYP26B1 suggests that the two enzymes have some distinct functional differences that result in the need for both enzymes.

For several decades, the CRABPs have been considered as important modulators of *at*RA metabolism and action (Napoli & Yoo, 2020; Napoli, 2017). It has been shown that CRABP2 localizes to the nucleus upon *at*RA binding and channels *at*RA to RARβ while CRABP1 does not (Budhu & Noy, 2002; Dong et al., 1999; Majumdar et al., 2011). CRABP1 also appeared to modulate *at*RA metabolism via facilitating *at*RA oxidation (Fiorella & Napoli, 1991, 1994). It was hypothesized that perhaps only one of the CYP26 enzymes interacts with CRABPs (CRABP1) (Nelson et al., 2016). Surprisingly this hypothesis regarding CRABP-CYP26 specificity turned out to be incorrect (Nelson et al., 2016; Yabut & Isoherranen, 2022; Zhong, Ortiz, Zelter, Nath, & Isoherranen, 2018). CRABP1 and CRABP2 had similar effects on the kinetics of *at*RA metabolism by CYP26A1 and CYP26B1 (Nelson et al., 2016; Yabut & Isoherranen, 2022). The kinetic data suggested that both CRABPs channel *at*RA for metabolism by CYP26A1 and CYP26B1 but the k_{cat} and K_m values for holo-CRABPs are decreased when compared to nominal *at*RA concentrations. The effect of CRABP1 and CRABP2 on recombinant CYP26 mediated *at*RA clearance was remarkably similar to the impact of CRABP1 on *at*RA hydroxylation in rat testes microsomes observed decades earlier (Fiorella & Napoli, 1991, 1994) demonstrating the reproducibility of the observation. What was surprising was the fact that both CRABPs appeared to directly interact with CYP26A1 and CYP26B1. In studies with CYP26C1, similar observations were made with CRABP1 and CRABP2 interacting with CYP26C1 (Zhong et al., 2018). In addition, the kinetic data suggested that apo-CRABPs inhibit CYP26 activity (Nelson et al., 2016; Yabut & Isoherranen, 2022).

The impact of CRABPs on *at*RA hydroxylation by CYP26 enzymes in vitro is established. However, the relevance of the CRABP-CYP interactions in vivo is unclear. Kinetic simulations can be used to explore the potential impact of CRABP-CYP26 interactions on *at*RA concentrations (Fig. 4). If CRABP1 is assumed to function as a simple binding sink (Fig. 4A) in a tissue like the skin where CYP26 is also expressed, simulations suggest that both CRABP1 and CYP26 expression levels impact cytosolic and total *at*RA concentrations and CYP26 expression also alters lipid partitioned *at*RA concentrations (Fig. 4B and C). The impact depends on a complex interplay of the kinetic constants, however. Based on the simulations, if CRABP1 concentration is kept constant, the higher CYP26 expression or higher CYP26 catalytic activity (CYP26A1 vs CYP26B1) results in lower free, cytosolic, lipid partitioned and total tissue *at*RA concentrations. If CYP26 activity is held constant, increased CRABP1 expression results in increased holo-CRABP1 and cytosolic *at*RA concentrations with the fraction of tissue *at*RA in the cytosol predicted to be increased (Fig. 4B and C). These results emphasize the importance of further studies to determine the expression levels of the CRABPs and CYP26 enzymes in the tissues of interest, and to measure the exact kinetic constants for *at*RA binding to CRABPs and CYP26s. In addition, these simulations show that changes in CRABP expression alone can result in increased cytosolic and total tissue *at*RA concentrations that may induce *at*RA signaling and RAR activation if the simple sink model applies.

The in vitro data suggest that holo-CRABP1 is a substrate of CYP26A1 and CYP26B1 with unique K_m and k_{cat} values (Fig. 4) and that apo-CRABP1 inhibits CYP26 enzymes (Nelson et al., 2016; Yabut & Isoherranen, 2022). If these protein-protein interactions are considered, the predicted impact of CRABP and CYP26 expression and the kinetic constants is more complicated (Fig. 4D,E and F) than shown for the simple binding sink model. Higher CYP26A1 or CYP26B1 expression still results in a lower free and total tissue *at*RA concentrations. However, higher CRABP1 expression is predicted to cause somewhat unpredictable effects on free and total tissue concentrations of *at*RA. When the kinetic constants

190   Nina Isoherranen and Yue Winnie Wen

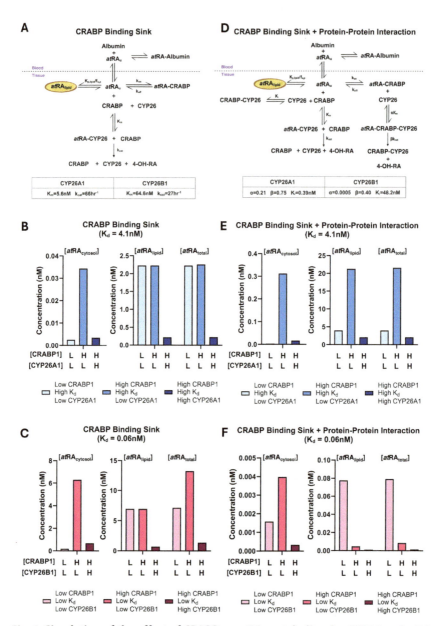

Fig. 4 Simulation of the effect of CRABP on *at*RA metabolism by CYP26 and *at*RA concentrations in different cellular compartments. Panel (A) shows the kinetic scheme for the conceptual model in which *at*RA binds to CRABP in the cytosol and acts as a binding sink without interactions with cellular lipids of CYP26. Panels (B) and (C) show the simulated cytosolic, lipid and total cellular *at*RA concentrations assuming the binding sink model shown in panel (A). Panel (D) shows the kinetic scheme for the

estimated with CYP26B1 are considered (Fig. 4F), higher CRABP1 expression results in lower lipid and total tissue atRA concentrations likely due to the high affinity predicted for holo-CRABP1 with CYP26B1 (Fig. 4F). Yet, the cytosolic concentration of atRA was predicted to be increased when CRABP1 concentrations increase and the fraction of total atRA in the cytosol was predicted to be higher. This is interesting as it suggests that even though total tissue atRA concentrations are lower the pharmacologically active atRA (cytosolic atRA) may be higher due to the network of protein-protein interactions. This possible discrepancy should be considered when interpreting tissue atRA concentrations. If the kinetic values estimated with CYP26A1 are considered (Fig. 4E), higher CRABP1 expression is predicted to result in higher lipid and total tissue atRA concentration and higher cytosolic atRA concentration. The higher lipid and total atRA concentrations are likely due to the decreased clearance of atRA when CRABP1 concentrations are higher. The apo-CRABP1 concentration is higher with higher CRABP1 expression which inhibits

---

conceptual model in which atRA binds to CRABP1 in the cytosol but apo-CRABP1 can interact with CYP26 enzymes and inhibit them and holo-CRABP1 can channel atRA to CYP26 for metabolism. Panels (E) and (F) show simulation results of atRA concentrations based on the protein-protein interaction model shown in panel (B). Panels B and E show simulation results of atRA concentrations based on the data obtained for CYP26A1 and Panels (C) and (F) show simulation results of atRA concentrations based on the data obtained for CYP26B1. The kinetic constants used in the simulations are listed below the scheme for CYP26A1 (B and E) (Yabut & Isoherranen, 2022) and CYP26B1 (C and F) mediated metabolism (Nelson et al., 2016) of atRA. For the simulations, the expression levels of CRABP in the simulations were set either as low (L), 2.4 nM, based on reported CRABP1 expression in rat skin (Ong et al., 1982) or as high (H), 84.7 nM, based on separate report of the CRABP1 expression in rat skin (Bailey & Siu, 1990). The binding constant of atRA to CRABP1 was set either as low K_d (L) of 60 pM based on stopped flow measurements reported (Dong et al., 1999) or as high K_d (H) of 4.1 nM based on binding experiments and fluorescence titrations reported (Yabut & Isoherranen, 2022). The expression level of the CYP26 enzymes is currently not known in different tissues and hence the CYP26 expression was set either low (L) at 10 nmols/tissue or high (H) at 100nmol/tissue. The atRA concentrations in the model were simulated to steady state with constant synthesis rate in a central compartment. atRA$_{cytosol}$: the sum of free atRA and holo-CRABP1 complex concentrations; atRA$_{lipid}$: concentration of atRA distributed into the lipids in the tissue; atRA$_{total}$: the total concentration of atRA (the sum of free atRA, holo-CRABP, and atRA partitioned into lipids) in the tissue. *Simulations were performed in MATLAB and Simulink (R2023b) and plotted using GraphPad Prism (V10.1.2). Panels (A) and (B) were created with Biorender.com.*

CYP26A1 mediated clearance of *at*RA. Due to the interactions between holo-CRABP1 and CYP26A1 the binding affinity of *at*RA to CRABP1 also affects the observed apparent clearance.

At present experimental data is insufficient to support the exact model structure for cellular CRABP-CYP interactions and there is considerable uncertainty in the individual kinetic constants. The modeling suggests that to interpret the impact of altered CRABP or CYP26 expression, the specific kinetics of the protein-protein interactions need to be established. It is interesting that all the organs affected in the *Cyp26* knockout have robust CRABP expression (skin: CRABP1 and CRABP2, spleen: CRABP1, adipose tissue: CRABP2, esophagus and GI tract: CRABP1 and CRABP2) while the liver hepatocytes do not express CRABPs nor was any toxicity observed in the liver. Together with the modeling shown, this suggests a coordinated function of CRABPs and CYP26s and further emphasize the necessity of CRABP expression in *at*RA signaling. These models support the role of the CRABPs in regulating *at*RA clearance via providing the cytosolic reservoir of *at*RA and via modulating *at*RA metabolism by CYP26 enzymes. The modeling also shows that depending on the specific kinetic constants for *at*RA binding to CRABPs and the protein-protein interactions, CYP26A1 and CYP26B1 may have different roles in regulating cellular *at*RA concentrations.

5. Conclusions

Retinoid signaling is essential for life and nuanced control of *at*RA concentrations in different cells is critical. Kinetic calculations can be used to define the cellular distribution of *at*RA and the lipid partitioning and metabolism of *at*RA. The kinetic calculations and modeling explain the long-held interpretation that *at*RA concentrations are regulated during post-natal life within the cell via autocrine mechanisms and not by central synthesis and elimination (endocrine). It is perhaps no surprise that *at*RA synthesis within a cell is by the aldehyde dehydrogenase 1A (ALDH1A) enzymes that are cytosolic. This cytosolic synthesis enables the CRABPs to bind the *at*RA synthesized before it can partition to the lipid and perhaps deliver *at*RA to the nucleus prior to overall lipid equilibrium is established. This may be of particular importance when dietary vitamin A is limiting resulting in less *at*RA synthesis. As shown by the simulations, observed changes, or lack of changes, in overall tissue *at*RA concentrations should be interpreted with caution as the cytosolic concentrations of *at*RA are a minority of overall tissue *at*RA. The data summarized here point to the need to better establish the absolute expression levels and kinetic constants for CRABPs and CYP26s in *at*RA target tissues to fully establish their role in retinoid homeostasis.

References

Abu-Abed, S., Dollé, P., Metzger, D., Beckett, B., Chambon, P., & Petkovich, M. (2001). The retinoic acid-metabolizing enzyme, CYP26A1, is essential for normal hindbrain patterning, vertebral identity, and development of posterior structures. *Genes & Development, 15*, 226–240.

Allegretto, E. A., McClurg, M. R., Lazarchik, S. B., Clemm, D. L., Kerner, S. A., Elgort, M. G., ... Heyman, R. A. (1993). Transactivation properties of retinoic acid and retinoid X receptors in mammalian cells and yeast. Correlation with hormone binding and effects of metabolism. *Journal of Biological Chemistry, 268*, 26625–26633. https://doi.org/10.1016/S0021-9258(19)74358-0.

Allenby, G., Janocha, R., Kazmer, S., Speck, J., Grippo, J. F., & Levin, A. A. (1994). Binding of 9-cis-retinoic acid and all-trans-retinoic acid to retinoic acid receptors alpha, beta, and gamma. Retinoic acid receptor gamma binds all-trans-retinoic acid preferentially over 9-cis-retinoic acid. *The Journal of Biological Chemistry, 269*, 16689–16695.

Amory, J. K., Ostrowski, K. A., Gannon, J. R., Berkseth, K., Stevison, F., Isoherranen, N., ... Walsh, T. (2017). Isotretinoin administration improves sperm production in men with infertility from oligoasthenozoospermia: A pilot study. *Andrology, 5*, 1115–1123. https://doi.org/10.1111/andr.12420.

Arnold, S. L. M., Amory, J. K., Walsh, T. J., & Isoherranen, N. (2012). A sensitive and specific method for measurement of multiple retinoids in human serum with UHPLC-MS/MS. *The Journal of Lipid Research, 53*, 587–598. https://doi.org/10.1194/jlr.D019745.

Arnold, S. L. M., Kent, T., Hogarth, C. A., Griswold, M. D., Amory, J. K., & Isoherranen, N. (2015). Pharmacological inhibition of ALDH1A in mice decreases all-trans retinoic acid concentrations in a tissue specific manner. *Biochemical Pharmacology, 95*, 177–192. https://doi.org/10.1016/j.bcp.2015.03.001.

Arnold, S. L. M., Kent, T., Hogarth, C. A., Schlatt, S., Prasad, B., Haenisch, M., ... Isoherranen, N. (2015). Importance of ALDH1A enzymes in determining human testicular retinoic acid concentrations. *Journal of Lipid Research, 56*, 342–357. https://doi.org/10.1194/jlr.M054718.

Bailey, J. S., & Siu, C. H. (1990). Unique tissue distribution of two distinct cellular retinoic acid binding proteins in neonatal and adult rat. *BBA—General Subjects, 1033*, 267–272. https://doi.org/10.1016/0304-4165(90)90131-F.

Bailey, J. S., & Siu, C. H. (1988). Purification and partial characterization of a novel binding protein for retinoic acid from neonatal rat. *Journal of Biological Chemistry, 263*, 9326–9332. https://doi.org/10.1016/s0021-9258(19)76544-2.

Belatik, A., Hotchandani, S., Bariyanga, J., & Tajmir-Riahi, H. A. (2012). Binding sites of retinol and retinoic acid with serum albumins. *European Journal of Medicinal Chemistry, 48*, 114–123. https://doi.org/10.1016/j.ejmech.2011.12.002.

Biesalski, H. K. (1989). Comparative assessment of the toxicology of vitamin A and retinoids in man. *Toxicology, 57*, 117–161.

Blomhoff, R., & Blomhoff, H. K. (2006). Overview of retinoid metabolism and function. *Journal of Neurobiology, 66*, 606–630. https://doi.org/10.1002/neu.20242.

Bradow, G., Kan, L., & Fenselau, C. (1989). Studies of intramolecular rearrangements of acyl-linked glucuronides using salicyclic acid, flufenamic acid, and (S)- and (R)-benoxaprofrn and confirmation of isomerization in acyl-linked .DELTA.9-11-carboxytetrahydrocannabinol glucuronide. *Chemical Research in Toxicology, 2*(5), 316–324.

Budhu, A. S., & Noy, N. (2002). Direct channeling of retinoic acid between cellular retinoic acid-binding protein II and retinoic acid receptor sensitizes mammary carcinoma cells to retinoic acid-induced growth arrest. *Molecular and Cellular Biology, 22*, 2632–2641.

Chen, H., Fantel, A. G., & Juchau, M. R. (2000). Catalysis of the 4-hydroxylation of retinoic acids by CYP3A7 in human fetal hepatic tissues. *Drug Metabolism and Disposition, 28*, 1051–1057. https://doi.org/10.1016/j.semnephrol.2007.10.007.

Chen, H., & Juchau, M. R. (1998). Recombinant human glutathione S-transferases catalyse enzymic isomerization of 13-cis-retinoic acid to all-trans-retinoic acid in vitro. *Biochemical Journal, 336*(Pt 1), 223–226.

Chen, H., & Juchau, M. R. (1997). Glutathione S-transferases act as isomerases in isomerization of 13-cis-retinoic acid to all-trans-retinoic acid in vitro. *Biochemical Journal, 327*(Pt 3), 721–726.

Chytil, F., & Ong, D. E. (1987). Intracellular vitamin A-binding proteins. *Annual Review of Nutrition, 7*, 321–335. https://doi.org/10.1146/annurev.nu.07.070187.001541.

Czuba, L. C., Fay, E. E., Lafrance, J., Smith, C. K., Shum, S., Moreni, S. L., ... Hebert, M. F. (2022). Plasma retinoid concentrations are altered in pregnant women. *Nutrients, 14*. https://doi.org/10.3390/nu14071365.

Czuba, L. C., & Isoherranen, N. (2024). LX-2 stellate cells are a model system for investigating the regulation of hepatic vitamin A metabolism and respond to tumor necrosis factor α and interleukin 1β. *Drug Metabolism and Disposition: The Biological Fate of Chemicals, 52*, 442–454. https://doi.org/10.1124/dmd.124.001679.

Czuba, L. C., Wu, X., Huang, W., Hollingshead, N., Roberto, J. B., Kenerson, H. L., ... Isoherranen, N. (2021). Altered vitamin A metabolism in human liver slices corresponds to fibrogenesis. *Clinical and Translational Science, 14*, 976–989. https://doi.org/10.1111/cts.12962.

Daruwalla, A., Choi, E. H., Palczewski, K., & Kiser, P. D. (2018). Structural biology of 11-cis-retinaldehyde production in the classical visual cycle. *Biochemical Journal, 475*, 3171–3188. https://doi.org/10.1042/BCJ20180193.

de Gerónimo, E., Rodriguez Sawicki, L., Bottasso Arias, N., Franchini, G. R., Zamarreño, F., Costabel, M. D., ... Falomir Lockhart, L. J. (2014). IFABP portal region insertion during membrane interaction depends on phospholipid composition. *Biochimica et Biophysica Acta (BBA)—Molecular and Cell Biology of Lipids, 1841*, 141–150. https://doi.org/10.1016/j.bbalip.2013.10.011.

Dong, D., Ruuska, S. E., Levinthal, D. J., & Noy, N. (1999). Distinct roles for cellular retinoic acid-binding proteins I and II in regulating signaling by retinoic acid. *The Journal of Biological Chemistry, 274*, 23695–23698.

Dranse, H. J., Sampaio, A. V., Petkovich, M., & Underhill, T. M. (2011). Genetic deletion of Cyp26b1 negatively impacts limb skeletogenesis by inhibiting chondrogenesis. *Journal of Cell Science, 124*, 2723–2734. https://doi.org/10.1242/jcs.084699.

Duester, G. (2008). Retinoic acid synthesis and signaling during early organogenesis. *Cell, 134*, 921–931. https://doi.org/10.1016/j.cell.2008.09.002.

Fagerberg, L., Hallström, B. M., Oksvold, P., Kampf, C., Djureinovic, D., Odeberg, J., ... Uhlén, M. (2014). Analysis of the human tissue-specific expression by genome-wide integration of transcriptomics and antibody-based proteomics. *Molecular & Cellular Proteomics: MCP, 13*, 397–406. https://doi.org/10.1074/mcp.M113.035600.

Fiorella, P. D., & Napoli, J. L. (1994). Microsomal retinoic acid metabolism. Effects of cellular retinoic acid-binding protein (type I) and C18-hydroxylation as an initial step. *The Journal of Biological Chemistry, 269*, 10538–10544.

Fiorella, P. D., & Napoli, J. L. (1991). Expression of cellular retinoic acid binding protein (CRABP) in Escherichia coli. Characterization and evidence that holo-CRABP is a substrate in retinoic acid metabolism. *Journal of Biological Chemistry, 266*, 16572–16579. https://doi.org/10.1016/s0021-9258(18)55339-4.

Gorry, P., Lufkin, T., Dierich, A., Rochette-Egly, C., Decimo, D., Dolle, P., ... Chambon, P. (1994). The cellular retinoic acid binding protein I is dispensable. *Proceedings of the National Academy of Sciences of the United States of America, 91*, 9032–9036.

Gudas, L. J. (2022). Retinoid metabolism: New insights. *Journal of Molecular Endocrinology.* https://doi.org/10.1530/JME-22-0082.

Hammond, T. G., Meng, X., Jenkins, R. E., Maggs, J. L., Castelazo, A. S., Regan, S. L., ... Williams, D. P. (2014). Mass spectrometric characterization of circulating covalent protein adducts derived from a drug Acyl glucuronide metabolite: Multiple albumin adductions in diclofenac patients. *Journal of Pharmacology and Experimental Therapeutics, 350*, 387–402. https://doi.org/10.1124/jpet.114.215079.

Hogarth, C. A., Arnold, S., Kent, T., Mitchell, D., Isoherranen, N., & Griswold, M. D. (2014). Processive pulses of retinoic acid propel asynchronous and continuous murine sperm production. *Biology of Reproduction, 92*, 37. https://doi.org/10.1095/biolreprod.114.126326.

Hogarth, C. A., Evans, E., Onken, J., Kent, T., Mitchell, D., Petkovich, M., & Griswold, M. D. (2015). CYP26 enzymes are necessary within the postnatal seminiferous epithelium for normal murine spermatogenesis. *Biology of Reproduction, 93*, 19. https://doi.org/10.1095/biolreprod.115.129718.

Hogarth, C. A., & Griswold, M. D. (2013). Retinoic acid regulation of male meiosis. *Current Opinion in Endocrinology, Diabetes, and Obesity, 20*, 217–223. https://doi.org/10.1097/MED.0b013e32836067cf.

Holloway, C., Zhong, G., Kim, Y., Ye, H., Sampath, H., Hammerling, U., ... Quadro, L. (2022). Retinoic acid regulates pyruvate dehydrogenase kinase 4 (*Pdk4*) to modulate fuel utilization in the adult heart: Insights from wild-type and β-carotene 9′,10′ oxygenase knockout mice. *The FASEB Journal, 36*. https://doi.org/10.1096/fj.202101910RR.

Holt, K., Nagar, S., & Korzekwa, K. (2019). Methods to predict volume of distribution. *Current Pharmacology Reports, 5*, 391–399. https://doi.org/10.1007/s40495-019-00186-5.

Honarbakhsh, M., Ericsson, A., Zhong, G., Isoherranen, N., Zhu, C., Bromberg, Y., ... Quadro, L. (2021). Impact of vitamin A transport and storage on intestinal retinoid homeostasis and functions. *Journal of Lipid Research, 62*, 100046. https://doi.org/10.1016/j.jlr.2021.100046.

Horowitz, B. Z. (2022). Did hypervitaminosis A have a role in Mawson's ill-fated Antarctic exploration? *Clinical Toxicology, 60*, 1277–1281. https://doi.org/10.1080/15563650.2022.2128366.

Hsu, K. T., & Storch, J. (1996). Fatty acid transfer from liver and intestinal fatty acid-binding proteins to membranes occurs by different mechanisms. *Journal of Biological Chemistry, 271*, 13317–13323. https://doi.org/10.1074/jbc.271.23.13317.

Isoherranen, N., & Zhong, G. (2019). Biochemical and physiological importance of the CYP26 retinoic acid hydroxylases. *Pharmacology & Therapeutics, 204*, 107400. https://doi.org/10.1016/J.PHARMTHERA.2019.107400.

Jeong, H., Armstrong, A. T., Isoherranen, N., Czuba, L., Yang, A., Zumpf, K., ... Wisner, K. L. (2023). Temporal changes in the systemic concentrations of retinoids in pregnant and postpartum women. *PLoS One, 18*, e0280424. https://doi.org/10.1371/journal.pone.0280424.

Jing, J., Nelson, C., Paik, J., Shirasaka, Y., Amory, J. K., & Isoherranen, N. (2017). Physiologically based pharmacokinetic model of all-*trans*-retinoic acid with application to cancer populations and drug interactions. *Journal of Pharmacology and Experimental Therapeutics, 361*, 246–258. https://doi.org/10.1124/jpet.117.240523.

Kedishvili, N. Y. (2013). Enzymology of retinoic acid biosynthesis and degradation. *Journal of Lipid Research, 54*, 1744–1760. https://doi.org/10.1194/jlr.R037028.
Kleywegt, G. J., Bergfors', T., Senn, H., Le Motte, P., Gsell, B., Shudo, K., & Jones, T. A. (n.d.). Crystal structures of cellular retinoic acid binding proteins I and II in complex with all-trans-retinoic acid and a synthetic retinoid.
Kretzschmar, R., & Leuschner, F. (1975). Biosynthesis and metabolism of retinoic acid. *Acta Dermato-Venereologica. Supplementum, 74*, 25–288.
Kumar, S., Chatzi, C., Brade, T., Cunningham, T. J., Zhao, X., & Duester, G. (2011). Sex-specific timing of meiotic initiation is regulated by Cyp26b1 independent of retinoic acid signalling. *Nature Communications, 2*, 151. https://doi.org/10.1038/ncomms1136.
Kurlandsky, S. B., Gamble, M. V., Ramakrishnan, R., & Blaner, W. S. (1995). Plasma delivery of retinoic acid to tissues in the rat. *The Journal of Biological Chemistry, 270*, 17850–17857.
Kurtz, P., Emmerling, D., & Donofrio, D. (1984). Subchronic toxicity of all-trans-retinoic acid and retinylidene dimedone in stprague-dawley rats. *Toxicology, 30*, 115–124.
Lampron, C., Rochette-Egly, C., Gorry, P., Dolle, P., Mark, M., Lufkin, T., ... Chambon, P. (1995). Mice deficient in cellular retinoic acid binding protein II (CRABPII) or in both CRABPI and CRABPII are essentially normal. *Development (Cambridge, England), 121*, 539–548.
Lansink, M., Van Bennekum, A. M., Blaner, W. S., & Kooistra, T. (1997). Differences in metabolism and isomerization of all-trans-retinoic acid and 9-cis-retinoic acid between human endothelial cells and hepatocytes. *European Journal of Biochemistry / FEBS, 247*, 596–604.
Lerner, U. H. (2024). Vitamin A—discovery, metabolism, receptor signaling and effects on bone mass and fracture susceptibility. *Frontiers in Endocrinology (Lausanne), 15*. https://doi.org/10.3389/fendo.2024.1298851.
Li, E., & Norris, A. W. (1996). Structure/function of cytoplasmic vitamin A-binding proteins. *Annual Review of Nutrition, 16*, 205–234. https://doi.org/10.1146/annurev.nu.16.070196.001225.
Lips, P. (2003). Hypervitaminosis A and fractures. *New England Journal of Medicine, 348*, 347–349. https://doi.org/10.1056/NEJMe020167.
Loudig, O., Babichuk, C., White, J., Abu-Abed, S., Mueller, C., & Petkovich, M. (2000). Cytochrome P450RAI(CYP26) promoter: A distinct composite retinoic acid response element underlies the complex regulation of retinoic acid metabolism. *Molecular Endocrinology, 14*, 1483–1497. https://doi.org/10.1210/me.14.9.1483.
Lutz, J. D., Dixit, V., Yeung, C. K., Dickmann, L. J., Zelter, A., Thatcher, J. E., ... Isoherranen, N. (2009). Expression and functional characterization of cytochrome P450 26A1, a retinoic acid hydroxylase. *Biochemical Pharmacology, 77*, 258–268. https://doi.org/10.1016/j.bcp.2008.10.012.
MacLean, G., Abu-Abed, S., Dolle, P., Tahayato, A., Chambon, P., & Petkovich, M. (2001). Cloning of a novel retinoic-acid metabolizing cytochrome P450, Cyp26B1, and comparative expression analysis with Cyp26A1 during early murine development. *Mechanisms of Development, 107*, 195–201.
Maclean, G., Dollé, P., & Petkovich, M. (2009). Genetic disruption of CYP26B1 severely affects development of neural crest derived head structures, but does not compromise hindbrain patterning. *Developmental Dynamics, 238*, 732–745. https://doi.org/10.1002/dvdy.21878.
Maden, M. (2007). Retinoic acid in the development, regeneration and maintenance of the nervous system. *Nature Reviews. Neuroscience, 8*, 755–765. https://doi.org/10.1038/nrn2212.
Majumdar, A., Petrescu, A. D., Xiong, Y., & Noy, N. (2011). Nuclear translocation of cellular retinoic acid-binding protein II is regulated by retinoic acid-controlled SUMOylation. *The Journal of Biological Chemistry, 286*, 42749–42757. https://doi.org/10.1074/jbc.M111.293464.

Molotkov, A., & Duester, G. (2003). Genetic evidence that retinaldehyde dehydrogenase Raldh1 (Aldh1a1) functions downstream of alcohol dehydrogenase Adh1 in metabolism of retinol to retinoic acid. *Journal of Biological Chemistry, 278*, 36085–36090. https://doi.org/10.1074/jbc.M303709200.

Muindi, J. R., Roth, M. D., Wise, R. A., Connett, J. E., O'Connor, G. T., Ramsdell, J. W., ... Sciurba, F. C. (2008). Pharmacokinetics and metabolism of all-trans-and 13-cis-retinoic acid in pulmonary emphysema patients. *Journal of Clinical Pharmacology, 48*, 96–107. https://doi.org/10.1177/0091270007309701.

Napoli, J. (2022). Retinoic acid: Sexually dimorphic, anti-insulin and concentration-dependent effects on energy. *Nutrients*.

Napoli, J. L. (2017). Cellular retinoid binding-proteins, CRBP, CRABP, FABP5: Effects on retinoid metabolism, function and related diseases. *Pharmacology & Therapeutics, 173*, 19–33. https://doi.org/10.1016/j.pharmthera.2017.01.004.

Napoli, J. L. (2012). Physiological insights into all-trans-retinoic acid biosynthesis. *Biochimica et Biophysica Acta—Molecular and Cell Biology of Lipids, 1821*, 152–167. https://doi.org/10.1016/j.bbalip.2011.05.004.

Napoli, J. L., & Yoo, H. S. (2020). Retinoid metabolism and functions mediated by retinoid binding-proteins. *Methods in enzymology* (pp. 55–75). Academic Press Inc. https://doi.org/10.1016/bs.mie.2020.02.004.

Nelson, C. H., Peng, C. C., Lutz, J. D., Yeung, C. K., Zelter, A., & Isoherranen, N. (2016). Direct protein–protein interactions and substrate channeling between cellular retinoic acid binding proteins and CYP26B1. *FEBS Letters*, 2527–2535. https://doi.org/10.1002/1873-3468.12303.

Nhieu, J., Lin, Y.-L., & Wei, L.-N. (2022). CRABP1 in non-canonical activities of retinoic acid in health and diseases. *Nutrients, 14*, 1528. https://doi.org/10.3390/nu14071528.

Norris, A. W., Rong, D., d'Avignon, D. A., Rosenberger, M., Tasaki, K., & Li, E. (1995). Nuclear magnetic resonance studies demonstrate differences in the interaction of retinoic acid with two highly homologous cellular retinoic acid binding proteins. *Biochemistry, 34*, 15564–15573. https://doi.org/10.1021/bi00047a023.

Noy, N. (2000). Retinoid-binding proteins: Mediators of retinoid action. *Biochemical Journal, 348*(Pt 3), 481–495.

N'soukpoé-Kossi, C. N., Sedaghat-Herati, R., Ragi, C., Hotchandani, S., & Tajmir-Riahi, H. A. (2007). Retinol and retinoic acid bind human serum albumin: Stability and structural features. *International Journal of Biological Macromolecules, 40*, 484–490. https://doi.org/10.1016/j.ijbiomac.2006.11.005.

Obrochta, K. M., Kane, M. A., & Napoli, J. L. (2014). Effects of diet and strain on mouse serum and tissue retinoid concentrations. *PLoS One, 9*, e99435. https://doi.org/10.1371/journal.pone.0099435.

O'Byrne, S. M., & Blaner, W. S. (2013). Retinol and retinyl esters: Biochemistry and physiology. *Journal of Lipid Research, 54*, 1731–1743. https://doi.org/10.1194/jlr.R037648.

Ong, D. E., & Chytil, F. (1975). Retinoic acid binding protein in rat tissue. Partial purification and comparison to rat tissue retinol binding protein. *Journal of Biological Chemistry, 250*, 6113–6117. https://doi.org/10.1016/s0021-9258(19)41166-6.

Ong, D. E., Crow, J. A., & Chytil, F. (1982). Radioimmunochemical determination of cellular retinol- and cellular retinoic acid-binding proteins in cytosols of rat tissues. *Journal of Biological Chemistry, 257*, 13385–13389. https://doi.org/10.1016/s0021-9258(18)33460-4.

Pennimpede, T., Cameron, D. a, MacLean, G. a, Li, H., Abu-Abed, S., & Petkovich, M. (2010). The role of CYP26 enzymes in defining appropriate retinoic acid exposure during embryogenesis. *Birth Defects Research. Part A, Clinical and Molecular Teratology, 88*, 883–894. https://doi.org/10.1002/bdra.20709.

Penniston, K. L., & Tanumihardjo, S. A. (2006). The acute and chronic toxic effects of vitamin A. *The American Journal of Clinical Nutrition, 83*, 191–201. https://doi.org/10.1093/ajcn/83.2.191.

Petkovich, M., & Chambon, P. (2022). Retinoic acid receptors at 35 years. *Journal of Molecular Endocrinology, 69*, T13–T24. https://doi.org/10.1530/JME-22-0097.

Raverdeau, M., Gely-Pernot, A., Feret, B., Dennefeld, C., Benoit, G., Davidson, I., ... Ghyselinck, N. B. (2012). Retinoic acid induces Sertoli cell paracrine signals for spermatogonia differentiation but cell autonomously drives spermatocyte meiosis. *Proceedings of the National Academy of Sciences of the United States of America, 109*, 16582–16587. https://doi.org/10.1073/pnas.1214936109.

Rodgers, T., Leahy, D., & Rowland, M. (2005). Physiologically based pharmacokinetic modeling 1: Predicting the tissue distribution of moderate-to-strong bases. *Journal of Pharmaceutical Sciences, 94*, 1259–1276. https://doi.org/10.1002/jps.20322.

Rodgers, T., & Rowland, M. (2007). Mechanistic Approaches to Volume of Distribution Predictions: Understanding the Processes, *24*, 918–933. https://doi.org/10.1007/s11095-006-9210-3.

Rodgers, T., & Rowland, M. (2006). Physiologically based pharmacokinetic modelling 2: Predicting the tissue distribution of acids, very weak bases, neutrals and zwitterions. *Journal of Pharmaceutical Sciences, 95*, 1238–1257. https://doi.org/10.1002/jps.20502.

Samokyszyn, V. M., Gall, W. E., Zawada, G., Freyaldenhoven, M. A., Chen, G., Mackenzie, P. I., ... Radominska-Pandya, A. (2000). 4-Hydroxyretinoic acid, a novel substrate for human liver microsomal UDP-glucuronosyltransferase(s) and recombinant UGT2B7. *The Journal of Biological Chemistry, 275*, 6908–6914.

Sass, J. O., Forster, A., Bock, K. W., & Nau, H. (1994). Glucuronidation and isomerization of all-trans- and 13-CIS-retinoic acid by liver microsomes of phenobarbital- or 3-methylcholanthrene-treated rats. *Biochemical Pharmacology, 47*, 485–492. https://doi.org/10.1016/0006-2952(94)90179-1.

Schug, T. T., Berry, D. C., Shaw, N. S., Travis, S. N., & Noy, N. (2007). Opposing effects of retinoic acid on cell growth result from alternate activation of two different nuclear receptors. *Cell, 129*, 723–733. https://doi.org/10.1016/j.cell.2007.02.050.

Senoo, H., Imai, K., Mezaki, Y., Miura, M., Morii, M., Fujiwara, M., & Blomhoff, R. (2012). Accumulation of vitamin A in the hepatic stellate cell of arctic top predators. *Anatomical Record, 295*, 1660–1668. https://doi.org/10.1002/ar.22555.

Sharma, M. K., Denovan-Wright, E. M., Boudreau, M. E. R., & Wright, J. M. (2003). A cellular retinoic acid-binding protein from zebrafish (Danio rerio): cDNA sequence, phylogenetic analysis, mRNA expression, and gene linkage mapping. *Gene, 311*, 119–128. https://doi.org/10.1016/S0378-1119(03)00580-8.

Sirbu, I. O., Gresh, L., Barra, J., & Duester, G. (2005). Shifting boundaries of retinoic acid activity control hindbrain segmental gene expression. *Development (Cambridge, England), 132*, 2611–2622. https://doi.org/10.1242/dev.01845.

Snyder, J. M., Zhong, G., Hogarth, C., Huang, W., Topping, T., LaFrance, J., ... Isoherranen, N. (2020). Knockout of Cyp26a1 and Cyp26b1 during postnatal life causes reduced lifespan, dermatitis, splenomegaly, and systemic inflammation in mice. *FASEB Journal, 34*, 15788–15804. https://doi.org/10.1096/fj.202001734R.

Stevison, F., Hogarth, C., Tripathy, S., Kent, T., & Isoherranen, N. (2017). Inhibition of the all trans-retinoic acid hydroxylases CYP26A1 and CYP26B1 results in dynamic, tissue-specific changes in endogenous atRA signaling. *Drug Metabolism and Disposition, 45*, 846–854. https://doi.org/10.1124/dmd.117.075341.

Stevison, F., Kosaka, M., Kenny, J. R., Wong, S., Hogarth, C., Amory, J. K., & Isoherranen, N. (2019). Does in vitro cytochrome P450 downregulation translate to in vivo drug-drug interactions? Preclinical and clinical studies with 13-cis-retinoic acid. *Clinical and Translational Science, 12*, 350–360. https://doi.org/10.1111/cts.12616.

Storch, J., & Corsico, B. (2008). The emerging functions and mechanisms of mammalian fatty acid–binding proteins. *Annual Review of Nutrition, 28*, 73–95. https://doi.org/10.1146/annurev.nutr.27.061406.093710.

Taimi, M., Helvig, C., Wisniewski, J., Ramshaw, H., White, J., Amad, M., ... Petkovich, M. (2004). A novel human cytochrome P450, CYP26C1, involved in metabolism of 9-cis and all-trans isomers of retinoic acid. *Journal of Biological Chemistry, 279*, 77–85. https://doi.org/10.1074/jbc.M308337200.

Tay, S., Dickmann, L., Dixit, V., & Isoherranen, N. (2010). A comparison of the roles of peroxisome proliferator-activated receptor and retinoic acid receptor on CYP26 regulation. *Molecular Pharmacology, 77*, 218–227. https://doi.org/10.1124/mol.109.059071.

Thatcher, J. E., Buttrick, B., Shaffer, S. A., Shimshoni, J. A., Goodlett, D. R., Nelson, W. L., & Isoherranen, N. (2011). Substrate specificity and ligand interactions of CYP26A1, the human liver retinoic acid hydroxylase. *Molecular Pharmacology, 80*, 228–239. https://doi.org/10.1124/mol.111.072413.

Thatcher, J. E., Zelter, A., & Isoherranen, N. (2010). The relative importance of CYP26A1 in hepatic clearance of all-trans retinoic acid. *Biochemical Pharmacology, 80*, 903–912. https://doi.org/10.1016/j.bcp.2010.05.023.

Topletz, A. R., Thatcher, J. E., Zelter, A., Lutz, J. D., Tay, S., Nelson, W. L., & Isoherranen, N. (2012). Comparison of the function and expression of CYP26A1 and CYP26B1, the two retinoic acid hydroxylases. *Biochemical Pharmacology, 83*, 149–163. https://doi.org/10.1016/j.bcp.2011.10.007.

Topletz, A. R., Tripathy, S., Foti, R. S., Shimshoni, J. A., Nelson, W. L., & Isoherranen, N. (2015). Induction of CYP26A1 by Metabolites of Retinoic Acid: Evidence That CYP26A1 Is an Important Enzyme in the Elimination of Active Retinoids. *87*, 430–441. https://doi.org/10.1124/mol.114.096784.

Topletz, A. R., Zhong, G., & Isoherranen, N. (2019). Scaling in vitro activity of CYP3A7 suggests human fetal livers do not clear retinoic acid entering from maternal circulation. *Scientific Reports, 9*, 4620. https://doi.org/10.1038/s41598-019-40995-8.

Tsukada, M., Schröder, M., Roos, T. C., Chandraratna, R. A. S., Reichert, U., Merk, H. F., ... Zouboulis, C. C. (2000). 13-Cis retinoic acid exerts its specific activity on human sebocytes through selective intracellular isomerization to all-trans Retinoic acid and binding to retinoid acid receptors. *Journal of Investigative Dermatology, 115*, 321–327. https://doi.org/10.1046/j.1523-1747.2000.00066.x.

Uehara, M., Yashiro, K., Mamiya, S., Nishino, J., Chambon, P., Dolle, P., & Sakai, Y. (2007). CYP26A1 and CYP26C1 cooperatively regulate anterior-posterior patterning of the developing brain and the production of migratory cranial neural crest cells in the mouse. *Developmental Biology, 302*, 399–411. https://doi.org/10.1016/j.ydbio.2006.09.045.

Urbach, J., & Rando, R. R. (1994). Thiol dependent isomerization of all-*trans*-retinoic acid to 9-*cis*-retinoic acid. *FEBS Letters, 351*, 429–432. https://doi.org/10.1016/0014-5793(94)00090-5.

Vaezeslami, S., Mathes, E., Vasileiou, C., Borhan, B., & Geiger, J. H. (2006). The structure of apo-wild-type cellular retinoic acid binding protein II at 1.4 Å and its relationship to ligand binding and nuclear translocation. *Journal of Molecular Biology, 363*, 687–701. https://doi.org/10.1016/j.jmb.2006.08.059.

Vane, F. M., Bugge, C. J., Rodriguez, L. C., Rosenberger, M., & Doran, T. I. (1990). Human biliary metabolites of isotretinoin: Identification, quantification, synthesis and biological activity. *Xenobiotica, 20*, 193–207. https://doi.org/10.3109/00498259009047155.

Wang, C. C., Campbell, S., Furner, R. L., & Hill, D. L. (1980). Disposition of all-trans- and 13-cis-retinoic acids and n-hydroxyethylretinamide in mice after intravenous administration. *Drug Metabolism and Disposition: The Biological Fate of Chemicals, 8*, 8–11.

White, J. A., Beckett-Jones, B., Guo, Y. D., Dilworth, F. J., Bonasoro, J., Jones, G., & Petkovich, M. (1997). cDNA cloning of human retinoic acid-metabolizing enzyme (hP450RAI) identifies a novel family of cytochromes p450 (CYP26). *Journal of Biological Chemistry, 272*, 18538–18541. https://doi.org/10.1074/jbc.272.30.18538.

White, J. A., Guo, Y., Baetz, K., Beckett-jones, B., Bonasoro, J., Hsu, K. E., ... Petkovich, M. (1996). Identification of the retinoic identification of the retinoic acid-inducible alltrans- retinoic acid 4-hydroxylase*. *Journal of Biological Chemistry, 271*, 29922–29927.

White, J. A., Ramshaw, H., Taimi, M., Stangle, W., Zhang, A., Everingham, S., ... Petkovich, M. (2000). Identification of the human cytochrome P450, P450RAI-2, which is predominantly expressed in the adult cerebellum and is responsible for all-transretinoic acid metabolism. *Proceedings of the National Academy of Sciences of the United States of America, 97*, 6403–6408. https://doi.org/10.1073/pnas.120161397.

Wolf, G. (1996). A history of vitamin A and retinoids. *The FASEB Journal, 10*, 1102–1107.

Yabut, K. C. B., & Isoherranen, N. (2022). CRABPs alter all-trans-retinoic acid metabolism by CYP26A1 via protein-protein interactions. *Nutrients, 14*, 1784. https://doi.org/10.3390/nu14091784.

Yabut, K. C. B., & Isoherranen, N. (2023). Impact of intracellular lipid binding proteins on endogenous and xenobiotic ligand metabolism and disposition. *Drug Metabolism and Disposition: The Biological Fate of Chemicals.* https://doi.org/10.1124/dmd.122.001010.

Yashiro, K., Zhao, X., Uehara, M., Yamashita, K., Nishijima, M., Nishino, J., ... Hamada, H. (2004). Regulation of retinoic acid distribution is required for proximodistal patterning and outgrowth of the developing mouse limb. *Developmental Cell, 6*, 411–422. https://doi.org/10.1016/S1534-5807(04)00062-0.

Yoo, H. S., Cockrum, M. A., & Napoli, J. L. (2023). Cyp26a1 supports postnatal retinoic acid homeostasis and glucoregulatory control. *Journal of Biological Chemistry, 299*, 104669. https://doi.org/10.1016/j.jbc.2023.104669.

Yoo, H. S., Rodriguez, A., You, D., Lee, R. A., Cockrum, M. A., Grimes, J. A., ... Napoli, J. L. (2022). The glucocorticoid receptor represses, whereas C/EBPβ can enhance or repress CYP26A1 transcription. *iScience, 25*, 104564. https://doi.org/10.1016/j.isci.2022.104564.

Yue, F., Cheng, Y., Breschi, A., Vierstra, J., Wu, W., Ryba, T., ... Mouse ENCODE Consortium (2014). A comparative encyclopedia of DNA elements in the mouse genome. *Nature, 515*, 355–364. https://doi.org/10.1038/nature13992.

Zhong, G., Hogarth, C., Snyder, J., Palau, L., Topping, T., Huang, W., ... Isoherranen, N. (2019). The retinoic acid hydroxylase Cyp26a1 has minor effects on postnatal vitamin A homeostasis, but is required for exogenous atRA clearance. *Journal of Biological Chemistry jbc, 009023*, RA119. https://doi.org/10.1074/jbc.RA119.009023.

Zhong, G., Kirkwood, J., Won, K. J., Tjota, N., Jeong, H., & Isoherranen, N. (2019). Characterization of Vitamin A metabolome in human livers with and without nonalcoholic fatty liver disease. *Journal of Pharmacology and Experimental Therapeutics, 370*, 92–103. https://doi.org/10.1124/jpet.119.258517.

Zhong, G., Ortiz, D., Zelter, A., Nath, A., & Isoherranen, N. (2018). CYP26C1 is a hydroxylase of multiple active retinoids and interacts with cellular retinoic acid binding proteins. *Molecular Pharmacology, 93*, 489–503. https://doi.org/10.1124/mol.117.111039.

Zhong, G., Seaman, C. J., Paragas, E. M., Xi, H., Herpoldt, K.-L., King, N. P., ... Isoherranen, N. (2021). Aldehyde oxidase contributes to all-trans-retinoic acid biosynthesis in human liver. *Drug Metabolism and Disposition: The Biological Fate of Chemicals, 49*, 202–211. https://doi.org/10.1124/dmd.120.000296.

Zile, M. H., Inhorn, R. C., & DeLuca, H. F. (1982). Metabolism in vivo of all-transretinoic acid. Biosynthesis of 13-cis-retinoic acid and all-trans- and 13-cis-retinoyl glucuronides in the intestinal mucosa of the rat. *The Journal of Biological Chemistry, 257*, 3544–3550.

CHAPTER EIGHT

Retinoic acid homeostasis and disease

Maureen A. Kane[*]

Department of Pharmaceutical Sciences, University of Maryland School of Pharmacy, Baltimore, MD, United States
[*]Corresponding author. e-mail address: mkane@rx.umaryland.edu

Contents

1. Introduction	202
1.1 ATRA biosynthetic pathway and homeostatic mechanisms	202
1.2 Binding proteins in ATRA homeostasis	205
2. ATRA homeostasis and disease	207
2.1 CYP26B1	209
2.2 POR	209
2.3 DHRS3	209
2.4 Cellular ROL-binding proteins in disease	210
2.5 Cellular ATRA binding proteins	215
2.6 Potential for gene-environment interaction	215
3. Other considerations	217
3.1 Direct quantification of ATRA is essential to understanding ATRA homeostasis	217
3.2 Reporter assays	222
3.3 Surrogates for ATRA and indirect measurement	222
4. Conclusions and future opportunities	224
Acknowledgments	224
References	224
Further reading	233

Abstract

Retinoids, particularly all-trans-retinoic acid (ATRA), play crucial roles in various physiological processes, including development, immune response, and reproduction, by regulating gene transcription through nuclear receptors. This review explores the biosynthetic pathways, homeostatic mechanisms, and the significance of retinoid-binding proteins in maintaining ATRA levels. It highlights the intricate balance required for ATRA homeostasis, emphasizing that both excess and deficiency can lead to severe developmental and health consequences. Furthermore, the associations are discussed between ATRA dysregulation and several diseases, including various genetic disorders, cancer, endometriosis, and heart failure, underscoring the role of retinoid-binding proteins like RBP1 in these conditions. The potential for gene-environment interactions in retinoid metabolism is also examined, suggesting that dietary factors may exacerbate genetic predispositions to ATRA-related pathologies. Methodological advancements in quantifying ATRA and its metabolites are reviewed, alongside the challenges inherent in studying retinoid dynamics. Future research directions are proposed to further elucidate the role of ATRA in health and disease, with the aim of identifying therapeutic targets for conditions linked to retinoid signaling dysregulation.

1. Introduction

Retinoids are isoprenoid lipids with a monocyclic ring and isoprenoid chain that can vary in respect to its conformation and/or polar end group at the terminus of the side chain (Sporn, Dunlop, Newton, & Henderson, 1976). All-*trans*-retinoic acid (ATRA), an active metabolite of vitamin A (also called retinol, ROL), mediates numerous systemic effects, including development, nervous system function, immune response, cell proliferation, cell differentiation, and reproduction. It acts by regulating transcription of hundreds of genes through binding to nuclear receptors (called RARA, RARB, RARG) and peroxisome proliferator-activated receptor delta (PPARD) (Balmer & Blomhoff, 2002; Chambon, 1996; Shaw, Elholm, & Noy, 2003). ATRA concentrations in vivo are precisely controlled in a temporally/spatially controlled manner by the expression of specific retinoid-binding proteins, enzymes, and receptors, which contribute to ATRA generation, signaling, and catabolism (Krezel, Kastner, & Chambon, 1999; Napoli, 1999a, 1999b; Ong, 1994; Ruberte, Friederich, Chambon, & Morriss-Kay, 1993; Yamagata et al., 1994). ATRA can function either cell autonomously or in a non-cell autonomous fashion as a morphogen (Dubey, Rose, Jones, & Saint-Jeannet, 2018; Shannon, Moise, & Trainor, 2017). Both excess and deficiency of ATRA have serious effects on embryonic development and have deleterious effect in the adult (Collins & Mao, 1999; Napoli, 2012, 2022; Shannon et al., 2017).

1.1 ATRA biosynthetic pathway and homeostatic mechanisms

The enzymes involved in ATRA biosynthesis are summarized in Fig. 1. Vitamin A (ROL) is diet derived from two major sources in the form of ROL or retinyl esters (RE) mainly from animal sources and in the form of carotenoids such as β-carotene from vegetable sources. Cellular ATRA production consists of a reversible and rate-limiting dehydrogenation of ROL to retinaldehyde (RAL) catalyzed by membrane-bound, short-chain retinol dehydrogenases (RDH). The forward reaction is catalyzed by several RDH including RDH10 (Ashique et al., 2012; Farjo et al., 2011; Sandell et al., 2007; Sandell, Lynn, Inman, McDowell, & Trainor, 2012).

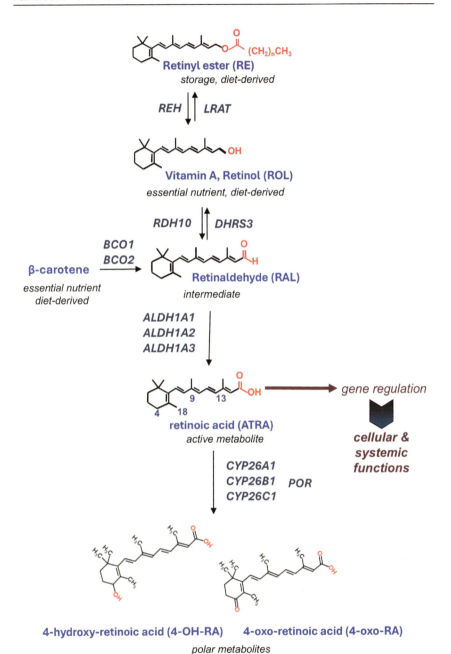

Fig. 1 **ATRA biosynthesis is regulated by enzymes.** Vitamin A is diet derived in the form of retinol (ROL), retinyl esters (RE) or β-carotene. All-*trans* retinoic acid (ATRA), the active metabolite of ROL, is biosynthesized by a two-step oxidation from ROL. The first reversible step of ROL to RAL is catalysed by ROL dehydrogenases, such as RDH10.

(Continued)

Dehydrogenase/reductase 3 (DHRS3; formerly named SDR16C1) is a major enzyme catalyzing the opposite reaction (RAL to ROL) (Billings et al., 2013; Napoli, 2012, 2020). RDH10 and DHRS3 have been shown to be critical during development (Ashique et al., 2012; Billings et al., 2013; Sandell et al., 2007). Additionally, RDH10 and DHRS3 function together as a oxidoreductase molecular complex inserted within the membrane of the endoplasmic reticulum (Adams et al., 2021). Additional RDH shown essential for efficient ATRA biosynthesis include RDH1 (RDH16 in human) and DHRS9 (Napoli, 2012, 2017). Additional RAL reductases include DHRS4 (formerly named RRD) (Napoli, 2017). In addition to being derived from ROL through the action of RDH, RAL can also be derived from the symmetric cleavage of β-carotene by BCO1 and BCO2. Intermediate RAL undergoes an irreversible dehydrogenation to form ATRA by cytosolic RAL dehydrogenases, such as ALDH1A1, ALDH1A2 and ALDH1A3 (formerly called RALDH1, RALDH2, and RALDH3). Additional RAL dehydrogenases include ALDH8A1 (formerly called RALDH4), which is unique in that it prefers 9-*cis* substrate over all-*trans* substrate. Storage of ROL is catalyzed by lecithin:retinol acyltransferase (LRAT) into RE, and ROL can be mobilized by stimulating RE hydrolases (REH). ATRA is catabolized as polar metabolites such as 4-oxo-retinoic acid (4-oxo-RA) or 4-hydroxy retinoic acid (4-OH-RA) by a number of cytochrome (CYP) P450 enzymes found in the microsomal membrane, with CYP26 demonstrating specificity for retinoids (Napoli, 2012). The cytochrome P450 oxidoreductase POR is an important factor regulating ATRA degradation. It is the universal electron donor for all microsomal P450 enzymes, including CYP26B1 (Fluck et al., 2004). Signaling by ATRA is mediated by ATRA binding to RARA, RARB or RARG dimerized with one of the three retinoid X receptors (RXRs), leading to altered transcription following binding to retinoic acid response elements (RAREs) located in regulatory regions of ATRA-target genes

---

Fig. 1—Cont'd RAL reductases, such as DHRS3, catalyse the reverse reaction of RAL to ROL. Intermediate RAL is then oxidized to ATRA irreversibly by RAL dehydrogenases, such as ALDH1A1, ALDH1A2, and ALDH1A3. The catabolism of ATRA is mediated by three cytochrome P450 enzymes CYP26A1, CYP26B1 and CYP26C1, which all require POR as a universal electron donor. They produce polar metabolites such as 4-OH-RA and 4-oxo-RA. RAL can also be derived from the symmetric cleavage of β-carotene by BCO1 or BCO2. Excess ROL is esterified and stored as RE where it can undergo hydrolysis to mobilize ROL in times of need. Additional ROL dehydrogenases (RDH) shown essential for efficient ATRA biosynthesis include RDH1 and DHRS9.

(Chambon, 1996; Gudas, 2022; Shaw et al., 2003). Similarly, ATRA binding to PPARD dimerized with a RXR leads to altered transcription following binding to PPAR response elements (PPREs) located in regulatory regions of target genes. Non-genomic actions of ATRA have also been reported (Chen & Napoli, 2008; Chen, Onisko, & Napoli, 2008).

1.2 Binding proteins in ATRA homeostasis

Retinoid-binding proteins play an important role in regulating ATRA homeostasis (Fig. 2). Bound to retinoid ligand (holo-protein), retinoid-binding proteins solubilize lipophilic retinoids that have limited solubility in aqueous media (e.g., ROL has aqueous solubility < 60 nM) (Napoli, 2016), protect their functional groups that would subject to spurious oxidative metabolism, and deliver the retinoids to specific enzymes for metabolism (Boerman & Napoli, 1996; Lapshina, Belyaeva, Chumakova, & Kedishvili, 2003; Napoli, 2012). Some apo-proteins (ligand-free) also have regulatory effects (Napoli, 2016, 2017). Circulating ROL is chaperoned by ROL-binding protein, type 4 (RBP4). Uptake of ROL from circulating holo-RBP4 is controlled by the membrane receptor STRA6. Intracellular chaperone RBP1 physically interacts with STRA6 to pick up ROL, and then delivers ROL to RDH (e.g., RDH10) which catalyze conversion of ROL to RAL. RAL reductase enzymes (e.g., DHRS3) catalyze the reverse reaction of RAL to ROL. RBP1 can also bind RAL and chaperones RAL to ALDH1A1, ALDH1A2, ALDH1A3, which irreversibly convert RAL to ATRA. RA is then chaperoned by a distinct set of ATRA-binding proteins (CRABP1, CRABP2 & FABP5). Once formed, ATRA can either (1) be transported by CRABP2 or FABP5 to the nucleus where ATRA binds to RARs or PPARD, respectively, that initiate gene transcription, or (2) be transported by CRABP1 to CYP26 enzymes to be degraded.

Of particular note, RBP1 is a widely expressed intracellular chaperone protein for ROL and RAL and is ubiquitously expressed in the body (Napoli, 2017). Both apo-RBP1 and holo-RBP1 participate in regulation of ATRA homeostasis (Fig. 2). Holo-RBP1 chaperones ROL to ATRA biosynthesizing enzymes while sequestering ROL (and RAL) from non-specific oxidation (Boerman & Napoli, 1996; Lapshina et al., 2003; Napoli, 2012). The RBP1 chaperone model has been discussed in detail in a series of reviews (Napoli, 2016, 2017) and postulates that RBP1 ensures efficient ROL metabolism, but also posits that RBP1 is not obligatory for ROL metabolism (Napoli, 2000, 2012). As such, enzymatic oxidation of ROL

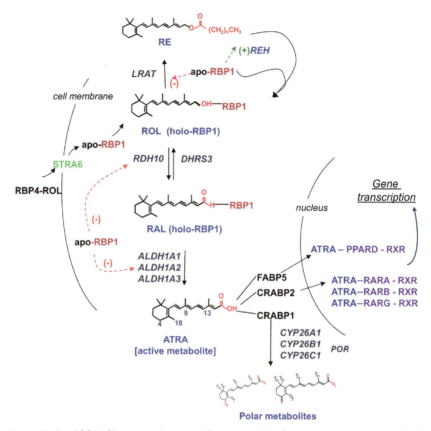

Fig. 2 Retinoid-binding proteins contribute to ATRA homeostasis. Uptake of ROL from circulating RBP4-bound ROL is controlled by the membrane receptor STRA6. Intracellular chaperone RBP1 physically interacts with STRA6 to pick up ROL, and then delivers ROL to RDH10, which catalyze conversion of ROL to RAL. DHRS3 catalyzes the reverse reaction of RAL to ROL. RBP1 can also bind RAL and chaperones RAL to ALDH1A1, ALDH1A2 and ALDH1A3, which irreversibly convert RAL to ATRA. ATRA is then chaperoned by a distinct set of ATRA-binding proteins (CRABP1, CRABP2, & FABP5). Once formed, ATRA can either (1) be transported by CRABP2 or FABP5 to the cell nucleus where ATRA binds to RAR or PPARD, respectively, and initiates gene transcription, or (2) be transported by CRABP1 to CYP26 enzymes to be degraded.

(and RAL) can proceed in the presence or absence of RBP1 chaperone. However, reactions that are normally restricted by RBP1 may occur in the absence of RBP1 chaperone, for example, metabolism by non-specific detoxifying enzymes that oxidize a wide range of substrates (Napoli, 2012). Apo-RBP1 has an additional regulatory influence on ATRA homeostasis as it is able to inhibit and stimulate specific ATRA biosynthesizing enzymes in order to control flux through the ROL pathway (Boerman & Napoli, 1991, 1996; Herr & Ong, 1992; Lapshina et al., 2003; Napoli, 2012). According to current reports, RBP1 loss affects ATRA homeostasis in $Rbp1^{-/-}$ mouse tissue, with some tissues having altered ATRA levels and some tissues possessing compensatory mechanisms to maintain ATRA levels (Kane, Chen, Sparks, & Napoli, 2005; Kane, Folias, Wang, & Napoli, 2010; Kane et al., 2011). The upregulation of homologous RBP2 and/or RBP3 chaperones (known as RBP7 in human) contributes to maintaining ATRA in the absence of RBP1 (Piantedosi, Ghyselinck, Blaner, & Vogel, 2005; Vogel et al., 2001). However, because RBP1 and its homologues (RBP2, RBP3) normally have distinct functions, compensatory upregulation does not usually result in full replacement of RBP1 functionality (Ghyselinck et al., 1999; Kane et al., 2011; Piantedosi et al., 2005; Vogel et al., 2001; Zizola, Schwartz, & Vogel, 2008). For example, in regards to ATRA homeostasis, when RBP2 or RBP3 is upregulated in the absence of RBP1, rates of ATRA biosynthesis are altered (Kane et al., 2011; Piantedosi et al., 2005).

2. ATRA homeostasis and disease

A number of retinoid pathway genes have known associations with diseases in the form of autosomal phenotypes including CYP26B1, POR, and DHRS3 (Table 1). These autosomal mutations are genetic disorders located on one of the numbered, non-sex chromosomes and can be passed from parent to offspring. Each of these enzymes also have similar phenotypes that are recapitulated in genetic loss-of-function mouse models. In mouse models, excess ATRA or deficiency of ATRA in the embryo can be lethal (Ashique et al., 2012; Billings et al., 2013; Sandell et al., 2007). In the case of ATRA homeostatic dysregulation, even hypomorphic (partial loss-of-function) variants can still be damaging (Zschocke, Byers, & Wilkie, 2023). In human, plasma ATRA has been suggested as a way to evaluate teratogenic potential of retinoid intake (van Vliet, Boelsma, de Vries, & van den Berg, 2001) and genetic mutations (Fukami et al., 2010). It has been postulated that even marginal increases in the levels of ATRA may be detrimental (Arnhold, Tzimas, Wittfoht, Plonait, & Nau, 1996). Here, we highlight key features of the human diseases and genetic mouse models used to dissect disease mechanisms.

Table 1 **Defects in ATRA biosynthesis in human disease.** CYP26B1 deficiency, POR deficiency, and DHRS3 deficiency all result in excess ATRA that has deleterious effects. *OMIM gene entry number, +Phenotype MIM number. From OMIM (Online Mendelian Inheritance in Man). OMIM is a comprehensive, authoritative compendium of human genes and genetic phenotypes on all known mendelian disorders and over 16,000 genes (https://www.omim.org).

*OMIM gene entry number	Gene	Phenotype MIM number+	RA status	Phenotype
*605207	CYP26B1 deficiency	MIM 614416	Excess ATRA	• Multiple skeletal defects • Severe craniofacial malformations
*124015	POR deficiency Antley-Bixler Syndrome	MIM 201750	Excess ATRA	• Craniosynostosis • Radiohumeral synostosis • Skeletal defects
*612830	DHRS3 deficiency		Excess ATRA	• Coronal craniosynostosis • Dysmorphic facial features • Congenital heart disease • Scoliosis

2.1 CYP26B1

Deficiency or loss-of-function of CYP26B1 results in excess ATRA due to insufficient degradation. In human, deficiency of CYP26B1 yields radio-humeral fusions with other skeletal and craniofacial anomalies (RHFCA) (Laue et al., 2011). RHFCA comprises a widely variable spectrum of craniofacial malformations including craniosynostosis, cranial ossification defects with occipital encephaloceles, characteristic facial dysmorphism, joint contractures and fusions, and digit abnormalities due, at least in part, to aberrant osteoblast-osteocyte transitioning (Grand et al., 2021; James, Levi, Xu, Carre, & Longaker, 2010; Laue et al., 2011; Li, Xie, & Jiang, 2018; Morton, Frentz, Morgan, Sutherland-Smith, & Robertson, 2016; Silveira et al., 2023; Yang, Lu et al., 2021; Yashiro et al., 2004). Studies in $Cyp26b1^{-/-}$ mice show excess ATRA signaling, defective limb patterning and delayed chondrocyte maturation (Yashiro et al., 2004). Wild-type embryos exposed to excess ATRA phenocopied the limb defects of $Cyp26b1^{-/-}$ mice. Cells derived from the $Cyp26b1^{-/-}$ mouse treated with ATRA had repressed proliferation and enhanced osteogenic differentiation of suture-derived mesenchymal cells (James et al., 2010). These observations are consistent with the notion that ATRA exposure may cause premature cranial suture fusion via enhanced osteogenesis (James et al., 2010; Laue et al., 2011).

2.2 POR

Deficiency in POR leads to a loss-of-function in ATRA catabolism, which yields excess ATRA. POR deficiency is known as Antley-Bixler syndrome, which has been described with genital anomalies and disordered steroidogenesis, as well as craniosynostosis, elbow anklyosis and other joint contractures. Excess ATRA levels in patients and mouse likely contribute to features consistent with ATRA embryopathy (Fukami et al., 2010; Schmidt et al., 2009).

2.3 DHRS3

DHRS3 mutations in human have recently been reported to be a ATRA embryopathy which includes craniosynostosis, cardiac malformations, and scoliosis (Hashimoto, 2017). Loss-of-function or ablation of DHRS3 results in excess ATRA which can be deleterious. Homozygosity for a 5'UTR deletion yielded elevated plasma ATRA in patients and pathogenic results (Hashimoto, 2017). Missense variants in DHRS3 have also been identified and may be pathogenic. The malformations observed in human recapitulated the phenotypes observed in mutant mice lacking DHRS3 (Billings et al., 2013). This includes axial skeletal anomalies such as cervical vertebrae fusion, and cranial skeletal anomalies that affect parietal ossification, with narrower coronal sutures and the onset of cransynostosis (Billings et al., 2013). Mouse embryos deficient in DHRS3 also had defects in palatogenesis, cardiac outflow tract formation, atrial and ventricular septation, vertebra formation, development of the coronary vasculature and die during late gestation (Billings et al., 2013; Wang et al., 2018). The elevated retinoids as a result of loss-of-function in human and ablation in mouse are similar. Patients harboring the 5′UTR deletion have increases in plasma ATRA between 66–78% accompanied by decreases in plasma ROL between 62–86%, as compared to WT. $Dhrs3^{-/-}$ mutant mouse embryos exhibited an increase in ATRA of 48% accompanied by a decrease in ROL of 61% at embryonic day 14.5 (Billings et al., 2013; Hashimoto, 2017).

2.4 Cellular ROL-binding proteins in disease

To date, dysregulation of ROL-binding proteins in disease has been linked to loss of expression and, thus, loss of regulatory control of ATRA homeostasis. Because RBP1 is a widely expressed intracellular chaperone protein for ROL and RAL and is ubiquitously expressed in the body (Napoli, 2017), this review will focus on this retinoid chaperone. Diseases with loss of RBP1 expression include cancer, endometriosis, and heart failure. In some cases the mechanism of loss is not known and in other cases loss of expression is due to epigenetic silencing (Esteller et al., 2002; Jeronimo et al., 2004; Kuppumbatti, Bleiweiss, Mandeli, Waxman, & Mira, 2000; Toki et al., 2010). Here, we highlight key features of the human diseases and genetic mouse models used to dissect disease mechanisms.

2.4.1 RBP1 and ATRA are reduced in breast cancer

Epigenetic disruption of RBP1 that results in silencing of RBP1 expression is a common alteration in at least ten different types of human cancer with the prevalence of this alteration ranging from 7% in head and neck cancer to 60% in lymphoma (Arapshian, Bertran, Kuppumbatti, Nakajo, & Mira-y-Lopez, 2004; Esteller et al., 2002; Kuppumbatti et al., 2000; Lotan, 2005; Wang, Shen, Wang, & Brooks, 2005). RBP1 downregulation in approximately 25% of breast cancers represents an epithelial cell loss-of-function that resembles ATRA deficiency (Kuppumbatti et al., 2000; Lotan, 2005; Napoli, 2012). In fact, breast cancer tumors have decreased RA levels (Yu et al., 2022). RBP1 loss has been reported in 10 of the 12 most common cancers indicating that many cancers may benefit from understanding RBP1 function (Colvin et al., 2011; Daxecker, Marth, & Daxenbichler, 1988; Esteller et al., 2002; Jeronimo et al., 2004; Kuppumbatti et al., 2000; Ong, Markert, & Chiu, 1978; Palan & Romney, 1980; Palan, Duttagupta, & Romney, 1980; Toki et al., 2010). These cancers with depleted RBP1 include (in order of frequency): prostate, breast, lung, colon & rectal, melanoma, bladder, non-Hodgkin lymphoma, leukemia, endometrial, and pancreatic cancers (Colvin et al., 2011; Daxecker et al., 1988; Esteller et al., 2002; Jeronimo et al., 2004; Kuppumbatti et al., 2000; Ong et al., 1978; Palan & Romney, 1980; Palan et al., 1980; Toki et al., 2010).

Both RBP1 and ATRA have been shown to be anti-tumorigenic in breast epithelial cells and the tumor suppressive effect of RBP1 has been attributed to its ability to regulate ATRA signaling (Lotan, 2005). Availability of ATRA in mammary gland is largely controlled by the metabolism of ROL to biosynthesize ATRA in epithelial cells. ATRA has been shown to maintain epithelial cells through regulating proliferation, apoptosis, and differentiation (Fields, Soprano, & Soprano, 2007). The importance of ATRA in breast tumorigenesis has been demonstrated through experiments that show inactivating ATRA signaling in mammary epithelial cells initiates tumors and fosters tumor progression and invasion (Farias et al., 2005; Kupumbati et al., 2006). Additionally, restoration of ATRA signaling in breast cancer cell lines reversed mammary tumor progression (Farias et al., 2005).

Downregulation of epithelial cell RBP1 has been linked to abrogated ATRA signaling in breast cancer, which has been postulated to result from diminished ATRA biosynthesis. Studies manipulating the expression of RBP1 in mammary epithelial cell lines have shown that biosynthesis of RA is proportional to the expression of CRBP1 (Yu et al., 2022). Lack of holo-RBP1 as substrate results in significantly less efficient metabolism of ROL to RAL (and subsequently metabolism of RAL to ATRA) in vitro (Napoli, 2012).

The loss of RBP1 expression has been reported to be an early event in cancer progression (Kuppumbatti et al., 2000) and RBP1 has been reported to be down regulated in hyperplastic (precancerous) tissue (Orlandi et al., 2006). Both $Rbp1^{-/-}$ mammary tissue and human hyperplastic lesions have ~35 % reduced endogenous ATRA levels (Pierzchalski, Yu, Norman, & Kane, 2013; Pierzchalski et al., 2014). Data show that RBP1 loss results in ATRA biosynthesis defects in vivo that precede tumor formation and ATRA biosynthesis defects become more pronounced as tumorigenesis progresses (Kuppumbatti et al., 2000; Yu et al., 2022). A reduction in ATRA available for signaling encourages microenvironmental remodeling, where reduced ATRA in the $Rbp1^{-/-}$ mouse mammary gland in vivo is sufficient to cause both epithelial and stromal hyperplasia. $Rbp1^{-/-}$ mammary gland has increased stromal collagen indicative of activated fibroblasts that secrete increased amounts of extracellular matrix components, including collagen (Pierzchalski et al., 2013; Yu et al., 2022). As such, the $Rbp1^{-/-}$ mouse is useful because it displays the hyperplasia and microenvironmental remodeling that precedes tumor formation but does not develop tumors under normal conditions (Ghyselinck et al., 1999), giving a unique model to study the initiating events in tumorigenesis that result from RBP1 loss.

Treatment with both natural and synthetic retinoids (compounds with ATRA activity) suppressed experimentally induced precancerous hyperplasia (Fields et al., 2007). Systemic ATRA treatment for cancer has exhibited significant toxicity in clinical trials resulting in diminished enthusiasm for its clinical potential (Fields et al., 2007). It is important to note that systemic treatment bypasses physiological mechanisms that regulate ATRA and partition ATRA to specific proteins that chaperone retinoids and control their action (Napoli, 2012; Noy, 2000). Each of the ligand-activated RARs or PPARD has distinct roles in physiology (Altucci, Leibowitz, Ogilvie, de Lera, & Gronemeyer, 2007). Treatment with nuclear receptor-specific agonists have reproduced some of the toxic, non-specific side effects of systemic ATRA, however, many of these compounds have not been fully tested in the clinic (Lu, Bertran, Samuels, Mira-y-Lopez, & Farias, 2010). Activation of endogenous ATRA production by re-expressing RBP1 has been proposed as a strategy to restore ATRA levels and ATRA signaling while using physiological regulatory mechanisms to avoid non-specific toxicity (Farias et al., 2010; Kwon et al., 2015; Yu et al., 2022).

2.4.2 RBP1 and ATRA are reduced in endometriosis

Endometriosis is a disease where endometrial cells translocate outside of the uterine cavity resulting a variety of symptoms, including chronic pelvic pain and infertility (Bulun et al., 2019; Nezhat et al., 2005). The pathogenesis of endometriosis involves loss of control of cell proliferation and is associated with local invasion and distant metastasis (Pierzchalski et al., 2014). Retinoids play fundamental roles in the normal maintenance of the endometrium (Kumarendran, Loughney, Prentice, Thomas, & Redfern, 1996; Loughney, Kumarendran, Thomas, & Redfern, 1995; Sharpe-Timms & Cox, 2002), where ATRA synthesis is most localized in endometrial stromal cells (Li, Kakkad, & Ong, 2004; Zheng, Sierra-Rivera, Luan, Osteen, & Ong, 2000). ATRA produced from metabolic conversion of ROL is necessary for normal endometrial cell differentiation and function (Bo & Smith, 1966). Numerous aspects of endometrial behavior are regulated by local ATRA production, including matrix metalloproteinase (MMP) secretion, gap junctional intracellular communication, and the expression of a variety of cytokines involved in stromal cell growth, adhesion, and differentiation (Bruner-Tran et al., 2002; Wu, Taylor, & Sidell, 2013). ATRA-regulated genes in the endometrium include IL6, CCL2, TNF, VEGFA, GJA1, various integrins, and FASL (Nozaki et al., 2006; Sago, Teitelbaum, Venstrom, Reichardt, & Ross, 1999; Sawatsri, Desai, Rock, & Sidell, 2000; Sidell et al., 2010; Wu et al., 2013). These ATRA-regulated genes are known to be aberrantly expressed in endometriotic lesions (Sharpe-Timms, 2001). Additionally, *RBP1* and other genes involved in ROL uptake and metabolism have been shown to be aberrantly expressed in endometriosis (Pavone et al., 2010; Pavone et al., 2011). Measurement of ATRA in endometriotic lesions from patients showed reduced ATRA as well as reduced RBP1 levels. Reductions in endometriotic lesion ATRA were correlated with reductions in RBP1 expression (Pierzchalski et al., 2014). Impaired ATRA synthesis was directly confirmed by treating primary endometrial stromal cell cultures with ROL and observing a significant reduction of ATRA production in lesion-derived cultures compared with those derived from patient-matched eutopic tissue (Pierzchalski et al., 2014).

Animal model systems of endometriosis and RBP1 deficiency support findings in human tissues. Reduced ATRA levels were found in peritoneal lesions that developed in a mouse model of endometriosis, in which recipient mice were inoculated intraperitoneally with syngeneic endometrial tissue to mimic a massive retrograde menses (Pierzchalski et al., 2014). Studies also showed in a mouse model of endometriosis that treatment with ATRA suppressed the establishment and growth of peritoneal implants, promoted macrophage differentiation, and inhibited peritoneal fluid accumulation of IL6 and CCL2 (Wieser, Wu, Shen, Taylor, & Sidell, 2012). Results in $Rbp1^{-/-}$ mice recapitulated observations in human, where there was a reduction of ATRA and an increase of ROL in the endometrium of the $Rbp1^{-/-}$ mice compared with wild-type animals. The myometrium and whole uteri were also evaluated in this model, showing significantly elevated ROL and RE but no differences in ATRA between $Rbp1^{-/-}$ and wild-type, indicating tissue specificity of the dysfunction in ATRA homeostasis. Absolute retinoid levels indicate that the endometrium appears to be the primary source of ATRA production in the uterus versus the myometrium (Pierzchalski et al., 2014).

Histochemical comparisons of the uteri of $Rbp1^{-/-}$ versus wild-type were able to establish that reduced RBP1 and reduced ATRA has a causative role in inducing phenotypic and functional changes in endometrial tissue consistent with endometriotic implants (Pierzchalski et al., 2014). $Rbp1^{-/-}$ mice display loss of definitive borders between endometrium and myometrium, stromal hypercellularity in endometrium, and accumulation of extracellular matrix in the endometrial stroma (Pierzchalski et al., 2014). Interestingly, hypercellularity, loss of tissue organization, and accumulation of stromal extracellular matrix was also observed in $Rbp1^{-/-}$ mammary gland.

In both endometriosis patients and $Rbp1^{-/-}$ mice, reduced levels of RBP1 and ATRA biosynthesis appear to play a role in the ability of endometrial cells to implant and grow at ectopic sites. However, these deficiencies may not be directly involved in the subfertility of endometriosis patients. Previous studies have demonstrated $Rbp1^{-/-}$ embryos do not exhibit abnormalities related to ATRA deficiency and the fertility of $Rbp1^{-/-}$ mice does not appear impaired (Ghyselinck et al., 1999; Matt et al., 2005). As such, it was concluded that the altered uterine features observed in $Rbp1^{-/-}$ mice do not seem to interfere with functions related to embryo implantation and early placentation, as well as later development of the fetus (Pierzchalski et al., 2014).

2.4.3 RBP1 and heart failure

RBP1 is ubiquitously expressed in the body, including the heart and has also been shown to be affected by heart disease (Neuville et al., 1997; Xu, Redard, Gabbiani, & Neuville, 1997; Yu et al., 2012). ATRA is critical to cardiac development, where both low and high levels of ATRA are teratogenic and lead to various defects (Soprano & Soprano, 1995). Low circulating levels of ATRA are associated with poor cardiovascular outcomes (Liu et al., 2016). Emerging literature shows ATRA is also essential in the adult heart and in heart disease (Bilbija et al., 2014; Choudhary et al., 2008; Yang et al., 2021; Da Silva et al., 2021) where intracardiac ATRA declines 40 % in patients with idiopathic dilated cardiomyopathy (IDCM) despite sufficient ROL levels (Yang, Parker et al., 2021).

A guinea pig model of heart failure showed a similar decline in ATRA and treatment with ATRA or CYP26 inhibition increased endogenous ATRA levels and mitigated cardiac dysfunction (Yang, Parker et al., 2021). The $Rbp1^{-/-}$ mouse model recapitulates the ATRA decline seen in human IDCM showing a similar decline in ATRA with sufficient ROL levels (Yang, Parker et al., 2021; Zalesak-Kravec et al., 2022). As such, the

$Rbp1^{-/-}$ mouse model provides a unique resource to evaluate the effect of cardiac ATRA insufficiency on cardiac function without having to induce ROL deficiency which would have significant confounding effects (Zalesak-Kravec et al., 2022).

2.5 Cellular ATRA binding proteins

It has been shown that CRABP2 and FABP5 allow partitioning of ATRA between RARs and PPARD, where expression of these binding proteins drives their signaling. For example, when CRABP2 expression is high and the ratio of CRABP2/FABP5 is high, ATRA functions through RARs and promotes cell death. When expression of FABP5 is high and the ratio of CRABP2/FABP5 is low, signals through PPARD and promotes cell survival. Thus, opposing effects of ATRA result from ATRA-binding protein influence on the alternate activation of two different nuclear receptors with distinct signaling (Schug, Berry, Shaw, Travis, & Noy, 2007). CRABP2, which delivers ATRA to RARA (Budhu, Gillilan, & Noy, 2001), is reduced in endometriosis and cancer and is likely significant to the persistence of reduced RBP1 (Pavone et al., 2010; Pavone et al., 2011). It is unknown whether CRABP2 loss precedes RBP1 reduction, however, loss of CRABP2 function can result in heritable chromatin repression of multiple loci downstream of RARA, including *RBP1* (Corlazzoli, Rossetti, Bistulfi, Ren, & Sacchi, 2009).

2.6 Potential for gene-environment interaction

As substrates for ATRA biosynthesis are nutritionally derived, there is a potential for a gene-environment interaction based upon nutritional status. Population studies have revealed that baseline levels of plasma retinoids, including ATRA, can vary according to retinoid intake (Arnhold et al., 1996; Siegel et al., 2004; van Vliet et al., 2001). Plasma ATRA can increase after the consumption of food or supplements rich in β-carotene or ROL and RE (Arnhold et al., 1996; Copper, Klaassen, Teerlink, Snow, & Braakhuis, 1999; Eckhoff & Nau, 1990; Eckhoff, Bailey, Collins, Slikker, & Nau, 1991; Eckhoff, Collins, & Nau, 1991; Sedjo et al., 2004; Siegel et al., 2004; van Vliet et al., 2001). Differing nutrient intake can yield unique subpopulations due to nutritional environment. For example, in the case of loss-of-function genetic defects that yield excess ATRA (e.g., loss of CYP26B1, POR, DHRS3), the resultant phenotype may be exacerbated by nutritional input. Differing nutritional environments have been evaluated in terms of nutritional adequacy of regional diets which varies globally. Dietary nutritional content and availability is not uniform, yielding unique nutritional subpopulations, such that high-income countries generally produce adequate amounts of dietary nutrients, whereas low-income countries often are deficient in dietary micronutrients (Chen, Chaudhary, & Mathys, 2021).

The potential for unique subpopulations due to nutritional environment is supported by the data from population studies of plasma retinoids in human. Population reference ranges, where a range of "normal" values are defined for an analyte across all persons, are evaluated by a clinical concept called the Index of Individuality (Ii). This index considers the within subject and between subject variability to determine the utility of a population-based reference value (Harris, 1974; Petersen, Fraser, Sandberg, & Goldschmidt, 1999). For population reference values to have utility in determining excess or deficiency from a single sample from an individual, the Ii should be >1.4 (Harris, 1974; Petersen et al., 1999). Studies of plasma retinoids have shown that the index of individuality for ATRA (0.86), and ROL (0.27) are all less than 1, indicating that population reference intervals are of limited utility for diagnosis of deficiency or excess in human (Soderlund, Sjoberg, Svard, Fex, & Nilsson-Ehle, 2002). In cases where the within subject variation is much less than the between subject variation, it is important to stratify populations and obtain separate reference intervals according to the subpopulation (Harris, 1974; Petersen et al., 1999; Soderlund et al., 2002).

Given the impact of retinoid intake on endogenous retinoid levels, nutritional environment is a variable to be considered in stratifying populations and defining a subpopulation. If populations come from likely distinct nutritional environments, comparison of baseline levels of control groups may reveal the need to stratify into unique sub-populations. In the case of genetic mutations, there is a necessity for similarly sourced controls for patient evaluation, with unaffected individuals from the same family as preferable and individuals from the same geographic-nutritional environment being acceptable. For subjects stratified into a unique subpopulation, thresholds for excess were defined according to the control population and the reported critical difference at the 95 % confidence level (Soderlund et al., 2002; Talwar et al., 2005). In the case where nutritional environment limits the available population and no wild-type is available in the subpopulation, comparison to heterozygotes is a valid concept to determine the effect of disease where the reference population may not be a healthy group, but is at steady-state (Murphy & Abbey, 1967; Petersen et al., 1999).

Animal models support the effect of retinoid intake on endogenous levels of ATRA and the existence of unique subpopulations according to genetic and/or dietary environment. Studies in non-human primates also showed a plasma ATRA dependence on intake of ROL where plasma ATRA and 13-*cis* retinoic acid (13-cis RA) increased after either RE or ROL dosing in a dose-dependent manner (Iida, Kawa, & Matsuda, 2009). Quantification of retinoids in various mouse models has demonstrated the effect of diet on endogenous retinoids and established retinoids varied by genetic (wild-type) strain and the amount of vitamin A in the diet in a strain specific manner (Obrochta, Kane, & Napoli, 2014).

3. Other considerations

3.1 Direct quantification of ATRA is essential to understanding ATRA homeostasis

Endogenous retinoids are produced through a series of enzymatic reactions from diet-derived substrates (Fig. 1). There are multiple isomers of retinoids that are detected in plasma and extra-ocular tissue including ATRA, 13-*cis* RA, 9-*cis* retinoic acid (9-*cis* RA), and 9,13-di-*cis* retinoic acid (9,13-di-*cis* RA). These isomers differ in their affinity for nuclear receptors and resultant biological activity. ATRA is a high affinity ligand for RAR and is the main active metabolite in plasma and tissue (Chambon, 1996; Napoli, 2012; Sucov & Evans, 1995). 13-*cis* RA has low affinity for RAR and is considered biologically inactive (Napoli, 2012). 13-*cis* RA has been postulated to induce dyslipidemia and insulin resistance, most likely through conversion into ATRA (Bershad et al., 1985; Heliovaara et al., 2007; McCormick, Kroll, & Napoli, 1983; Rodondi et al., 2002; Staels, 2001). 9-*cis* RA is a high affinity ligand for both RAR and RXR, however its measured concentrations are limited in location and amount (Germain et al., 2006; Kane, 2012). 9,13-di-*cis* RA has low affinity for RAR and is considered biologically inactive. 9,13-di-*cis* RA has been reported to vary during pregnancy and increase after dosing with 9-*cis* RA and may possibly arise due to conversion from 13-*cis* RA and/or 9-*cis* RA (Arnhold et al., 1996; Horst, Reinhardt, Goff, Koszewski, & Napoli, 1995; Horst, Nonnecke et al., 1995). Most tissues and fluids have a characteristic isomer distribution, for example, the endogenous isomer distribution of ATRA in human plasma has been described (Arnhold et al., 1996; Arnold, Amory, Walsh, & Isoherranen, 2012; Jones, Pierzchalski, Yu, & Kane, 2015). Retinoic acid levels are typically in the low nM range (Kane & Napoli, 2010). Polar metabolites as 4-oxo-RA, 4-OH-RA, and 13-cis-4-oxo-RA, among others, are typically increased in times of ATRA excess (Arnhold et al., 1996; van Vliet et al., 2001). Plasma levels retinoid isomers (13-*cis* RA, 9-*cis* RA, 9,13-di-*cis* RA), and ROL are not affected by food intake (Arnold et al., 2012; Eckhoff, Collins et al., 1991; Jones et al., 2015; Kane et al., 2005; Kane, Folias, & Napoli, 2008; Kane, Folias, Wang, & Napoli, 2008), however, postprandial RE varies significantly (Krasinski et al., 1989; Krasinski, Cohn, Russell, & Schaefer, 1990; Krasinski, Cohn, Schaefer, & Russell, 1990; Relas, Gylling, & Miettinen, 2000). Population values for RE in plasma are typically reported for the fasting state (Krasinski et al., 1989; Krasinski, Cohn et al., 1990; Tanumihardjo et al., 2016).

9-*cis*-13,14-dihydro-RA, recently identified as an RXR ligand and named vitamin A_5, has been discussed in a number of reviews (Banati et al., 2024; Krezel, Ruhl, & de Lera, 2019; Ruhl, Krezel, & de Lera, 2018). Pharmacological treatment with 9-*cis*-13,14-dihydro-RA or its presumed precursor, 9-*cis*-13,14-dihydro-ROL, indicated that enhancement of RXR-mediated signaling protects against the effects of social defeat stress by displaying anti-depressant effects in mice (Krzyzosiak et al., 2021). While this finding has generated enthusiasm for a new ligand-activated mechanism of retinoid-regulated nuclear receptor signaling in the brain, the endogenous nature of 9-*cis*-13,14-dihydro-RA remains to be elucidated. Unfortunately, the studies purporting to establish 9-*cis*-13,14-dihydro-RA as an endogenous retinoid of potential nutritional origin are not accompanied by a sufficient description of the analytical methodology that is required to establish these compounds' endogenous existence (Krezel et al., 2021). Throughout literature (Banati et al., 2024; Krezel et al., 2021; Krezel, Ruhl, & de Lera, 2019; Krzyzosiak et al., 2021; Ruhl, Krezel, & de Lera, 2018) the authors repeatedly cite a publication as a "validated method" for the measurement of 9-*cis*-13,14-dihydro-RA, its precursors, and other endogenous retinoids (Ruhl, 2006), but the publication does not describe the analytical methodology. It is both insufficient and inappropriate to be cited for any detection and quantification of retinoids, as it has many elements that are woefully inadequate methodologically and is certainly not a validated method for any retinoid according to accepted criteria (for example, (FDA, 2018)). In addition to the unfortunate analytical shortcomings of the vitamin A_5 literature, the endogenous nature of 9-*cis*-13,14-dihydro-RA also remains unclear given that there are no enzymes identified yet that mediate the chemical conversions from nutritional precursors or enzymes that regulate degradation/excretion. No binding proteins have been identified to chaperone 9-*cis*-13,14-dihydro-ROL or 9-*cis*-13,14-dihydro-RA in order to protect them from non-specific oxidation and direct them to specific enzymes like other retinoids. There also is a lack of pharmacokinetic data to describe the metabolism of 9-*cis*-13,14-dihydro-ROL or 9-*cis*-13, 14-dihydro-RA. Careful enzymatic studies and use of biochemical and genetic models are needed to elucidate these points in the future. In contrast, 9-*cis*-RA, an endogenous retinoid with RXR activity, has analytical methodology described in detail, as well as relevant biosynthetic enzymes that have been characterized for their activity towards 9-*cis*-retinoids as substrates and analyzed for their relevant in vivo expression. 9-*cis*-ROL and 9-*cis*-RA have retinoid-binding proteins that bind with high affinity and mechanisms for converting 9-*cis* retinoids to soluble derivatives able to be excreted have been described. The literature describing these enzymatic mechanisms, as well as an in-depth assessment of analytical considerations for retinoid analysis, have been reviewed in detail (Kane & Napoli, 2010; Kane, 2012).

3.1.1 Methods for quantification

Endogenous retinoids have disparate abundance that demands attention to analytical methodology. The concentration of different retinoid species can differ by as much as six orders of magnitude, making not technically practical methods that simultaneously detect all retinoids. For example, RE levels can reach high micromolar levels whereas endogenous ATRA is typically low nanomolar in concentration (Kane & Napoli, 2010).

The gold-standard for direct detection and quantification of nanomolar levels of ATRA is liquid chromatography-tandem mass spectrometry (LC-MS/MS)-based approaches due to their sensitivity and specificity (Arnold et al., 2012; Czuba, Zhong, Yabut, & Isoherranen, 2020; Jones et al., 2015; Kane & Napoli, 2010; Kane et al., 2005; Kane, Wang et al., 2008). LC-MS/MS-based approaches rely on a unique precursor to product ion m/z transition, termed multiple reaction monitoring (MRM) or selected reaction monitoring (SRM), to yield specificity (Arnold et al., 2012; Czuba et al., 2020; Kane & Napoli, 2010; Kane et al., 2005; Kane, Wang et al., 2008). Tandem quadrupole instruments are used for this type of quantitative analysis because they yield the most robust and reproducible analytical detection, with typical linear ranges of 3–5 orders of magnitude and typical instrument coefficients of variation of < 5 % (Kane et al., 2005; Kane, Wang et al., 2008). Quantitation is derived from calibration curves constructed from authentic standards. Stable isotope-labeled retinoids or nonendogenous retinoid derivatives are typically used as internal standards to correct for extraction efficiency and variability during data acquisition (Kane & Napoli, 2010).

Although MRM detection in LC-MS/MS methods is one of the most selective detection modalities, the tandem quadrupole instruments that are typically employed are low mass resolution and abundant lipid species of similar nominal mass can coelute with the analyte of interest and interfere with quantification (Jones et al., 2015; Kane, 2012). Because most extraction methods for retinoids also extract lipids, the potential for interferences is significant and requires additional caution and careful method validation. Current strategies to identify and mitigate interference is to use of chromatographic separation to move interfering species to retention times away from ATRA and/or the inclusion of additional mass transitions to impart additional selectivity (Jones et al., 2015; Kane, 2012). Liquid chromatography-multiple reaction monitoring cubed (LC-MRM^3, also known as liquid chromatography-multistage tandem mass spectrometry) is a similar tandem mass spectrometry detection scheme conducted on a hybrid tandem quadrupole-linear ion trap instrument that includes an additional mass transition during detection that yields additional specificity for ATRA quantification (Jones et al., 2015).

ATRA quantification poses several significant analytical challenges including low endogenous concentrations, often spatially localized occurrence, and existence of endogenous geometric isomers with distinct biological functions (Kane & Napoli, 2010). Endogenous retinoid isomers are isobaric and need to be separated chromatographically before mass spectrometric detection in order to enable their individual quantification since they do not have the same biological activity or function. This is particularly important in tissues where *cis*-retinoid isomers have significant biological roles, such as the eye and pancreas. Chromatographic methods for separation of isomers and isomeric retinoid detection has been previously reviewed (Kane & Napoli, 2010; Kane, 2012).

More abundant retinoids (e.g., ROL and RE) can be feasibly detected by liquid chromatography with UV detection (LC-UV) due to their favorable molar absorptivity (typically $\varepsilon = \sim 30{,}000$–$60{,}000 \text{ M}^{-1} \text{ cm}^{-1}$), and unique absorption wavelengths, which are red-shifted from many other biological molecules (Kane & Napoli, 2010; Kane, Wang et al., 2008). LC-MS/MS methodology also exists for quantification of ROL and RE, which have less of a requirement for the sensitivity of LC-MS/MS-based approaches (Wang, Yoo, Obrochta, Huang, & Napoli, 2015).

RAL has a reactive aldehyde group that requires derivatization for accurate quantification. Derivatization with hydroxylamine or O-ethylhydroxylamine converts RAL to the respective *syn* and *anti*-retinaloxime adducts. Ther derivatization process releases RAL that has formed covalent Schiff base adducts with amine groups of proteins, phospholipids and other compounds (Golczak, Bereta, Maeda, & Palczewski, 2010; Kane & Napoli, 2010; Kane, Wang et al., 2008; Wang et al., 2015). Whereas LC-UV methods have been popular (Kane & Napoli, 2010; Kane, Wang et al., 2008), recent LC-MS/MS methodology provides more sensitive detection of RAL for samples of limited quantity (Golczak et al., 2010; Wang et al., 2015).

Particular experimental challenges that are critical to control when working with retinoids include the susceptibility of retinoids to light-induced isomerization, oxidation, and adherence to plastic surfaces. Yellow or red laboratory lights (blocking UV wavelengths <~500 nm) are recommended for experimental and sample preparation work. Sample should be protected from non-UV blocking laboratory settings light by cover and/or amber vials. Samples should be kept cold (on ice) during sample preparation to reduce spurious oxidation and glass pipets, vials and containers, should be used as much as possible to prevent sample loss by adsorption to plastic (Kane & Napoli, 2010).

LC-MS/MS-based approaches offer the advantage of direct, absolute quantitation of ATRA that is applicable to diverse sample types including cell lines and tissues. Direct quantification of ATRA (or other retinoids) has temporal accuracy representative of retinoid levels at the time of collection and is highly reproducible for validated assays (Jones et al., 2015; Kane et al., 2005; Kane, Wang et al., 2008). Limitations of LC-MS/MS-based approaches include the requirement for expensive, specialized instrumentation and minimum tissue requirements that may necessitate pooling of samples of limited amount. LC-MS/MS approaches do not provide spatial visualization, and measurement of specific spatial regions is limited by the ability to dissect those regions.

3.1.2 Supra-physiological dosing to aid in measurement

Administration of a supra-physiological doses of ROL to raise ATRA levels to an amount detectable by UV absorbance is a strategy that has been employed to facilitate quantification of low abundance retinoids when LC-MS/MS-based approaches were not available (Molotkov & Duester, 2002). However, this approach is problematic for studying ATRA homeostasis such that it induces an artificial environment where ATRA levels are raised far above steady-state likely overwhelming normal metabolism and homeostatic mechanisms as well as eliciting retinoid toxicity responses (Biesalski, 1989; Molotkov & Duester, 2002; Napoli, 2012, 2016).

3.2 Reporter assays

Reporter screens that provide crucial information about the spatial resolution or extent of ATRA signaling within cells or tissues have been reviewed recently (Shannon et al., 2020). These reporters are generally based on the expression of a readily assayable enzyme (beta-galactosidase or luciferase) or a fluorescent protein driven by a promoter controlled by ATRA signaling (Shannon et al., 2020). Reporter systems offer valuable spatial information and ATRA signaling read-outs, that currently cannot be obtained through other techniques. However, interpretation of reporter assay response requires caution in terms of concentration and temporal relationships.

Reporter-based methods are nonquantitative and do not provide a linear response proportional to the level of ATRA in tissues. Thus, reporter assays provide a qualitative assessment of ATRA signaling which is correlated with RA levels. Caution and consideration are warranted by reports that high levels of ATRA can turn off reporter response in some cases (McCaffery & Drager, 1994). Valuable qualitative relationships can be assessed under comparable conditions, however, specific, absolute quantitation for ATRA cannot be achieved via reporter assays.

Reporter assays reflect RAR activation, and therefore, represent the longer-term consequences of receptor activation with a delay in the reporter signal appearance often hours after a change in ATRA concentration (McCaffery & Drager, 1994; Wagner, Han, & Jessell, 1992; Wagner, 1997; Zetterstrom et al., 1999). Additionally, the half-life of the reporters employed generally outlasts the ATRA input. In many cases ATRA may even be significantly catabolized by the time the reporter signal is visualized and reporter assays may not accurately correlate with ATRA concentration via ATRA signaling in real-time (Shannon et al., 2020). As a result, temporal relationships should always be cautiously and carefully interpreted.

Lack of specificity may confound results for ATRA where some reporter assays have cited that endogenous retinoids other than ATRA can produce signals to varying extents including 3,4-didehydro-retinoic acid, 9-*cis* RA, 4-oxo-RA, 4-hydroxy-RA, and 4-hydroxy-ROL (McCaffery & Drager, 1994; Wagner et al., 1992; Wagner, 1997; Zetterstrom et al., 1999).

3.3 Surrogates for ATRA and indirect measurement

3.3.1 Surrogate metabolites

There are no surrogate metabolites that have a direct correlation with ATRA. ROL typically exceeds ATRA by 100–1000x such that this mismatch in concentration does not make ROL a reasonable reporter of ATRA levels. Additionally, ROL is separated by two enzymatic steps and controlled by mechanisms independent of ATRA that make it neither a good indicator of ROL status nor a viable surrogate for ATRA levels (Eckhoff, Collins et al., 1991). For example, the clinical threshold for ROL deficiency is 0.7 nmol/mL and low ROL status is defined as 0.7–1.05 nmol/mL (West, 2002). Even at very reduced plasma ROL levels, plasma ROL can be present at 100x greater levels than ATRA indicating there is still ample substrate for ATRA synthesis. Thus, changes in ROL do not necessarily correlate with ATRA status. RAL is a tightly regulated intermediate present at 10–100x greater abundance than ATRA, where there is no correlation between RAL levels and ATRA levels. RE has an extremely large dynamic range (Kane & Napoli, 2010) making it a poor reporter of ROL status and ATRA homeostasis (Eckhoff, Collins et al., 1991). Polar metabolites tend to increase after elevated ATRA levels, as a result of ATRA-induced upregulation of CYP26 enzymes. The polar metabolite profile can vary in abundance, identity, and temporal occurrence and has no demonstrated direct correlation with RA levels (Arnhold et al., 1996; Eckhoff, Collins et al., 1991).

3.3.2 Surrogate gene expression

Gene expression of ATRA-inducible genes is sometimes used as a correlate of RA levels and/or used to infer changes in ATRA levels. However, levels of mRNA and levels of ATRA do not have a quantitative correlation, are temporally separated, and gene expression can be subject to other, sometime complex, regulatory influences. Induction of such genes often reflects a previous elevation in ATRA that occurred in a time period that is impossible to identify without direct measurement of ATRA. ATRA can be back to baseline levels at the time of gene induction, making it a poor reporter of ATRA status.

3.3.3 ELISA

Some studies have used a commercially available enzyme-linked immunosorbent assay (ELISA) to quantify ATRA. In one publication, the authors state: "This kit could not distinguish all-*trans* ROL, all-*trans* RAL or ATRA. Thus, the concentration of ATRA was the concentration of all forms." (Shibata et al., 2021). Such a detection and quantification scheme is not valid for ATRA quantification or for any retinoid quantification.

4. Conclusions and future opportunities

ATRA homeostasis is complex and context-dependent with many mechanistic details yet to be fully elucidated. Using modern experimental tool such as genetic mouse models and advance analysis methods, human diseases will be further understood and potentially targeted by therapeutics. In addition to cancer, endometriosis, and heart disease, many opportunities remain for investigation of disease states with dysregulated ATRA or ATRA signaling including, but not limited to, neurodegeneration, radiation-induced tissue damage, myopia, asthma, fibrosis, infection and inflammatory diseases.

Acknowledgments

Support was provided by NIH/NHLBI: R01HL164478, NIH/NEI: R01EY033361, and NIH/NIAID: 75N93020D00011. Additional support was provided by the University of Maryland School of Pharmacy Mass Spectrometry Center (SOP1841-IQB2014).

References

Adams, M. K., Belyaeva, O. V., Wu, L., Chaple, I. F., Dunigan-Russell, K., Popov, K. M., & Kedishvili, N. Y. (2021). Characterization of subunit interactions in the hetero-oligomeric retinoid oxidoreductase complex. *The Biochemical Journal, 478*, 3597–3611.

Altucci, L., Leibowitz, M. D., Ogilvie, K. M., de Lera, A. R., & Gronemeyer, H. (2007). RAR and RXR modulation in cancer and metabolic disease. *Nature Reviews. Drug Discovery, 6*, 793–810.

Arapshian, A., Bertran, S., Kuppumbatti, Y. S., Nakajo, S., & Mira-y-Lopez, R. (2004). Epigenetic CRBP downregulation appears to be an evolutionarily conserved (human and mouse) and oncogene-specific phenomenon in breast cancer. *Molecular Cancer, 3*, 13.

Arnhold, T., Tzimas, G., Wittfoht, W., Plonait, S., & Nau, H. (1996). Identification of 9-cis-retinoic acid, 9,13-di-cis-retinoic acid, and 14-hydroxy-4,14-retro-retinol in human plasma after liver consumption. *Life Sciences, 59*, PL169–PL177.

Arnold, S. L., Amory, J. K., Walsh, T. J., & Isoherranen, N. (2012). A sensitive and specific method for measurement of multiple retinoids in human serum with UHPLC-MS/MS. *Journal of Lipid Research, 53*, 587–598.

Ashique, A. M., May, S. R., Kane, M. A., Folias, A. E., Phamluong, K., Choe, Y., … Peterson, A. S. (2012). Morphological defects in a novel Rdh10 mutant that has reduced retinoic acid biosynthesis and signaling. *Genesis (New York, N. Y.: 2000), 50*, 415–423.

Balmer, J. E., & Blomhoff, R. (2002). Gene expression regulation by retinoic acid. *Journal of Lipid Research, 43*, 1773–1808.

Banati, D., Hellman-Regen, J., Mack, I., Young, H. A., Benton, D., Eggersdorfer, M., … Ruhl, R. (2024). Defining a vitamin A5/X specific deficiency - vitamin A5/X as a critical dietary factor for mental health. *International Journal for Vitamin and Nutrition Research, 94*, 443–475.

Bershad, S., Rubinstein, A., Paterniti, J. R., Le, N. A., Poliak, S. C., Heller, B., … Brown, W. V. (1985). Changes in plasma lipids and lipoproteins during isotretinoin therapy for acne. *The New England Journal of Medicine, 313*, 981–985.

Biesalski, H. K. (1989). Comparative assessment of the toxicology of vitamin A and retinoids in man. *Toxicology, 57*, 117–161.

Bilbija, D., Elmabsout, A. A., Sagave, J., Haugen, F., Bastani, N., Dahl, C. P., ... Valen, G. (2014). Expression of retinoic acid target genes in coronary artery disease. *International Journal of Molecular Medicine, 33*, 677–686.

Billings, S. E., Pierzchalski, K., Butler Tjaden, N. E., Pang, X. Y., Trainor, P. A., Kane, M. A., & Moise, A. R. (2013). The retinaldehyde reductase DHRS3 is essential for preventing the formation of excess retinoic acid during embryonic development. *The FASEB Journal, 27*, 4877–4889.

Bo, W. J., & Smith, M. S. (1966). The effect of retinol and retinoic acid on the morphology of the rat uterus. *The Anatomical Record, 156*, 5–9.

Boerman, M. H., & Napoli, J. L. (1991). Cholate-independent retinyl ester hydrolysis. Stimulation by Apo-cellular retinol-binding protein. *The Journal of Biological Chemistry, 266*, 22273–22278.

Boerman, M. H., & Napoli, J. L. (1996). Cellular retinol-binding protein-supported retinoic acid synthesis. Relative roles of microsomes and cytosol. *The Journal of Biological Chemistry, 271*, 5610–5616.

Bruner-Tran, K. L., Eisenberg, E., Yeaman, G. R., Anderson, T. A., McBean, J., & Osteen, K. G. (2002). Steroid and cytokine regulation of matrix metalloproteinase expression in endometriosis and the establishment of experimental endometriosis in nude mice. *The Journal of Clinical Endocrinology and Metabolism, 87*, 4782–4791.

Budhu, A., Gillilan, R., & Noy, N. (2001). Localization of the RAR interaction domain of cellular retinoic acid binding protein-II. *Journal of Molecular Biology, 305*, 939–949.

Bulun, S. E., Yilmaz, B. D., Sison, C., Miyazaki, K., Bernardi, L., Liu, S., ... Wei, J. (2019). Endometriosis. *Endocrine Reviews, 40*, 1048–1079.

Chambon, P. (1996). A decade of molecular biology of retinoic acid receptors. *The FASEB Journal, 10*, 940–954.

Chen, C., Chaudhary, A., & Mathys, A. (2021). Nutrient adequacy of global food production. *Frontiers in Nutrition, 8*, 739755.

Chen, N., & Napoli, J. L. (2008). All-trans-retinoic acid stimulates translation and induces spine formation in hippocampal neurons through a membrane-associated RARalpha. *The FASEB Journal, 22*, 236–245.

Chen, N., Onisko, B., & Napoli, J. L. (2008). The nuclear transcription factor RARalpha associates with neuronal RNA granules and suppresses translation. *The Journal of Biological Chemistry, 283*, 20841–20847.

Choudhary, R., Palm-Leis, A., Scott, R. C., 3rd, Guleria, R. S., Rachut, E., ... Pan, J. (2008). All-trans retinoic acid prevents development of cardiac remodeling in aortic banded rats by inhibiting the renin-angiotensin system. *American Journal of Physiology. Heart and Circulatory Physiology, 294*, H633–H644.

Collins, M. D., & Mao, G. E. (1999). Teratology of retinoids. *Annual Review of Pharmacology and Toxicology, 39*, 399–430.

Colvin, E. K., Susanto, J. M., Kench, J. G., Ong, V. N., Mawson, A., Pinese, M., ... Biankin, A. V. (2011). Retinoid signaling in pancreatic cancer, injury and regeneration. *PLoS One, 6*, e29075.

Copper, M. P., Klaassen, I., Teerlink, T., Snow, G. B., & Braakhuis, B. J. (1999). Plasma retinoid levels in head and neck cancer patients: a comparison with healthy controls and the effect of retinyl palmitate treatment. *Oral Oncology, 35*, 40–44.

Corlazzoli, F., Rossetti, S., Bistulfi, G., Ren, M., & Sacchi, N. (2009). Derangement of a factor upstream of RARalpha triggers the repression of a pleiotropic epigenetic network. *PLoS One, 4*, e4305.

Czuba, L. C., Zhong, G., Yabut, K. C., & Isoherranen, N. (2020). Analysis of vitamin A and retinoids in biological matrices. *Methods in Enzymology, 637*, 309–340.

Da Silva, F., Jian Motamedi, F., Weerasinghe Arachchige, L. C., Tison, A., Bradford, S. T., Lefebvre, J., ... Schedl, A. (2021). Retinoic acid signaling is directly activated in cardiomyocytes and protects mouse hearts from apoptosis after myocardial infarction: *Elife, 10*.

Daxecker, F., Marth, C., & Daxenbichler, G. (1988). Retinoic binding in melanomas of the eye and in normal choroid. *Cancer Letters, 41*, 119–122.

Dubey, A., Rose, R. E., Jones, D. R., & Saint-Jeannet, J. P. (2018). Generating retinoic acid gradients by local degradation during craniofacial development: One cell's cue is another cell's poison. *Genesis (New York, N. Y.: 2000), 56*.

Eckhoff, C., Bailey, J. R., Collins, M. D., Slikker, W., Jr., & Nau, H. (1991). Influence of dose and pharmaceutical formulation of vitamin A on plasma levels of retinyl esters and retinol and metabolic generation of retinoic acid compounds and beta-glucuronides in the cynomolgus monkey. *Toxicology and Applied Pharmacology, 111*, 116–127.

Eckhoff, C., Collins, M. D., & Nau, H. (1991). Human plasma all-trans-, 13-cis- and 13-cis-4-oxoretinoic acid profiles during subchronic vitamin A supplementation: Comparison to retinol and retinyl ester plasma levels. *The Journal of Nutrition, 121*, 1016–1025.

Eckhoff, C., & Nau, H. (1990). Identification and quantitation of all-trans- and 13-cis-retinoic acid and 13-cis-4-oxoretinoic acid in human plasma. *Journal of Lipid Research, 31*, 1445–1454.

Esteller, M., Guo, M., Moreno, V., Peinado, M. A., Capella, G., Galm, O., ... Herman, J. G. (2002). Hypermethylation-associated inactivation of the cellular retinol-binding-protein 1 gene in human cancer. *Cancer Research, 62*, 5902–5905.

Farias, E. F., Ong, D. E., Ghyselinck, N. B., Nakajo, S., Kuppumbatti, Y. S., & Mira y Lopez, R. (2005). Cellular retinol-binding protein I, a regulator of breast epithelial retinoic acid receptor activity, cell differentiation, and tumorigenicity. *Journal of the National Cancer Institute, 97*, 21–29.

Farias, E. F., Petrie, K., Leibovitch, B., Murtagh, J., Chornet, M. B., Schenk, T., ... Waxman, S. (2010). Interference with Sin3 function induces epigenetic reprogramming and differentiation in breast cancer cells. *Proceedings of the National Academy of Sciences of the United States of America, 107*, 11811–11816.

Farjo, K. M., Moiseyev, G., Nikolaeva, O., Sandell, L. L., Trainor, P. A., & Ma, J. X. (2011). RDH10 is the primary enzyme responsible for the first step of embryonic Vitamin A metabolism and retinoic acid synthesis. *Developmental Biology, 357*, 347–355.

FDA (2018). Bioanalytical Method Validation: Guidance for Industry. *Food and Drug Administration*.

Fields, A. L., Soprano, D. R., & Soprano, K. J. (2007). Retinoids in biological control and cancer. *Journal of Cellular Biochemistry, 102*, 886–898.

Fluck, C. E., Tajima, T., Pandey, A. V., Arlt, W., Okuhara, K., Verge, C. F., ... Miller, W. L. (2004). Mutant P450 oxidoreductase causes disordered steroidogenesis with and without Antley-Bixler syndrome. *Nature Genetics, 36*, 228–230.

Fukami, M., Nagai, T., Mochizuki, H., Muroya, K., Yamada, G., Takitani, K., & Ogata, T. (2010). Anorectal and urinary anomalies and aberrant retinoic acid metabolism in cytochrome P450 oxidoreductase deficiency. *Molecular Genetics and Metabolism, 100*, 269–273.

Germain, P., Chambon, P., Eichele, G., Evans, R. M., Lazar, M. A., Leid, M., ... Gronemeyer, H. (2006). International union of pharmacology. LXIII. Retinoid X receptors. *Pharmacological Reviews, 58*, 760–772.

Ghyselinck, N. B., Bavik, C., Sapin, V., Mark, M., Bonnier, D., Hindelang, C., ... Chambon, P. (1999). Cellular retinol-binding protein I is essential for vitamin A homeostasis. *The EMBO Journal, 18*, 4903–4914.

Golczak, M., Bereta, G., Maeda, A., & Palczewski, K. (2010). Molecular biology and analytical chemistry methods used to probe the retinoid cycle. *Methods in Molecular Biology, 652*, 229–245.

Grand, K., Skraban, C. M., Cohen, J. L., Dowsett, L., Mazzola, S., Tarpinian, J., ... Deardorff, M. A. (2021). Nonlethal presentations of CYP26B1-related skeletal anomalies and multiple synostoses syndrome. *American Journal of Medical Genetics. Part A, 185*, 2766–2775.

Gudas, L. J. (2022). Retinoid metabolism: new insights. *Journal of Molecular Endocrinology, 69,* T37–T49.

Harris, E. K. (1974). Effects of intra- and interindividual variation on the appropriate use of normal ranges. *Clinical Chemistry, 20,* 1535–1542.

Hashimoto, A. (2017). *Next generation sequencing to identify new genetic causes of familial craniosynostosis.* University of Oxford.

Heliovaara, M. K., Remitz, A., Reitamo, S., Teppo, A. M., Karonen, S. L., & Ebeling, P. (2007). 13-cis-Retinoic acid therapy induces insulin resistance, regulates inflammatory parameters, and paradoxically increases serum adiponectin concentration. *Metabolism: Clinical and Experimental, 56,* 786–791.

Herr, F. M., & Ong, D. E. (1992). Differential interaction of lecithin-retinol acyltransferase with cellular retinol binding proteins. *Biochemistry, 31,* 6748–6755.

Horst, R. L., Reinhardt, T. A., Goff, J. P., Koszewski, N. J., & Napoli, J. L. (1995). 9,13-Di-cis-retinoic acid is the major circulating geometric isomer of retinoic acid in the periparturient period. *Archives of Biochemistry and Biophysics, 322,* 235–239.

Horst, R. L., Reinhardt, T. A., Goff, J. P., Nonnecke, B. J., Gambhir, V. K., Fiorella, P. D., & Napoli, J. L. (1995). Identification of 9-cis,13-cis-retinoic acid as a major circulating retinoid in plasma. *Biochemistry, 34,* 1203–1209.

Iida, T., Kawa, G., & Matsuda, T. (2009). A case of preserving renal function by renal autotransplantation for bilateral urothelial carcinoma of the ureter. *International Journal of Urology: Official Journal of the Japanese Urological Association, 16,* 587.

James, A. W., Levi, B., Xu, Y., Carre, A. L., & Longaker, M. T. (2010). Retinoic acid enhances osteogenesis in cranial suture-derived mesenchymal cells: Potential mechanisms of retinoid-induced craniosynostosis. *Plastic and Reconstructive Surgery, 125,* 1352–1361.

Jeronimo, C., Henrique, R., Oliveira, J., Lobo, F., Pais, I., Teixeira, M. R., & Lopes, C. (2004). Aberrant cellular retinol binding protein 1 (CRBP1) gene expression and promoter methylation in prostate cancer. *Journal of Clinical Pathology, 57,* 872–876.

Jones, J. W., Pierzchalski, K., Yu, J., & Kane, M. A. (2015). Use of fast HPLC multiple reaction monitoring cubed for endogenous retinoic acid quantification in complex matrices. *Analytical Chemistry, 87,* 3222–3230.

Kane, M. A. (2012). Analysis, occurrence, and function of 9-cis-retinoic acid. *Biochimica et Biophysica Acta, 1821,* 10–20.

Kane, M. A., Chen, N., Sparks, S., & Napoli, J. L. (2005). Quantification of endogenous retinoic acid in limited biological samples by LC/MS/MS. *The Biochemical Journal, 388,* 363–369.

Kane, M. A., Folias, A. E., & Napoli, J. L. (2008). HPLC/UV quantitation of retinal, retinol, and retinyl esters in serum and tissues. *Analytical Biochemistry, 378,* 71–79.

Kane, M. A., Folias, A. E., Pingitore, A., Perri, M., Krois, C. R., Ryu, J. Y., ... Napoli, J. L. (2011). CrbpI modulates glucose homeostasis and pancreas 9-cis-retinoic acid concentrations. *Molecular and Cellular Biology, 31,* 3277–3285.

Kane, M. A., Folias, A. E., Wang, C., & Napoli, J. L. (2008). Quantitative profiling of endogenous retinoic acid in vivo and in vitro by tandem mass spectrometry. *Analytical Chemistry, 80,* 1702–1708.

Kane, M. A., Folias, A. E., Wang, C., & Napoli, J. L. (2010). Ethanol elevates physiological all-trans-retinoic acid levels in select loci through altering retinoid metabolism in multiple loci: A potential mechanism of ethanol toxicity. *The FASEB Journal, 24,* 823–832.

Kane, M. A., & Napoli, J. L. (2010). Quantification of endogenous retinoids. *Methods in Molecular Biology, 652,* 1–54.

Krasinski, S. D., Cohn, J. S., Russell, R. M., & Schaefer, E. J. (1990). Postprandial plasma vitamin A metabolism in humans: A reassessment of the use of plasma retinyl esters as markers for intestinally derived chylomicrons and their remnants. *Metabolism: Clinical and Experimental, 39,* 357–365.

Krasinski, S. D., Cohn, J. S., Schaefer, E. J., & Russell, R. M. (1990). Postprandial plasma retinyl ester response is greater in older subjects compared with younger subjects. Evidence for delayed plasma clearance of intestinal lipoproteins. *The Journal of Clinical Investigation, 85*, 883–892.

Krasinski, S. D., Russell, R. M., Otradovec, C. L., Sadowski, J. A., Hartz, S. C., Jacob, R. A., & McGandy, R. B. (1989). Relationship of vitamin A and vitamin E intake to fasting plasma retinol, retinol-binding protein, retinyl esters, carotene, alpha-tocopherol, and cholesterol among elderly people and young adults: increased plasma retinyl esters among vitamin A-supplement users. *The American Journal of Clinical Nutrition, 49*, 112–120.

Krezel, W., Kastner, P., & Chambon, P. (1999). Differential expression of retinoid receptors in the adult mouse central nervous system. *Neuroscience, 89*, 1291–1300.

Krezel, W., Rivas, A., Szklenar, M., Ciancia, M., Alvarez, R., de Lera, A. R., & Ruhl, R. (2021). Vitamin A5/X, a New Food to Lipid Hormone Concept for a Nutritional Ligand to Control RXR-Mediated Signaling. *Nutrients, 13*.

Krezel, W., Ruhl, R., & de Lera, A. R. (2019). Alternative retinoid X receptor (RXR) ligands. *Molecular & Cellular Endocrinology, 491*, 110436.

Krzyzosiak, A., Podlesny-Drabiniok, A., Vaz, B., Alvarez, R., Ruhl, R., de Lera, A. R., & Krezel, W. (2021). Vitamin A5/X controls stress-adaptation and prevents depressive-like behaviors in a mouse model of chronic stress. *Neurobiology of Stress, 15*, 100375.

Kumarendran, M. K., Loughney, A. D., Prentice, A., Thomas, E. J., & Redfern, C. P. (1996). Nuclear retinoid receptor expression in normal human endometrium throughout the menstrual cycle. *Molecular Human Reproduction, 2*, 123–129.

Kuppumbatti, Y. S., Bleiweiss, I. J., Mandeli, J. P., Waxman, S., & Mira, Y. L. R. (2000). Cellular retinol-binding protein expression and breast cancer. *Journal of the National Cancer Institute, 92*, 475–480.

Kupumbati, T. S., Cattoretti, G., Marzan, C., Farias, E. F., Taneja, R., & Mira-y-Lopez, R. (2006). Dominant negative retinoic acid receptor initiates tumor formation in mice. *Molecular Cancer, 5*, 12.

Kwon, Y. J., Petrie, K., Leibovitch, B. A., Zeng, L., Mezei, M., Howell, L., ... Waxman, S. (2015). Selective inhibition of SIN3 corepressor with avermectins as a novel therapeutic strategy in triple-negative breast cancer. *Molecular Cancer Therapeutics, 14*, 1824–1836.

Lapshina, E. A., Belyaeva, O. V., Chumakova, O. V., & Kedishvili, N. Y. (2003). Differential recognition of the free versus bound retinol by human microsomal retinol/sterol dehydrogenases: Characterization of the holo-CRBP dehydrogenase activity of RoDH-4. *Biochemistry, 42*, 776–784.

Laue, K., Pogoda, H. M., Daniel, P. B., van Haeringen, A., Alanay, Y., von Ameln, S., ... Robertson, S. P. (2011). Craniosynostosis and multiple skeletal anomalies in humans and zebrafish result from a defect in the localized degradation of retinoic acid. *American Journal of Human Genetics, 89*, 595–606.

Li, X. H., Kakkad, B., & Ong, D. E. (2004). Estrogen directly induces expression of retinoic acid biosynthetic enzymes, compartmentalized between the epithelium and underlying stromal cells in rat uterus. *Endocrinology, 145*, 4756–4762.

Li, J., Xie, H., & Jiang, Y. (2018). Mucopolysaccharidosis IIIB and mild skeletal anomalies: Coexistence of NAGLU and CYP26B1 missense variations in the same patient in a Chinese family. *BMC Medical Genetics, 19*, 51.

Liu, Y., Chen, H., Mu, D., Li, D., Zhong, Y., Jiang, N., ... Xia, M. (2016). Association of serum retinoic acid with risk of mortality in patients with coronary artery disease. *Circulation Research, 119*, 557–563.

Lotan, R. (2005). A crucial role for cellular retinol-binding protein I in retinoid signaling. *Journal of the National Cancer Institute, 97*, 3–4.

Loughney, A. D., Kumarendran, M. K., Thomas, E. J., & Redfern, C. P. (1995). Variation in the expression of cellular retinoid binding proteins in human endometrium throughout the menstrual cycle. *Human Reproduction (Oxford, England), 10*, 1297–1304.

Lu, Y., Bertran, S., Samuels, T. A., Mira-y-Lopez, R., & Farias, E. F. (2010). Mechanism of inhibition of MMTV-neu and MMTV-wnt1 induced mammary oncogenesis by RARalpha agonist AM580. *Oncogene, 29*, 3665–3676.

Matt, N., Schmidt, C. K., Dupe, V., Dennefeld, C., Nau, H., Chambon, P., ... Ghyselinck, N. B. (2005). Contribution of cellular retinol-binding protein type 1 to retinol metabolism during mouse development. *Developmental Dynamics: An Official Publication of the American Association of Anatomists, 233*, 167–176.

McCaffery, P., & Drager, U. C. (1994). Hot spots of retinoic acid synthesis in the developing spinal cord. *Proceedings of the National Academy of Sciences of the United States of America, 91*, 7194–7197.

McCormick, A. M., Kroll, K. D., & Napoli, J. L. (1983). 13-cis-retinoic acid metabolism in vivo. The major tissue metabolites in the rat have the all-trans configuration. *Biochemistry, 22*, 3933–3940.

Molotkov, A., & Duester, G. (2002). Retinol/ethanol drug interaction during acute alcohol intoxication in mice involves inhibition of retinol metabolism to retinoic acid by alcohol dehydrogenase. *The Journal of Biological Chemistry, 277*, 22553–22557.

Morton, J. E., Frentz, S., Morgan, T., Sutherland-Smith, A. J., & Robertson, S. P. (2016). Biallelic mutations in CYP26B1: A differential diagnosis for Pfeiffer and Antley-Bixler syndromes. *American Journal of Medical Genetics Part A, 170*, 2706–2710.

Murphy, E. A., & Abbey, H. (1967). The normal range—A common misuse. *Journal of Chronic Diseases, 20*, 79–88.

Napoli, J. L. (1999a). Interactions of retinoid binding proteins and enzymes in retinoid metabolism. *Biochimica et Biophysica Acta, 1440*, 139–162.

Napoli, J. L. (1999b). Retinoic acid: Its biosynthesis and metabolism. *Progress in Nucleic Acid Research and Molecular Biology, 63*, 139–188.

Napoli, J. L. (2000). A gene knockout corroborates the integral function of cellular retinol-binding protein in retinoid metabolism. *Nutrition Reviews, 58*, 230–236.

Napoli, J. L. (2012). Physiological insights into all-trans-retinoic acid biosynthesis. *Biochimica et Biophysica Acta, 1821*, 152–167.

Napoli, J. L. (2016). Functions of intracellular retinoid binding-proteins. *Sub-Cellular Biochemistry, 81*, 21–76.

Napoli, J. L. (2017). Cellular retinoid binding-proteins, CRBP, CRABP, FABP5: Effects on retinoid metabolism, function and related diseases. *Pharmacology & Therapeutics, 173*, 19–33.

Napoli, J. L. (2020). Post-natal all-trans-retinoic acid biosynthesis. *Methods in Enzymology, 637*, 27–54.

Napoli, J. L. (2022). Retinoic acid: Sexually dimorphic, anti-insulin and concentration-dependent effects on energy. *Nutrients, 14*.

Neuville, P., Geinoz, A., Benzonana, G., Redard, M., Gabbiani, F., Ropraz, P., & Gabbiani, G. (1997). Cellular retinol-binding protein-1 is expressed by distinct subsets of rat arterial smooth muscle cells in vitro and in vivo. *The American Journal of Pathology, 150*, 509–521.

Nezhat, C., Littman, E. D., Lathi, R. B., Berker, B., Westphal, L. M., Giudice, L. C., & Milki, A. A. (2005). The dilemma of endometriosis: Is consensus possible with an enigma? *Fertility and Sterility, 84*, 1587–1588.

Noy, N. (2000). Retinoid-binding proteins: Mediators of retinoid action. *The Biochemical Journal, 348*(Pt 3), 481–495.

Nozaki, Y., Yamagata, T., Sugiyama, M., Ikoma, S., Kinoshita, K., & Funauchi, M. (2006). Anti-inflammatory effect of all-trans-retinoic acid in inflammatory arthritis. *Clinical Immunology (Orlando, Fla.), 119*, 272–279.

Obrochta, K. M., Kane, M. A., & Napoli, J. L. (2014). Effects of diet and strain on mouse serum and tissue retinoid concentrations. *PLoS One, 9*, e99435.

Ong, D. E. (1994). Cellular transport and metabolism of vitamin A: Roles of the cellular retinoid-binding proteins. *Nutrition Reviews, 52*, S24–S31.

Ong, D. E., Markert, C., & Chiu, J. F. (1978). Cellular binding proteins for vitamin A in colorectal adenocarcinoma of rat. *Cancer Research, 38*, 4422–4426.

Orlandi, A., Ferlosio, A., Ciucci, A., Francesconi, A., Lifschitz-Mercer, B., Gabbiani, G., ... Czernobilsky, B. (2006). Cellular retinol binding protein-1 expression in endometrial hyperplasia and carcinoma: Diagnostic and possible therapeutic implications. *Modern Pathology: An Official Journal of the United States and Canadian Academy of Pathology, Inc, 19*, 797–803.

Palan, P. R., Duttagupta, C., & Romney, S. L. (1980). Sex difference in cellular retinol- and retinoic acid-binding proteins in human colon adenocarcinomas. *Cancer Letters, 11*, 97–101.

Palan, P. R., & Romney, S. L. (1980). Cellular binding proteins for vitamin A in human carcinomas and in normal tissues. *Cancer Research, 40*, 4221–4224.

Pavone, M. E., Dyson, M., Reirstad, S., Pearson, E., Ishikawa, H., Cheng, Y. H., & Bulun, S. E. (2011). Endometriosis expresses a molecular pattern consistent with decreased retinoid uptake, metabolism and action. *Human Reproduction (Oxford, England), 26*, 2157–2164.

Pavone, M. E., Reierstad, S., Sun, H., Milad, M., Bulun, S. E., & Cheng, Y. H. (2010). Altered retinoid uptake and action contributes to cell survival in endometriosis. *The Journal of Clinical Endocrinology and Metabolism, 95*, E300–E309.

Petersen, P. H., Fraser, C. G., Sandberg, S., & Goldschmidt, H. (1999). The index of individuality is often a misinterpreted quantity characteristic. *Clinical Chemistry and Laboratory Medicine: CCLM / FESCC, 37*, 655–661.

Piantedosi, R., Ghyselinck, N., Blaner, W. S., & Vogel, S. (2005). Cellular retinol-binding protein type III is needed for retinoid incorporation into milk. *The Journal of Biological Chemistry, 280*, 24286–24292.

Pierzchalski, K., Taylor, R. N., Nezhat, C., Jones, J. W., Napoli, J. L., Yang, G., ... Sidell, N. (2014). Retinoic acid biosynthesis is impaired in human and murine endometriosis. *Biology of Reproduction, 91*, 84.

Pierzchalski, K., Yu, J., Norman, V., & Kane, M. A. (2013). CrbpI regulates mammary retinoic acid homeostasis and the mammary microenvironment. *The FASEB Journal, 27*, 1904–1916.

Relas, H., Gylling, H., & Miettinen, T. A. (2000). Effect of stanol ester on postabsorptive squalene and retinyl palmitate. *Metabolism: Clinical and Experimental, 49*, 473–478.

Rodondi, N., Darioli, R., Ramelet, A. A., Hohl, D., Lenain, V., Perdrix, J., ... Mooser, V. (2002). High risk for hyperlipidemia and the metabolic syndrome after an episode of hypertriglyceridemia during 13-cis retinoic acid therapy for acne: A pharmacogenetic study. *Annals of Internal Medicine, 136*, 582–589.

Ruberte, E., Friederich, V., Chambon, P., & Morriss-Kay, G. (1993). Retinoic acid receptors and cellular retinoid binding proteins. III. Their differential transcript distribution during mouse nervous system development. *Development (Cambridge, England), 118*, 267–282.

Ruhl, R. (2006). Method to determine 4-oxo-retinoic acids, retinoic acids and retinol in serum and cell extracts by liquid chromatography/diode-array detection atmospheric pressure chemical ionisation tandem mass spectrometry. *Rapid Communications in Mass Spectrometry, 20*, 2497–2504.

Ruhl, R., Krezel, W., & de Lera, A. R. (2018). 9-Cis-13,14-dihydroretinoic acid, a new endogenous mammalian ligand of retinoid X receptor and the active ligand of a potential new vitamin A category: vitamin A5. *Nutrition Reviews, 76*, 929–941.

Sago, K., Teitelbaum, S. L., Venstrom, K., Reichardt, L. F., & Ross, F. P. (1999). The integrin alphavbeta5 is expressed on avian osteoclast precursors and regulated by retinoic acid. *Journal of Bone and Mineral Research: The Official Journal of the American Society for Bone and Mineral Research, 14,* 32–38.

Sandell, L. L., Lynn, M. L., Inman, K. E., McDowell, W., & Trainor, P. A. (2012). RDH10 oxidation of Vitamin A is a critical control step in synthesis of retinoic acid during mouse embryogenesis. *PLoS One, 7,* e30698.

Sandell, L. L., Sanderson, B. W., Moiseyev, G., Johnson, T., Mushegian, A., Young, K., ... Trainor, P. A. (2007). RDH10 is essential for synthesis of embryonic retinoic acid and is required for limb, craniofacial, and organ development. *Genes & Development, 21,* 1113–1124.

Sawatsri, S., Desai, N., Rock, J. A., & Sidell, N. (2000). Retinoic acid suppresses interleukin-6 production in human endometrial cells. *Fertility and Sterility, 73,* 1012–1019.

Schmidt, K., Hughes, C., Chudek, J. A., Goodyear, S. R., Aspden, R. M., Talbot, R., ... Tickle, C. (2009). Cholesterol metabolism: The main pathway acting downstream of cytochrome P450 oxidoreductase in skeletal development of the limb. *Molecular and Cellular Biology, 29,* 2716–2729.

Schug, T. T., Berry, D. C., Shaw, N. S., Travis, S. N., & Noy, N. (2007). Opposing effects of retinoic acid on cell growth result from alternate activation of two different nuclear receptors. *Cell, 129,* 723–733.

Sedjo, R. L., Ranger-Moore, J., Foote, J., Craft, N. E., Alberts, D. S., Xu, M. J., & Giuliano, A. R. (2004). Circulating endogenous retinoic acid concentrations among participants enrolled in a randomized placebo-controlled clinical trial of retinyl palmitate. *Cancer Epidemiology, Biomarkers & Prevention: A Publication of the American Association for Cancer Research, Cosponsored by the American Society of Preventive Oncology, 13,* 1687–1692.

Shannon, S. R., Moise, A. R., & Trainor, P. A. (2017). New insights and changing paradigms in the regulation of vitamin A metabolism in development. *Wiley Interdisciplinary Reviews: Developmental Biology, 6.*

Shannon, S. R., Yu, J., Defnet, A. E., Bongfeldt, D., Moise, A. R., Kane, M. A., & Trainor, P. A. (2020). Identifying vitamin A signaling by visualizing gene and protein activity, and by quantification of vitamin A metabolites. *Methods in Enzymology, 637,* 367–418.

Sharpe-Timms, K. L. (2001). Endometrial anomalies in women with endometriosis. *Annals of the New York Academy of Sciences, 943,* 131–147.

Sharpe-Timms, K. L., & Cox, K. E. (2002). Paracrine regulation of matrix metalloproteinase expression in endometriosis. *Annals of the New York Academy of Sciences, 955,* 396–406 147-156; discussion 157-148.

Shaw, N., Elholm, M., & Noy, N. (2003). Retinoic acid is a high affinity selective ligand for the peroxisome proliferator-activated receptor beta/delta. *The Journal of Biological Chemistry, 278,* 41589–41592.

Shibata, M., Pattabiraman, K., Lorente-Galdos, B., Andrijevic, D., Kim, S. K., Kaur, N., ... Sestan, N. (2021). Regulation of prefrontal patterning and connectivity by retinoic acid. *Nature, 598,* 483–488.

Sidell, N., Feng, Y., Hao, L., Wu, J., Yu, J., Kane, M. A., ... Taylor, R. N. (2010). Retinoic acid is a cofactor for translational regulation of vascular endothelial growth factor in human endometrial stromal cells. *Molecular Endocrinology (Baltimore, Md.), 24,* 148–160.

Siegel, E. M., Craft, N. E., Roe, D. J., Duarte-Franco, E., Villa, L. L., Franco, E. L., & Giuliano, A. R. (2004). Temporal variation and identification of factors associated with endogenous retinoic acid isomers in serum from Brazilian women. *Cancer Epidemiology, Biomarkers & Prevention: A Publication of the American Association for Cancer Research, Cosponsored by the American Society of Preventive Oncology, 13,* 1693–1703.

Silveira, K. C., Fonseca, I. C., Oborn, C., Wengryn, P., Ghafoor, S., Beke, A., ... Kannu, P. (2023). CYP26B1-related disorder: Expanding the ends of the spectrum through clinical and molecular evidence. *Human Genetics, 142*, 1571–1586.

Soderlund, M. B., Sjoberg, A., Svard, G., Fex, G., & Nilsson-Ehle, P. (2002). Biological variation of retinoids in man. *Scandinavian Journal of Clinical and Laboratory Investigation, 62*, 511–519.

Soprano, D. R., & Soprano, K. J. (1995). Retinoids as teratogens. *Annual Review of Nutrition, 15*, 111–132.

Sporn, M. B., Dunlop, N. M., Newton, D. L., & Henderson, W. R. (1976). Relationships between structure and activity of retinoids. *Nature, 263*, 110–113.

Staels, B. (2001). Regulation of lipid and lipoprotein metabolism by retinoids. *Journal of the American Academy of Dermatology, 45*, S158–S167.

Sucov, H. M., & Evans, R. M. (1995). Retinoic acid and retinoic acid receptors in development. *Molecular Neurobiology, 10*, 169–184.

Talwar, D. K., Azharuddin, M. K., Williamson, C., Teoh, Y. P., McMillan, D. C., & St, J. O. R. D. (2005). Biological variation of vitamins in blood of healthy individuals. *Clinical Chemistry, 51*, 2145–2150.

Tanumihardjo, S. A., Russell, R. M., Stephensen, C. B., Gannon, B. M., Craft, N. E., Haskell, M. J., ... Raiten, D. J. (2016). Biomarkers of nutrition for development (BOND)-vitamin A review. *The Journal of Nutrition, 146*, 1816S–1848S.

Toki, K., Enokida, H., Kawakami, K., Chiyomaru, T., Tataroano, S., Yoshino, H., ... Nakagawa, M. (2010). CpG hypermethylation of cellular retinol-binding protein 1 contributes to cell proliferation and migration in bladder cancer. *International Journal of Oncology, 37*, 1379–1388.

van Vliet, T., Boelsma, E., de Vries, A. J., & van den Berg, H. (2001). Retinoic acid metabolites in plasma are higher after intake of liver paste compared with a vitamin A supplement in women. *The Journal of Nutrition, 131*, 3197–3203.

Vogel, S., Mendelsohn, C. L., Mertz, J. R., Piantedosi, R., Waldburger, C., Gottesman, M. E., & Blaner, W. S. (2001). Characterization of a new member of the fatty acid-binding protein family that binds all-trans-retinol. *The Journal of Biological Chemistry, 276*, 1353–1360.

Wagner, M. A. (1997). Use of reporter cells to study endogenous retinoid sources in embryonic tissues. *Methods in Enzymology, 282*, 98–107.

Wagner, M., Han, B., & Jessell, T. M. (1992). Regional differences in retinoid release from embryonic neural tissue detected by an in vitro reporter assay. *Development (Cambridge, England), 116*, 55–66.

Wang, S., Huang, W., Castillo, H. A., Kane, M. A., Xavier-Neto, J., Trainor, P. A., & Moise, A. R. (2018). Alterations in retinoic acid signaling affect the development of the mouse coronary vasculature. *Developmental Dynamics: An Official Publication of the American Association of Anatomists, 247*, 976–991.

Wang, Y. A., Shen, K., Wang, Y., & Brooks, S. C. (2005). Retinoic acid signaling is required for proper morphogenesis of mammary gland. *Developmental Dynamics: An Official Publication of the American Association of Anatomists, 234*, 892–899.

Wang, J., Yoo, H. S., Obrochta, K. M., Huang, P., & Napoli, J. L. (2015). Quantitation of retinaldehyde in small biological samples using ultrahigh-performance liquid chromatography tandem mass spectrometry. *Analytical Biochemistry, 484*, 162–168.

West, K. P., Jr. (2002). Extent of vitamin A deficiency among preschool children and women of reproductive age. *The Journal of Nutrition, 132*, 2857S–2866S.

Wieser, F., Wu, J., Shen, Z., Taylor, R. N., & Sidell, N. (2012). Retinoic acid suppresses growth of lesions, inhibits peritoneal cytokine secretion, and promotes macrophage differentiation in an immunocompetent mouse model of endometriosis. *Fertility and Sterility, 97*, 1430–1437.

Wu, J., Taylor, R. N., & Sidell, N. (2013). Retinoic acid regulates gap junction intercellular communication in human endometrial stromal cells through modulation of the phosphorylation status of connexin 43. *Journal of Cellular Physiology, 228,* 903–910.

Xu, G., Redard, M., Gabbiani, G., & Neuville, P. (1997). Cellular retinol-binding protein-1 is transiently expressed in granulation tissue fibroblasts and differentially expressed in fibroblasts cultured from different organs. *The American Journal of Pathology, 151,* 1741–1749.

Yamagata, T., Momoi, M. Y., Yanagisawa, M., Kumagai, H., Yamakado, M., & Momoi, T. (1994). Changes of the expression and distribution of retinoic acid receptors during neurogenesis in mouse embryos. *Brain Research. Developmental Brain Research, 77,* 163–176.

Yang, M., Lu, X., Zhang, Y., Wang, C., Cai, Z., Li, Z., ... Jiang, H. (2021). Whole-exome sequencing analysis in 10 families of sporadic microtia with thoracic deformities. *Molecular Genetics & Genomic Medicine, 9,* e1657.

Yang, N., Parker, L. E., Yu, J., Jones, J. W., Liu, T., Papanicolaou, K. N., ... Foster, D. B. (2021). Cardiac retinoic acid levels decline in heart failure. *JCI Insight, 6.*

Yashiro, K., Zhao, X., Uehara, M., Yamashita, K., Nishijima, M., Nishino, J., ... Hamada, H. (2004). Regulation of retinoic acid distribution is required for proximodistal patterning and outgrowth of the developing mouse limb. *Developmental Cell, 6,* 411–422.

Yu, M., Ishibashi-Ueda, H., Ohta-Ogo, K., Gabbiani, G., Yamagishi, M., Hayashi, K., ... Hao, H. (2012). Transient expression of cellular retinol-binding protein-1 during cardiac repair after myocardial infarction. *Pathology International, 62,* 246–253.

Yu, J., Perri, M., Jones, J. W., Pierzchalski, K., Ceaicovscaia, N., Cione, E., & Kane, M. A. (2022). Altered RBP1 gene expression impacts epithelial cell retinoic acid, proliferation, and microenvironment. *Cells, 11.*

Zalesak-Kravec, S., Huang, W., Jones, J. W., Yu, J., Alloush, J., Defnet, A. E., ... Kane, M. A. (2022). Role of cellular retinol-binding protein, type 1 and retinoid homeostasis in the adult mouse heart: A multi-omic approach. *The FASEB Journal, 36,* e22242.

Zetterstrom, R. H., Lindqvist, E., Mata de Urquiza, A., Tomac, A., Eriksson, U., Perlmann, T., & Olson, L. (1999). Role of retinoids in the CNS: Differential expression of retinoid binding proteins and receptors and evidence for presence of retinoic acid. *The European Journal of Neuroscience, 11,* 407–416.

Zheng, W. L., Sierra-Rivera, E., Luan, J., Osteen, K. G., & Ong, D. E. (2000). Retinoic acid synthesis and expression of cellular retinol-binding protein and cellular retinoic acid-binding protein type II are concurrent with decidualization of rat uterine stromal cells. *Endocrinology, 141,* 802–808.

Zizola, C. F., Schwartz, G. J., & Vogel, S. (2008). Cellular retinol-binding protein type III is a PPARgamma target gene and plays a role in lipid metabolism. *American Journal of Physiology. Endocrinology and Metabolism, 295,* E1358–E1368.

Zschocke, J., Byers, P. H., & Wilkie, A. O. M. (2023). Mendelian inheritance revisited: Dominance and recessiveness in medical genetics. *Nature Reviews. Genetics, 24,* 442–463.

Further reading

Noy, N. (2016). Non-classical transcriptional activity of retinoic acid. *Sub-Cellular Biochemistry, 81,* 179–199.

Yabut, K. C. B., & Isoherranen, N. (2022). CRABPs alter all-trans-retinoic acid metabolism by CYP26A1 via protein-protein interactions. *Nutrients, 14.*

CHAPTER NINE

The multifaceted roles of retinoids in eye development, vision, and retinal degenerative diseases

Zachary J. Engfer[a,b,*] and Krzysztof Palczewski[a,b,c,d,*]
[a]Center for Translational Vision Research, Department of Ophthalmology, Gavin Herbert Eye Institute, University of California, Irvine, Irvine, CA, United States
[b]Department of Physiology and Biophysics, University of California, Irvine, Irvine, CA, United States
[c]Department of Chemistry, University of California Irvine, Irvine, CA, United States
[d]Department of Molecular Biology and Biochemistry, University of California, Irvine, Irvine, CA, United States
*Corresponding authors. e-mail address: zengfer@uci.edu; kpalczew@uci.edu

Contents

1. Introduction	236
2. Vitamin A: The root of all retinoids	238
3. Retinoic acid signaling in the developing and mature retina	246
4. Retinaldehydes, retinyl esters, and retinols in vision	256
5. Conclusion	277
Acknowledgments	278
Author Contributions	278
Competing Interest Statement	278
References	278

Abstract

Vitamin A (all-*trans*-retinol; *at*-Rol) and its derivatives, known as retinoids, have been adopted by vertebrates to serve as visual chromophores and signaling molecules, particularly in the eye/retina. Few tissues rely on retinoids as heavily as the retina, and the study of genetically modified mouse models with deficiencies in specific retinoid-metabolizing proteins has allowed us to gain insight into the unique or redundant roles of these proteins in *at*-Rol uptake and storage, or their downstream roles in retinal development and function. These processes occur during embryogenesis and continue throughout life. This review delves into the role of these genes in supporting retinal function and maps the impact that genetically modified mouse models have had in studying retinoid-related genes. These models display distinct perturbations in retinoid biochemistry, physiology, and metabolic flux, mirroring human ocular diseases.

Abbreviations

Abca4	ATP-binding cassette, subfamily A, member 4.
Aldh1a1	Aldehyde dehydrogenase family 1, subfamily a1.
Aldh1a2	Aldehyde dehydrogenase family 1, subfamily a2.
Aldh1a3	Aldehyde dehydrogenase family 1, subfamily a3.
at-RA	all-*trans*-retinoic acid.
at-Ral	all-*trans*-retinal.
at-RDHs	all-*trans*-retinol dehydrogenases (predominantly RDH8).
at-REs	all-*trans*-retinyl esters.
at-Rol	all-*trans*-retinol.
9*cis*-Rol	9-*cis*-retinol.
9*cis*-RA	9-*cis* retinoic acid.
11*cis*-Ral	11-*cis*-retinal.
11*cis*-RDHs	11-*cis*-retinol dehydrogenases (predominantly RDH5).
11*cis*-Rol	11-*cis*-retinol.
Crabp1	Cellular retinoic acid-binding protein 1.
Crabp2	Cellular retinoic acid-binding protein 2.
Cyp26a	Cytochrome P450 family 26, subfamily A member 1.
Cyp26c	Cytochrome P450 family 26, subfamily C member 1.
LRAT	lecithin-retinol acyltransferase.
Rara	Retinoic acid receptor, alpha.
Rarb	Retinoic acid receptor, beta.
RAREs	Retinoic acid response elements.
Rarg	Retinoic acid receptor, gamma.
Rars	Retinoic acid receptors (encompasses alpha, beta, and gamma isotypes).
Rbp1	retinol-binding protein 1 (alias: cellular retinol-binding protein 1).
Rbp3	retinol-binding protein 3 (alias interphotoreceptor-binding protein).
Rbp4	retinol-binding protein 4 (alias: serum retinol-binding protein).
RDH10	retinol dehydrogenase 10.
Rgr	retinal G-protein coupled receptor.
Rho	rhodopsin.
Rlbp1	Retinaldehyde-binding protein 1 (alias: cellular retinaldehyde-binding protein).
Rpe65	Retinoid isomerohydrolase RPE65.
Rxra	Retinoid X receptor, alpha.
Rxrb	Retinoid X receptor, beta.
Rxrg	Retinoid X receptor, gamma.
Rxrs	Retinoid X receptors (encompasses alpha, beta, and gamma isotypes).
Stra6	Signaling receptor and transporter of retinol STRA6.
Ttr	Transthyretin.

1. Introduction

Retinoids constitute a diverse class of both natural and synthetic derivatives of vitamin A. Several naturally occurring retinoids, including vitamin A itself, serve various biological roles in mammalian systems and lower organisms (Alvarez et al., 2014; Andre, Ruivo, Gesto, Castro, & Santos, 2014). All-*trans*-retinoic acid (a*t*-RA) and its derivatives are crucial drivers of numerous signaling pathways involved in cell growth, immune homeostasis, reproduction, and development (Clagett-Dame & Knutson, 2011; Duester, 2008; Ghyselinck & Duester, 2019; Hall, Grainger, Spencer, & Belkaid, 2011; Wang et al., 1998). Similarly, a*t*-Rol derivatives are implicated in light perception within the retina (Kiser, Golczak, & Palczewski, 2014; Palczewski & Kiser, 2020). There has been a significant interest in utilizing retinoids and retinoid mimetics as therapeutics for a range of human diseases, including vitamin A deficiencies, retinitis pigmentosa (RP), psoriasis, and cancers (Bavik et al., 2015; Carazo et al., 2021; Fritsch, 1992). Therefore, understanding how retinoids support normal biological functions in mammalian tissues and how fluctuations in retinoids maintain functional homeostasis across various tissues is critical for the betterment of human health.

Few organs rely on retinoids as consistently as the retina. Here, a*t*-RA signaling is indispensable for proper tissue patterning during embryonic development, while retinaldehyde isomerization and reduction/oxidation play critical roles in sustaining the perception and transduction of light stimuli. Ensuring effective spatial and temporal control of retinoid levels within the retina is vital for its proper function, a task accomplished by various retinoid-binding/transport proteins and enzymes that interconvert retinoid species. Genetic deficiencies in these pathways have been associated with a broad spectrum of diseases affecting retinal development and function, many of which have been molecularly characterized using genetically modified mouse models. The earliest mouse models of retinal degenerative disorders emerged from spontaneous mutations in inbred mouse colonies, some of which are still utilized today to simulate conditions such as RP (Chang et al., 2002). The advent of mammalian gene-editing technologies has enabled targeted disruption of specific genomic loci, revealing intriguing functional redundancies in certain retinoid-associated genes, while other genetic alterations have unveiled the critical roles of individual genes in normal eye development and retinal function. These foundational discoveries in modified mice have contributed to understanding human diseases linked to defects in the identical retinoid-associated genes. Significant functional conservation in retinoid-associated genes combined with a diverse library of murine genetic tools positions mice as compelling preclinical models for testing novel genetic and small-molecule therapeutics targeting pathogenic defects in retinoid-associated genes (*e.g.* RP, Leber Congenital Amaurosis, Stargardt disease) (McBee, Palczewski, Baehr, & Pepperberg, 2001; Palczewska et al., 2014; Palczewska et al., 2010; Sears et al., 2017; Zhang et al., 2015).

This review focuses on the role of various retinoids in the retina/eye, highlighting the enzymatic, molecular, and metabolic overlaps among enzymes crucial for at-RA signaling in the developing retina and those essential for visual-cycle function in the mature retina. Additionally, this chapter underscores the importance of modified mouse models in ongoing studies of at-RA signaling and visual-cycle disease pathogenesis. Understanding retinoid flux within the retina facilitates the optimization of small-molecule therapeutics capable of limiting the rate of visual chromophore (11-*cis*-retinaldehyde, 11*cis*-Ral) regeneration in the visual cycle, thereby extending the function of photoreceptors harboring pathogenic mutations in visual-cycle genes.

2. Vitamin A: The root of all retinoids

At-Rol is a small, lipophilic vitamin that is essential for a wide range of biological processes in vertebrates. At-Rol cannot be synthesized endogenously within vertebrates, and thus must be obtained from dietary sources (in the case of mature mammals) or maternal sources (in the case of embryonic, fetal, and neonatal mammals) (Clagett-Dame & Knutson, 2011). To respond to fluctuating demands for retinoids throughout the body, at-Rol and its retinyl esters (at-REs) derivatives are stored in the liver and adipose tissue. They are poised to release at-Rol into circulation when tissue at-RA levels are low. At-Rol also serves as the core metabolic source of the 11*cis*-Ral chromophore used in vision. The synthesis of 11*cis*-Ral is imperative for light perception within the retina (Kiser & Palczewski, 2016, 2021; Palczewski, 2010; Palczewski & Kiser, 2020). Systemic at-Rol deficiency remains a global human health problem; as of 2013, an estimated 29% of children aged 6–59 months of age in low- and middle-income countries were found to be Vitamin A-deficient (Stevens et al., 2015). Extreme vitamin A deficiency has severe effects on eye health and vision, causing nyctalopia (progressive night-blindness), xerophthalmia (complex dry eye syndrome), and corneal ulceration (Sommer, 2008). Early observations of pregnant mice and rats fed diets lacking vitamin A demonstrated that vitamin-A deprivation reduced fertility, caused an increase in embryonic and fetal abnormalities including fetal death/resorption, and greatly impacted the health of severely-deprived, mature animals (McCarthy & Cerecedo, 1952; White, Highland, Kaiser, & Clagett-Dame, 2000). In a rat model of vitamin A deficiency where pregnant dams were fed different levels of at-Rol, their offspring had increasing rates of caudal hindbrain malformations as a function of decreasing levels of maternal at-Rol supplementation (White et al., 2000). On the contrary, over-supplementation with at-Rol has teratogenic effects on developing rat embryos. High doses of at-Rol administered during pregnancy results in pups with congenital defects such as hydrocephalus, spina bifida, and eye malformations (Cohlan, 1954).

An early study on rats conducted by Dowling and Wald recorded certain systemic effects resulting from prolonged vitamin A deprivation, such as weight loss, growth arrest, compromised epithelial barriers, and neurological and respiratory defects, could be mitigated by supplementing vitamin A-deficient rats with at-RA (Dowling & Wald, 1960). Despite the reduction in systemic morbidities observed in the vitamin A-deprived rats due to at-RA supplementation, it failed to prevent progressive thinning of the outer nuclear layer and the loss of photoreceptors in the retina. This early observation anticipated subsequent research that established a mechanistic distinction between at-RA signaling and metabolism necessary for general cellular functions, as well as retinol/retinaldehyde metabolism essential for the survival and function of photoreceptors.

Both at-Rol and at-REs are obtainable from dietary sources and are absorbed by the enterocytes of the small intestine. Free at-Rol enters the enterocytes of the small intestine *via* passive diffusion, facilitated by its lipophilic properties (During & Harrison, 2007). Conversely, at-REs undergo hydrolysis to at-Rol by at-RE-hydrolases in the gut lumen before absorption by the enterocytes (Schreiber et al., 2012). Alongside at-Rol, pro-vitamin A carotenoids are absorbed through scavenger receptor class B member 1 (SCARB1, also known as SR-BI), CD36, and other class 2 scavenger receptors responsible for the uptake of lipophilic vitamins and cholesterol esters (Borel et al., 2013; von Lintig, Moon, Lee, & Ramkumar, 2020; Werder et al., 2001). Carotenoids such as β,β-carotene undergo oxidative cleavage into all-*trans*-retinaldehyde (at-Ral), which is then further reduced to at-Rol. In the case of β,β-carotene, the β,β-carotene oxygenase 1 (BCO1) enzyme symmetrically cleaves it into two molecules of at-Ral. Other carotenoids undergo asymmetric cleavage by β,β-carotene oxygenase 2 (BCO2) in conjunction with BCO1, yielding a single molecule of at-Ral and apo-carotenal metabolites (Shete & Quadro,

2013; von Lintig, 2012). Notably, SCARB1 expression is regulated by the *at*-RA-inducible transcription factor intestine-specific homeobox (ISX) (Lobo et al., 2010). β,β-carotene accumulates in *Bco1*-knockout mice due to upregulated SCARB1 expression, which could be prevented through dietary retinoid supplementation. Retinoid supplementation induced *at*-RA production, leading to increased ISX expression levels. This upregulation of ISX resulted in decreased β,β-carotene uptake by suppressing SCARB1 expression (Lobo et al., 2010). ISX was also found to repress *Bco1* gene expression in an *at*-RA-dependent manner (Widjaja-Adhi et al., 2017). This negative feedback loop exemplifies how retinoid homeostasis is maintained by regulating dietary levels of absorption.

Following the cleavage of β,β-carotene in enterocytes, the *at*-Ral is reduced to *at*-Rol by class I, II, IV, and VII alcohol dehydrogenases (ADHs), and by short-chain dehydrogenases/reductases (SDRs), such as microsomal retinol dehydrogenase (RDH) 11 (Duester, 2000; von Lintig et al., 2020). The reversible reduction of retinaldehyde to retinol is dependent on a high $NADH/NAD^+$ cofactor ratio in the case of ADHs and a high $NADPH/NADP^+$ cofactor ratio in the case of SDRs (Duester, 2000). This dependence on NADH or NADPH for the reduction of *at*-Ral to *at*-Rol links retinoid metabolism to the overarching redox environment within the cells. Dependence on $NAD^+/NADH$ and $NADP^+/NADPH$ cofactors in the redox chemistry of *at*-Rol/*at*-Ral is likely coupled in the putative retinoid oxidoreductase complex (ROC) (Adams et al., 2021; Adams, Belyaeva, Wu, & Kedishvili, 2014). This hetero-oligomeric complex consists of dehydrogenase/reductase superfamily member 3 (DHRS3), which utilizes $NADP^+/NADPH$ as a cofactor, and RDH10, which utilizes $NAD^+/NADH$ as a cofactor. This complex is proposed to interconvert *at*-Ral and *at*-Rol in a steady-state manner. At-Rol molecules that are directly absorbed or produced *via* the cleavage of carotenoids in the enterocytes of the small intestine are bound by cellular retinol-binding protein type II (RBP2) and shuttled to lecithin-retinol acyltransferase (LRAT) (Ong, 1993; Ong & Page, 1987; von Lintig et al., 2020). LRAT re-esterifies the *at*-Rol to produce *at*-REs, which are incorporated into chylomicrons for secretion into the lacteals and subsequent transport through the bloodstream to the liver and other tissues (Blomhoff, Helgerud, Rasmussen, Berg, & Norum, 1982; Tso & Balint, 1986). LRAT plays a key role in the regulation of intestinal retinoid production from carotenoids *via* the ISX-dependent feedback loop (Ramkumar et al., 2021). Mice with a genetic ablation of LRAT expression demonstrated a 40–50 % decrease in retinoid-absorption efficiency when retinol was administered, and an increase in the ratio of free retinol to a*t*-REs within chylomicrons (O'Byrne et al., 2005). Along with its key role in intestinal retinoid homeostasis, LRAT also has critical roles in a*t*-RE storage within hepatic stellate cells (HSCs) and the retinal pigment epithelium (RPE), both of which are discussed in later sections. In instances of high intracellular free (*i.e.* unbound to RBP2) retinol concentrations, it is hypothesized that a second enzyme, diacylglycerol O-acyltransferase 1 (DGAT1), assists in the esterification of retinol in an acyl-CoA-dependent manner (Wongsiriroj et al., 2008; Yen, Monetti, Burri, & Farese, 2005). It is still unclear, however, whether DGAT1 plays a significant role in enterocyte/liver retinoid homeostasis. DGAT1 expression in mouse skin epidermis is essential for preventing retinoid toxicity and alopecia, but when researchers examined total a*t*-RE content in liver preparations of the *Dgat1* knockouts, they saw no significant differences in a*t*-RE content when compared to wild-type controls under both a*t*-Rol-sufficient and -deficient conditions (Shih et al., 2009).

After being secreted by enterocytes, the a*t*-REs circulating in the remnants of chylomicrons are absorbed by parenchymal hepatocytes in the liver through two distinct receptors, the low-density lipoprotein receptor (LDLR) and, in the case of APOE-enriched chylomicron remnants, the LDL receptor-related protein (LRP1) (Blaner et al., 1987; Blaner et al., 1985; Blaner et al., 2016; Cooper, 1997). Following absorption of a*t*-REs into the hepatocytes, they are hydrolyzed to a*t*-Rol for relay to the HSCs, which are the predominant retinoid-storage cells in the liver. Several enzymes display a*t*-RE-hydrolase activity, though it remains unclear which enzymes are the predominant a*t*-RE hydrolases under physiological conditions in the hepatocytes and HSCs (Blaner et al., 1985; Blaner et al., 2016; Grumet, Taschler, & Lass, 2016; Schreiber et al., 2012; Taschler et al., 2015).

The hydrolyzed a*t*-Rol then freely diffuses from the hepatocytes to HSCs or is ferried bound to RBP1 (a.k.a. CRBP1) (Blaner et al., 2016). Despite RBP1's purported role in the shuttling of retinoids between hepatocytes and the HSCs, mice lacking RBP1 were found to only have a 50 % decrease in hepatic content of all-*trans*-retinyl palmitate (the dominant hepatic a*t*-RE species) (Ghyselinck et al., 1999). These RBP1-deficient mice also displayed an increased sensitivity to systemic a*t*-Rol deficiency following a switch to a vitamin A-deficient diet. RBP1's function in shuttling retinoids to LRAT for re-esterification is counterbalanced by that of ADH1, which acts as a sink for free a*t*-Rol, buffering a*t*-Rol and a*t*-RE levels. Genetic deletion of *Adh1* in mice resulted in increased hepatic a*t*-RE formation, while co-deletion of

Adh1 and *Rbp1* resulted in mice that were resistant to the rapid loss of liver *at*-REs exhibited by *Rbp1* knock-out mice during conditions of vitamin A-deficiency (Molotkov, Ghyselinck, Chambon, & Duester, 2004). Another study on *Adh1* and *Adh4* single knock-out and double knock-out mice, studied under *at*-Rol-sufficient and -deficient conditions, revealed that the two enzymes are important for establishing a balance between resisting *at*-Rol toxicity during periods of excess (a function attributable to ADH1) and maintaining *at*-Rol levels and storage during periods of vitamin-A deficiency (a function attributable to ADH4) (Molotkov, Deltour, Foglio, Cuenca, & Duester, 2002). It was initially thought that retinol-binding protein 4 (RBP4) might participate in the transfer of retinol from hepatocytes to the HSCs, but genetic ablation of *Rbp4* in mice did not significantly affect retinoid storage in HSCs (Quadro et al., 1999; Quadro et al., 2005). Though its activity is low, the ubiquitously-expressed ADH3 is also hypothesized to play a general role in curtailing *at*-Rol levels during periods of excess (Pares, Farres, Kedishvili, & Duester, 2008). Human and mouse ADH2 homologs, also expressed in the liver, exhibit differential activity, with the human enzyme displaying high affinity for 9-*cis*-retinol (9*cis*-Rol) and lower affinity for *at*-Rol, and the mouse Adh2 displaying little affinity for 9*cis*-Rol or *at*-Rol. This dichotomy suggests that ADH2 plays a specific role in hepatic retinoid metabolism in humans, while it plays little to no significant role in murine hepatic retinoid metabolism (Hellgren et al., 2007). The role of ADHs in retinoid metabolism has come into question, with mounting evidence to suggest that ADHs are incapable of accessing and oxidizing *at*-Rol under physiological conditions (for review, see (Napoli, 2012)).

Once *at*-Rol is delivered to the HSCs, it binds to RBP1 and is shuttled to LRAT for re-esterification. The resulting *at*-REs are stored in large lipid droplets. Liver-localized retinoids account for approximately 70 % of the total retinoid species in the adult human, and 80–90 % of the total retinoid species in the adult mouse (Blaner et al., 2016). In *Lrat* knock-out mice, there was an almost complete elimination of total liver *at*-RE content, indicating that LRAT is almost exclusively responsible for the esterification of *at*-Rol in the mouse liver (O'Byrne et al., 2005; Ruiz et al., 2007). This deficiency in liver *at*-RE content in the *Lrat*-KO mice corresponded with a dramatic decrease in total liver retinol content, suggesting that LRAT-dependent esterification also has a significant role in the delivery of retinol to the liver for storage. *at*-REs serve as a means of stable retinoid storage in the HSCs and buffer circulating *at*-Rol levels in response to fluctuating demand from peripheral tissues such as the retina. To re-enter circulation from the HSCs, a*t*-REs are hydrolyzed to a*t*-Rol by a*t*-REs hydrolases. The a*t*-Rol then travels back to the parenchymal hepatocytes and forms complexes with RBP4 and transthyretin (TTR). RBP4 is serum-localized and responsible for shuttling a*t*-Rol to peripheral tissues. After assembling in hepatocytes, the *at*-Rol-RBP4-TTR complex re-enters the bloodstream and distributes a*t*-Rol to peripheral tissues such as the eye (Monaco, 2000). Transthyretin, which normally tetramerizes to form a 55-kilodalton (kDa) complex during its biosynthesis, protects two molecules of the 21-kDa RBP4 from premature glomerular filtration by the kidneys. The *in vitro* kinetics of the dissociation of apo-RBP4 (a*t*-Rol-free RBP4) from TTR is faster than the kinetics of dissociation of holo-RBP4 (a*t*-Rol-bound) from TTR, providing a mechanistic explanation for the specific filtration of apo-RBP4 from circulation by the kidneys following delivery of a*t*-Rol to peripheral tissues (Monaco, 2000).

In mice, hepatocytes are responsible for synthesizing the majority of RBP4 secreted into the circulation, as documented in hepatocyte-specific *Rbp4*-knockout mice (Thompson et al., 2017). Mice with global *Rbp4* deletion displayed an increased sensitivity to vitamin A deficiency, decreased serum a*t*-Rol levels, as well as a rapid retinal-degenerative phenotype when their dietary intake of vitamin A was restricted. Hepatic storage of a*t*-Rol was found to be uncompromised, though mobilization of retinol into the bloodstream was decreased, indicating a reliance on RBP4 for efficient release of a*t*-Rol into the bloodstream (Quadro et al., 1999). Studies in rats and rat hepatocyte cultures demonstrated that RBP4 secretion into the circulation was dependent on a*t*-Rol levels, with a dramatic increase in circulating/secreted RBP4 levels following administration of a*t*-Rol (Muto, Smith, Milch, & Goodman, 1972; Ronne et al., 1983). Furthermore, examination of embryonic mice homozygous or heterozygous for a *Rbp4* knock-out allele revealed that fetal liver retinoid accumulation is dependent on maternal retinoid status and RBP4 function rather than on fetal expression of functional RBP4, highlighting the importance of maternal retinoid delivery to support normal fetal development and retinoid accumulation (Quadro et al., 2005).

Pathogenic mutations in RBP4 cause a variety of ocular dysfunctions in humans, such as RP, cataracts, and iris and retinal colobomas, along with other non-ocular phenotypes such as acne (Plaisancie et al., 2023). Mutant RBP4 variants derived from human patients with varying levels of ocular pathology have been recombinantly expressed, purified, and biochemically profiled (Chou et al., 2015). This profiling revealed that the mutant, patient-derived RBP4s were unable to bind free at-Rol while maintaining stable secretion, TTR-association, and STRA6-binding characteristics found in wild-type RBP4. These findings suggest that the mutant RBP4 variants, which were found circulating in the sera of human patients, are able to compete with the wild-type retinol-bound RBP4 variants for binding to STRA6, the essential transporter responsible for at-Rol uptake in many of the body's cell types (see below). This competition for STRA6 binding impedes at-Rol import into cells, thus explaining the dominant nature of the RBP4 mutations that were characterized. The decreased delivery of at-Rol to a fetus carrying one of the characterized RBP4 mutations would be compounded if the mother is also a carrier of the same mutation; maternally-derived retinol would be delivered with decreased efficiency to the placenta for transport to the fetus and then be further impaired by inefficient delivery of at-Rol in the fetal circulation to the developing eye. Ocular defects from change-in-function mutations in RBP4 are recapitulated in knock-out mice with complete elimination of RBP4 expression, though the severity of ocular phenotypes seems to be dependent on the background strain of the knock-outs (Shen et al., 2016). These studies of genetically modified mice emphasize the critical dependence that fetuses have on maternal sources of at-Rol during both prenatal development and postnatal lactation.

As mentioned above, STRA6 is a critical, membrane-bound receptor that facilitates the efficient transport of at-Rol from circulation to the cytosol (Kawaguchi et al., 2007). Testing the association of radio-iodine-labeled holo-RBP4-TTR (bovine and human) with isolated bovine RPE cells in the presence and absence of unlabeled holo-RBP4-TTR revealed a saturable level of specific, preferential binding of holo-RBP4 over apo-RBP4 (Heller, 1975). These findings suggested the presence of a receptor on the surface of RPE cells responsible for the specific uptake of at-Rol *via* binding to holo-RBP4-TTR. In 1995, the first *Stra6* transcript was cloned from mouse embryonic carcinoma cells and characterized as a retinoic-acid responsive gene (Bouillet et al., 1995). Follow-up studies in cell cultures demonstrated that induction of STRA6 transcription was decreased when RXRA or RARG were specifically ablated, suggesting that RXRA/RARG heterodimers are essential for upregulating STRA6 expression in response to at-RA (Bouillet et al., 1997; Taneja et al., 1995). RXRA and RARG were later confirmed to be present at the *at*-RA response element (RARE) that drives expression of STRA6 (Laursen, Kashyap, Scandura, & Gudas, 2015). Another related study with RARA-knock-out cell cultures demonstrated that STRA6 transcription was increased in response to at-RA when compared to WT cell cultures, suggesting a potential role for RARA in suppressing at-RA -induced STRA6 transcription (Taneja et al., 1995). Widespread profiling of *Stra6* transcription *via in situ* hybridization at various stages of mouse embryonic development has implicated STRA6 in the transport of at-Rol across the yolk sac membrane to supply embryos with at-Rol at early stages of development. As the placenta matures and the yolk sac atrophies, STRA6 expression was shown to increase in the labyrinthine zone of the placental barrier, which denotes a gradual transference of fetal retinoid sources from the retinoid-rich yolk sac to maternal circulation. *Stra6* transcription was also seen in many other embryonic mouse tissues such as the nervous system, face, eye, muscle, skeletal system, gut, and genitourinary tract, implying a widespread reliance on STRA6 for efficient import of at-Rol throughout development, likely for use in the synthesis of at-RA (Bouillet et al., 1997). Adult mouse tissues profiled in the same manner, including the choroid plexus of the brain, the RPE in the eye, spleen, kidney, and male/female reproductive systems, displayed a more restricted transcription pattern. The connection between STRA6 induction and at-Rol transport, however, would not be confirmed until 2007 (Kawaguchi et al., 2007). In 2007, STRA6 was shown to be a binding partner for RBP4 and recombinantly-expressed STRA6 was shown to be an efficient importer of at-Rol from holo-RBP4. The same study demonstrated that this at-Rol could be intracellularly sequestered as at-REs when STRA6 was co-expressed with LRAT. Furthermore, they demonstrated that deleterious point mutations in STRA6 impaired at-Rol uptake in cultured cells (Kawaguchi et al., 2007).

Initial studies on *Stra6*-null mice with exons 5, 6, and 7 deleted *via* recombination revealed the knockout mice had hypertrophic vitreous and a vascularized structure in the vitreous humor, implicating STRA6 function in normal development of the eye, while the loss of cone photoreceptors and shortened photoreceptor inner/outer segments implicate STRA6 in the maintenance of photoreceptor layer morphology and function. This first study noted that photoreceptor electroretinogram (ERG) responses to light were not completely abolished in these mice, nor was there a total elimination of retinoids, as shown by HPLC-based retinoid analyses, implying a vestigial function of the *Stra6*-null gene (Ruiz et al., 2012). A different *Stra6*-null mouse with a deletion of exon 2 exhibited additional morphological changes to the retina, including discolored regions of the RPE and disordered areas of choroidal thickness and vascularization (Amengual et al., 2014). These *Stra6*-null mice displayed a more prominent reduction in retinoid content and both photopic (cone-based) and scotopic (rod-based) electroretinograms with increasing intensities of light stimulus, demonstrating a range of pathological severity in models of *Stra6* deficiency. This range of ocular defects are recapitulated in human patients with Matthew-Wood Syndrome, a rare condition caused by truncations in STRA6. These deficiencies in STRA6 manifest in microphthalmia/anophthalmia, with additional congenital heart defects and diaphragmatic hernia (Golzio et al., 2007; Pasutto et al., 2007). A regulatory mechanism for STRA6-mediated import of *at*-Rol was identified in zebrafish, where increased intracellular [Ca^{2+}] promoted calmodulin binding to STRA6 and preferential binding of apo-RBP4 to STRA6 (Zhong et al., 2020). These results align with the solved cryo-electron microscopy structure of zebrafish STRA6, which included calmodulin bound to the cytosolic portions of STRA6 (Chen et al., 2016). This mechanism of regulating *at*-Rol influx through STRA6 has yet to be validated in mammalian systems.

The overarching mechanism of *at*-Rol uptake from dietary sources or maternal circulation, coupled with hepatic storage and release into the bloodstream, serves as the core framework for controlling the *at*-RA signaling required for retinal development and synthesis of the key visual pigment in mature retina.

3. Retinoic acid signaling in the developing and mature retina

The biosynthesis of *at*-RA is largely dictated by levels of free *at*-Rol within a cell, which is first converted to *at*-Ral by cytosolic ADHs and/or RDHs. Expression patterns of some cytosolic ADHs and RDHs are tissue- and cell-type dependent while others are ubiquitous; many isotypes also have distinct spatiotemporal roles in *at*-Rol metabolism, depending on their involvement in development or adult tissue functions (Duester, 2000; Kumar, Sandell, Trainor, Koentgen, & Duester, 2012; Molotkov, Fan, et al., 2002). RBPs act as stable reservoirs of *at*-Rol and help to regulate the quantities of free (*i.e.*, unbound) *at*-Rol available for the reversible oxidation to *at*-Ral by RDHs and/or ADHs. The chemical equilibrium between *at*-Rol and *at*-Ral is defined by the overarching redox state of the cell; oxidation of *at*-Rol to *at*-Ral is driven by NAD^+ or $NADP^+$ cofactors, depending on the dehydrogenase participating in the oxidation reaction. Once at-Rol is oxidized to at-Ral, it can be further oxidized to at-RA by retinaldehyde dehydrogenases (RALDHs) for downstream signaling (see Fig. 1A).

In early mouse embryonic development, ADH- and RDH-isotypes display unique expression patterns that underlie their involvement in the first stages of at-RA synthesis (Pares et al., 2008). An early study using a combination of at-RA-dependent reporter assays and *in situ* hybridization revealed that little-to-no at-RA synthesis occurs until the late primitive-streak stage of mouse embryogenesis (Rossant, Zirngibl, Cado, Shago, & Giguere, 1991); at-RA synthesis was found to coincide with an increase in ADH4 transcription levels in the posterior mesoderm at 7.5 days postcoitum (d.p.c.). At 8.5–9.5 d.p.c., ADH4 transcripts were detected in anterior neural folds and the caudal neuropore prior to the close of the neural tube, corresponding with early at-RA production in the trunk (Ang, Deltour, Hayamizu, Zgombic-Knight, & Duester, 1996). Genetic disruption of ADH4 with sufficient maternal at-Rol levels, however, had little effect on the viability of newborn mouse pups, highlighting the partially redundant functions of ADH isotypes in the embryonic synthesis of at-RA, which were validated with the generation of compound ADH-null mice (Deltour, Foglio, & Duester, 1999a, 1999b). According to *in situ* hybridization, ADH3 transcription is almost ubiquitous in the developing mouse from 6.5 d.p.c. onwards, but its widespread expression precedes the production of at-RA at 7.5–8.5 d.p.c. Nevertheless, ADH3 has limited involvement in the earliest at-RA synthesis, likely due to its poor affinity for at-Rol compared to ADH4 (Ang et al., 1996; Balkan, Colbert, Bock, & Linney, 1992). Alternatively, the lack of at-RA synthesis at earlier stages of development could be dependent on the bioavailability of unbound at-Rol for the synthesis of at-Ral and subsequent synthesis of at-RA (Ang et al., 1996; Pares et al., 2008). Despite the increase in ADH3 transcription prior to the onset of at-RA synthesis and the contentious affinity of ADH3 for at-Rol, *Adh3*-null mice were found to have a diminution in at-RA synthesis, as well as congenital growth defects that were corrected by maternal at-Rol supplementation. *Adh3*-null neonate mice also exhibited postnatal lethality under maternal at-Rol deficiency (Boleda, Saubi, Farres, & Pares, 1993; Molotkov, Fan, et al., 2002). These observations make a strong argument for the involvement of ADH3 in mammalian embryonic at-RA biosynthesis and development, though its intrinsic activity as a RDH is much lower than those of ADH1 and ADH4 (Yang et al., 1994). In contrast to the expression patterns for ADH3 and ADH4, ADH1 transcripts are

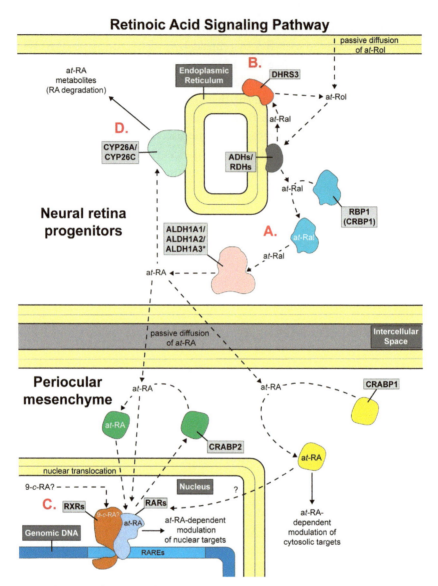

Fig. 1 Schematic diagrams of at-RA signaling in the developing retina. (A) Conversion of at-Rol to at-Ral by ADHs/ RDHs followed by RBP1 binding and shuttling at-Ral to ALDH1 isotypes for irreversible oxidation to at-RA. (B) Reduction of at-Ral to at-Rol by DHRS3. (C) Transcriptional modulation of repressive RAR-RXR heterodimers by at-RA to initiate downstream transcription of at-RA-responsive genes. (D) Degradations of at-RA by CYP26 isotypes to inactive metabolites.

localized to metanephric tissues at 10.5 d.p.c., suggesting its involvement in genito-urinary tract development (Ang et al., 1996). As mentioned previously, the roles of ADHs in metabolizing at-Rol to at-Ral is contentious, and more studies will need to be performed to validate the role that ADH plays in at-RA synthesis *in vivo* (for review, see (Napoli, 2012)).

RDH10 also plays a key role in early embryonic at-RA synthesis and development. RDH10 was proven to be essential for mouse embryonic development, as *Rdh10*-null embryos did not survive past 13 d.p.c. (Sandell et al., 2007). This embryonically-lethal phenotype was attributed to a lack of RDH10-dependent at-Rol-oxidation activity, which impaired embryonic at-RA synthesis and caused stunted forelimb development, lung agenesis, craniofacial and optic vesicle defects, and abnormal somitogenesis. Defects were localized to tissues that normally express RDH10, and residual at-RA production was observed at embryonic day 9.5 (E9.5) in the neural ectoderm. *Rdh10*-null embryos were salvaged until birth with maternal at-RA supplementation, implying that functional RDH10 is not necessary for guiding the spatiotemporal distribution of at-RA that drives normal embryonic development (Sandell et al., 2007). It was later proposed that RDH10 is the primary driver of early embryonic at-RA production, as RDH10 can access latent pools of at-Rol in cellular membranes that are inaccessible to cytosolic ADHs (Farjo et al., 2011). It was initially thought that RDH1 also participates role in the oxidation of at-Rol to at-Ral, as RDH1 expression also occurs early in mouse embryonic development. The study of *Rdh1*-knockout mice revealed that RDH1 is dispensable for normal mouse embryonic development. *Rdh1*-null mice demonstrated no ostensible embryonic defects and had normal fertility, though they had increased adiposity and increased at-Rol and at-RE content in the liver when compared to age-matched WT mice (Zhang, Hu, Krois, Kane, & Napoli, 2007). Ablation of RDH1 was later demonstrated to lower mouse body temperature and decrease at-RA synthesis in brown adipose tissue that normally follows refeeding after a period of fasting; this decrease in at-RA synthesis led to metabolic dysregulation and hypertrophy of both brown adipose tissue and white adipose tissue in the *Rdh1*-null mice (Krois et al., 2019). Collectively, these studies in mutant mice have shed light on the complex series of redundancies and unique enzymatic functions in the rate-limiting oxidation of at-Rol to at-Ral, which is the first step in the generation of *at*-RA. With the advent of new single-cell-multiomic and cell lineage-tracing technologies, additional experiments will need to be done to understand why certain classes of ADHs acquired seemingly redundant *at*-Rol-oxidation functionalities, while other enzymes such as RDH10 acquired critical, non-redundant functionalities in development and beyond to adulthood.

Another enzyme, dehydrogenase/reductase superfamily member 3 (DHRS3), was shown to reduce *at*-Ral back to *at*-Rol as an indispensable means of preventing excess *at*-RA synthesis during embryonic development and into maturity, potentially in concert with RDH10 (Adams et al., 2014; Haeseleer, Huang, Lebioda, Saari, & Palczewski, 1998) (Fig. 1B). Generation and characterization of *Dhrs3*-null mice demonstrated that DHRS3 is crucial for normal embryonic development. *Dhrs3*-null mice displayed late embryonic and perinatal lethality due to cardiovascular, skeletal, and palatal malformations (Billings et al., 2013). These defects were attributed to a 40% increase in *at*-RA levels, with concurrent decreases in *at*-Rol/*at*-RA levels and a compensatory upregulation of *at*-RA-degrading CYP26A1. DHRS3 was also shown to be *at*-RA-inducible, suggesting that a negative feedback loop exists to limit the timeframe of *at*-RA production (Cerignoli et al., 2002; Zolfaghari, Chen, & Ross, 2012).

Following the production of *at*-Ral by RDHs and ADHs, *at*-Ral is further oxidized to *at*-RA by aldehyde dehydrogenases 1A1 (ALDH1A1, a.k.a. RALDH1), 1A2 (ALDH1A2, a.k.a. RALDH2), and 1A3 (ALDH1A3, a.k.a. RALDH3). These ALDH1A isotypes exhibit distinct and specific spatiotemporal expression patterns during embryonic development onward to adulthood, as determined using systemic immunolocalization of the three isotypes in sectioned mouse tissues at various embryonic and postnatal timepoints (Niederreither, Fraulob, Garnier, Chambon, & Dolle, 2002). In early mouse embryonic development, ALDH1A2 is the first of the three aldehyde dehydrogenases to be transcribed at E7.5 in the paraxial mesoderm, corresponding to the earliest detection of *at*-RA production, as seen with *at*-RA reporter assays (Balkan et al., 1992; Duester, 2008; Niederreither, Fraulob, et al., 2002; Niederreither, McCaffery, Drager, Chambon, & Dolle, 1997). *Aldh1a2*-null mice display embryonic lethality at E9.5–E10.5, with impaired axial rotation, shortening of the posterior region, no limb buds, anatomical heart defects, and a lack of neural tube closure (Niederreither, Subbarayan, Dolle, & Chambon, 1999). These defects corresponded to a widespread lack of embryonic *at*-RA synthesis and downstream *at*-RA-responsive gene transcription. Maternal administration of *at*-RA had variable degrees of impact on the development of *Aldh1a2*-null embryos, with some embryos displaying a near-WT developmental phenotype at 10 d.p.c., while others displayed little-to-no rescue of the *Aldh1a2*-null phenotype. Crossing *Aldh1a2*-null mice with a transgenic *at*-RA-dependent reporter (*RARE-lacZ*) mouse line revealed that embryonic *at*-RA production remained in the regions of embryos expressing ALDH1A1 or ALDH1A3 isotypes, independent of ALDH1A2 activity (Niederreither, Vermot, Fraulob, Chambon, & Dolle, 2002). More extensive characterization of *Aldh1a2*-null and *Aldh1a1/Aldh1a2*-null embryos rescued with *at*-RA supplementation revealed that ALDH1A2 normally generates the *at*-RA necessary to drive retinal invagination in the optic vesicle, which creates the optic cup during embryonic eye development (Mic, Molotkov, Fan, Cuenca, & Duester, 2000).

ALDH1A1 and ALDH1A3 display distinct expression patterns throughout mouse embryonic development to maturity and have an interesting pattern of expression in the fetal eye. Thus, ALDH1A1 is expressed in the developing dorsal retina, while ALDH1A3 is expressed in the developing ventral retina (Li et al., 2000; Mic et al., 2000; Niederreither, Fraulob, et al., 2002). These findings initially suggested that independent ALDH1A drivers of *at*-RA production were necessary for proper dorsoventral patterning of the fetal retina, but genetic ablation of *Aldh1a1* and *Aldh1a3* revealed that neither were essential for the dorsoventral patterning of the fetal retina (Dupe et al., 2003; Fan et al., 2003; Matt et al., 2005). Phenotyping of *Aldh1a3*-null mice revealed that the mice do not survive past birth due to blockage of nasal passages (choanal atresia), along with minor ocular-developmental defects, implicating ALDH1A3 as a critical factor in the fetal development of nasal passages but not as a critical driver of retinal development (Dupe et al., 2003; Molotkov, Molotkova, & Duester, 2006). This phenotype was rescued with maternal *at*-RA supplementation, suggesting that spatial localization of *at*-RA production is not necessary for fetal craniofacial/eye development. *Aldh1a1*-null mice, however, survived to adulthood with no congenital defects. Adult *Aldh1a1*-null mice had few phenotypic traits that differed from WT mice, aside from a dramatic reduction in liver *at*-RA production, which suggests that ALDH1A1 is more important in *at*-RA signaling or metabolism of *at*-Ral within the adult liver than for the fetal production of *at*-RA that is essential for development (Fan et al., 2003). Generation of *Aldh1a1/Aldh1a3* double-knockout mutant embryos revealed few phenotypic differences from *Aldh1a3*-null mice aside from an over-proliferation of the perioptic mesenchyme, which was determined to be a paracrine target of *at*-RA produced in the developing retina (Matt et al., 2005; Molotkov et al., 2006). Additional observations in double- and triple-null *Aldh1a1*, *Aldh1a2*, and *Aldh1a3* mice led to the conclusion that *at*-RA is necessary for eye-morphogenetic movements *via* paracrine signaling to the periocular mesenchyme during optic development, but is not necessary for dorsoventral patterning of the retina (Molotkov et al., 2006). Salvage of these mouse models of ALDH1A1, ALDH1A2, and ALDH1A3 deficiencies with maternal at-RA is consistent with observations of *Rdh10*-null embryos supplemented with at-RA, emphasizing the limited importance of spatial at-RA localization and delivery to fetal tissues during development. It's likely that other signaling pathways help to regulate which downstream at-RA response elements are activated to guide tissue-specific development, especially in the context of the embryonic/fetal eye.

Once at-RA is synthesized in the developing retina, it diffuses across cell membranes to the periocular mesenchyme, which is present from E10.5-E14.5 (Cvekl & Wang, 2009). In target tissues such as the periocular mesenchyme, at-RA binds to cellular retinoic acid-binding proteins (CRABP1, CRABP2). These two isotypes display unique expression patterns in embryonic and mature tissues, and purportedly serve to shuttle at-RA to downstream cytosolic and nuclear targets; in the case of CRABP2, binding of at-RA to apo-CRABP2 was shown to expose a nuclear localization sequence, thus facilitating import of the holo-CRABP2 into the nucleus to the nuclear at-RA receptors (RARs) (Sessler & Noy, 2005; Wei, 2016). It is hypothesized that CRABP1 primarily serves as a cytosolic sink for at-RA, while CRABP2 primarily serves as a cytosolic-nuclear shuttle for at-RA (Donovan, Olofsson, Gustafson, Dencker, & Eriksson, 1995). CRABP1 and CRABP2 transcripts were localized *via in situ* hybridization to the developing neural retina at E10.5, as RBP1 expression begins to emerge in both the neural retina and surrounding RPE progenitors (Perez-Castro, Toth-Rogler, Wei, & Nguyen-Huu, 1989). Neither CRABP1 nor CRABP2 were found to be required for embryonic development or maturation to adulthood in mice, implying that passive diffusion of at-RA across cell membranes (*e.g.*, from cytosol to nucleus) is sufficient to drive embryogenesis (Gorry et al., 1994; Lampron et al., 1995). Researchers studying CRABP-mutant mice did note minor congenital limb abnormalities due to CRABP1 and CRABP2 deletions, suggesting that CRABP isotypes have a minor role in controlling spatiotemporal localization of at-RA in the developing limbs, though additional studies need to be performed to clarify the mechanistic basis of such defects. Despite their apparent lack of necessity in mouse development, dysregulated CRABP expression and/or endogenous at-RA levels leads to downstream disruptions in cell cycle arrest- and cell survival-pathways in cancers, leading to increased resistance to *at*-RA-based chemotherapy, metastasis, and proliferation (Napoli, 2017; Tang & Gudas, 2011; Won et al., 2004; Wu, Lin, Tseng, Chen, & Wang, 2019). These findings establish CRABPs as important regulators of *at*-RA-dependent cell signaling in mature, actively dividing cells. Due to the post-mitotic nature of many cell types in the mature retina, it is unlikely that CRABPs play significant roles in regulating the cell cycle within photoreceptors or other neuronal cell types. These proteins, however, serve as subtle regulators of endogenous *at*-RA levels, thus controlling downstream transcription of *at*-RA-inducible genes.

In the absence of *at*-RA, retinoic-acid receptors (RARs) assemble into transcriptionally-repressive heterodimers with retinoid-X receptors (RXRs) at specific RAREs within the genome. Upon binding of *at*-RA to high-affinity RARs, the RAR-RXR heterodimer undergoes a conformational change, leading to the dissociation of affiliated nuclear receptor corepressor proteins (NCOR1 and NCOR2), and recruitment of transcriptional coactivators (NCOA1, NCOA2 and NCOA3), which initiates the downstream transcription of *at*-RA-responsive genes (Bastien & Rochette-Egly, 2004) (Fig. 1C). Unlike RARs, RXRs have the capacity to homodimerize at specific RAREs, and heterodimerize with other classes of nuclear receptors (*e.g.*, PPARs). RXRs have no affinity for *at*-RA (Blaner et al., 2016; Lefebvre, Benomar, & Staels, 2010; Vivat-Hannah, Bourguet, Gottardis, & Gronemeyer, 2003; Yu et al., 1991). Interestingly, they do have a high affinity for 9-*cis*-retinoic acid (9*cis*-RA), though this retinoid is only detectable *via* mass spectrometry in the mature pancreas (Heyman et al., 1992; Kane, 2012).

Single and combinatorial RAR- and RXR-null mice have been studied extensively. Genetic disruption of single RARA, RARB, and RARG isotypes in mice have indicated some functional redundancies in embryonic retinoic acid signaling. Though all single RAR-null mice are viable and can develop to adulthood, growth and anatomical defects were noted in some mice from each genotype (for review see (Mark, Ghyselinck, & Chambon, 2006, 2009)). Eliminating the entire *Rara* gene within mice led to testis degeneration and early postnatal lethality (Lufkin et al., 1993), while many *Rarb*-null mice exhibit persistence and hyperplasia of the primary vitreous body, implicating RARB in lens development within the eye (Ghyselinck et al., 1997). Several human subjects with change-of-function mutations in *RARB* manifested microphthalmia or anoptalmia with diaphragmatic hernia, suggesting that retinoic acid signaling through RARB has an even more prominent role in guiding eye development within humans (Srour et al., 2013). *Rarg*-null mice had growth defects, early postnatal lethality, and male sterility similar to *Rara*-null mice (Lohnes et al., 1993). Importantly, *Rarb* and *Rarg* are not expressed in the developing mouse retina (a neuroectoderm-derived cell population), but rather in the periocular mesenchyme, a neural-crest-derived cell population (Mori, Ghyselinck, Chambon, & Mark, 2001). A study using a neural-crest-specific ablation of *Rarb* and *Rarg* proved that the ocular defects observed in germline *Rarb-Rarg*-null mice indeed stemmed from *Rarb* and *Rarg* signaling insufficiency in the periocular mesenchyme (Matt et al., 2005). Combinatorial RAR knockout fetuses have much more severe abnormalities and do not survive past fetal development or birth (Mark et al., 2006, 2009). Generation and phenotyping of RXRA-null mutant mice revealed that disruption of RXRA resulted in embryonic lethality, with fetuses displaying consistent, severe disruptions in ocular and heart development along with poor tissue vascularization and edema, solidifying RXRA as an essential receptor for proper embryogenesis (Kastner et al., 1994). Morphological defects in the developing mouse eye are further exacerbated in *Rarb-Rxra*-null and *Rarg-Rxra*-null mice (Mark et al., 2006). All of these observations in RAR- and RXR-null mice emphasize the critical role that retinoic acid signaling through RAR-RXR receptor pairs plays in governing overall embryonic development and more specific patterning of ocular tissues.

As stated previously, *at*-RA production and downstream signaling is necessary for proper embryonic/fetal eye development; following production of *at*-RA in the periocular mesenchyme, first by ALDH1A2 and then by spatially-defined expression of ALDH1A1 (expressed in the dorsal retina) and ALDH1A3 (expressed in the ventral retina). The synthesized *at*-RA diffuses to the periocular mesenchyme where it signals through RXRA/RARB and RXRA/RARG heterodimers to control genes that dictate ventral retinal growth, anterior eye segment development, and periocular mesenchyme proliferation (Matt et al., 2005). Engagement of *at*-RA signaling pathways is tightly controlled both spatially and temporally, and there are *at*-RA-inducible cytochrome P450 family 26 (CYP26) genes that efficiently degrade *at*-RA to inactive metabolites such as 4-OH-*at*-RA, thereby halting downstream signaling. For example, localized CYP26A1 expression separates distinct regions of ALDH1A1 (dorsal)- and ALDH1A3 (ventral)-based *at*-RA production in the developing retina (Chithalen, Luu, Petkovich, & Jones, 2002; Sakai, Luo, McCaffery, Hamada, & Drager, 2004; Thatcher & Isoherranen, 2009). The horizontal stripe of CYP26A1 expression limits at-RA diffusion between the dorsal and ventral portions of the retina, and is further enhanced by expression of CYP26C1, which was found to emerge at later timepoints of fetal retinal development to further aid in the elimination of at-RA (Sakai et al., 2004). Despite this pronounced control of at-RA diffusion in the developing retina, generation and phenotyping of *Cyp26a1*-null mouse embryos revealed that CYP26A1 expression was not required for expression of dorsal and ventral retinal markers at E10.5, implying that it is not essential for the spatial patterning of the retina. The use of an at-RA-dependent reporter in the *Cyp26a1*-knockout embryos verified that there is a lack of a clear border between dorsal and ventral at-RA production in the *Cyp26a1*-null mice (Sakai et al., 2004). The authors of the study suggested that the zone lacking at-RA within the center of the retina is important in defining high-acuity areas of vision, though follow-up functional studies on mature *Cyp26a1*-null retinas are needed to test this hypothesis. This could be another example of functional redundancy in retinoid metabolism, where other CYP26 isotypes are capable of compensating for a lack of CYP26A1 activity during retinal development (Fig. 1D).

There is also a more recent, growing interest in the postnatal roles of at-RA in the mature retina. At-RA signaling in peripheral Müller glia was shown to prolong peripheral cone survival in the *rd1* mutant-mouse model of RP, which has deleterious mutations in PDE6B, a gene involved in the photoreceptor phototransduction cascade (Amamoto, Wallick, & Cepko, 2022). When at-RA production was downregulated in *rd1* mice through conditional knockout of the three ALDH1A isotypes, there was a significant diminution of peripheral cones, suggesting that at-RA signaling is essential for peripheral cone survival. To cross-validate these findings, they also employed the use of a constitutively active RARA-VP16-transactivator fusion construct to show that at-RA signaling independent of at-RA production was sufficient to increase cone survival in the central retina of *rd1* mice. The source of the at-RA driving signaling and downstream cone survival in the *rd1* mutant mice was then narrowed down to the peripheral Müller glia. These Müller glia cells express ALDH1A1, the likely source of the at-RA promoting peripheral cone survival. Comparison of *rd1* and *rd10* mice (which have a different pathogenic mutation in *Pde6b*) revealed that at-RA signaling through RARs induces photoreceptor hyperactivity in degenerating retinas (Telias et al., 2019). This hyperactivity was shown to reduce photoreceptor responses to light in the *rd10* mice and could be pharmacologically remedied by intravitreal injection with a RAR inverse agonist BMS 204493, or genetically remedied by substituting RARA with a truncated dominant-negative version of RARA that is constitutively repressed. These findings along with others that implicate at-RA signaling in scleral shaping of the eye and myopia highlight the importance of persistent at-RA production and signaling in the mature eye (Brown et al., 2023).

Overall, at-RA signaling has been shown to be critical for the normal development of the mammalian retina and other tissues of the eye, though it remains unclear how at-RA signaling interfaces with other developmental-signaling pathways during different stages of development to guide eye/retina morphogenesis, especially in the case of mice with specific defects in genes responsible for at-RA synthesis or downstream nuclear signaling. It is also unclear why there seems to be precise spatial and temporal distribution of these proteins in the pre- and post-natal retina, when broad maternal supplementation with at-RA appears to correct many of the physiological defects caused by null mutations in the at-RA-producing ALDH1A isotypes (Dupe et al., 2003; Molotkov et al., 2006). Precise control of at-RA degradation or at-RA signaling through RARs/RXRs in peripheral tissues (*e.g.* the periocular mesenchyme) appear to be more important in dictating stages of eye/ retinal development and morphogenesis than at-RA production, which does not seem to require tight control for proper eye development.

4. Retinaldehydes, retinyl esters, and retinols in vision

In the mature mammalian retina, at-Rol serves as a precursor to 11*cis*-Ral, which is the essential visual pigment that isomerizes to at-Ral in response to light stimuli, initiating phototransduction in both rod and cone photoreceptors. The process of generating 11*cis*-Ral from at-Rol begins when at-Rol is transported to the retina in the holo-RBP4-TTR complex in the bloodstream. Delivery of at-Rol to the retina is almost exclusively through the choroidal vasculature, which forms a dense network of blood vessels between the outer retina (*i.e.*, the RPE and photoreceptors) and the sclera (Lejoyeux et al., 2022; Nickla & Wallman, 2010). In the posterior pole of the eye, the choroidal vasculature is stratified into three distinct layers, with the innermost layer (*i.e.*, closest to the RPE) being a fenestrated capillary bed called the choriocapillaris (Morse, 1983). In the anterior pole of the eye, however, the choriocapillaris sits underneath only one layer of larger blood vessels (Hasegawa et al., 2007). The choriocapillaris delivers oxygen and nutrients, including at-Rol, from the bloodstream to the retina, which is one of the most metabolically demanding tissues of the human body (Hurley, 2021). Additionally, the choriocapillaris siphons waste products from the outer retina and acts as a heat sink to maintain the temperature of the retina (Lejoyeux et al., 2022; Nickla & Wallman, 2010). The holo-RBP4-TTR complex diffuses across the choriocapillaris endothelial cells and the extracellular matrix-rich Bruch's membrane to the basal side of the RPE monolayer in the retina. Together, these three layers form the blood-retinal barrier, which tightly controls the influx and egress of gasses, waste products, nutrients, and ions to-and-from the retina. Following diffusion to the basal side of the RPE, at-Rol dissociates from the holo-RBP4-TTR complex and is transported across the basal plasma membrane of RPE cells through STRA6, the essential at-Rol transporter that is highly expressed in mammalian RPE cells (see Fig. 2A).

Following entry of at-Rol into the RPE, it binds to RBP1, which serves as a cytosolic at-Rol reservoir and a shuttle to LRAT, another enzyme that is highly enriched within the endoplasmic reticulum (ER) membranes of the RPE (Herr & Ong, 1992; Sears & Palczewski, 2016) (Fig. 2B). Initially, it was unclear whether at-Rol must dissociate from holo-RBP1 prior to entry into the LRAT active site for esterification. Studies of liver microsomes containing LRAT clarified that the dissociation of at-Rol from RBP1 was not required for binding to LRAT, suggesting that holo-RBP1 directly associates with LRAT to transfer at-Rol and drive the formation of at-REs (Herr & Ong, 1992). The same comparative study established apo-RBP1 as an inhibitor of LRAT activity, revealing an additional facet of retinoid homeostasis that exists in both the liver and RPE. The esterification of at-Rol to at-REs in the RPE is essential for the entry of bloodstream-derived at-Rol into the visual cycle, which exists to generate the 11cis-Ral chromophore for light perception and to regenerate it from at-Ral following photoisomerization in the photoreceptors. Rbp1-null mice display decreased at-REs formation in the RPE, though they do not exhibit significant changes in the rate of 11cis-Ral regeneration following bleaching of photoreceptors with light. This lack of a change in visual cycle kinetics is expected, as retinoid diffusion likely serves as a rate-limiting step of the visual cycle (Kiser et al., 2014; Saari et al., 2002; Sears & Palczewski, 2016). In contrast, Lrat-null mice are almost completely devoid of retinyl ester formation in the retina and do not generate visual chromophore, exhibiting a retinal-degenerative phenotype characterized by progressive death of the photoreceptors and RPE cells (Batten et al., 2004).

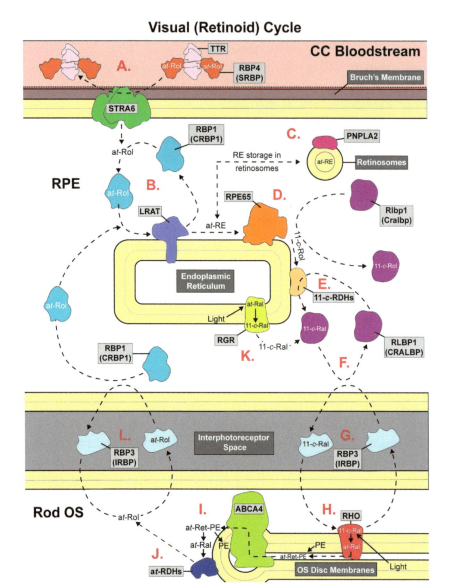

Fig. 2 Schematic diagram of the visual cycle and related retinoid transport/isomerization pathways in the mature retina. (A) Entry of RBP4-bound at-Rol into the RPE from the choriocapillaris (CC) *via* the STRA6 transporter. (B) Shuttling of at-Rol to LRAT for esterification by RBP1. (C) Stable storage of at-REs within self-aggregating retinosomes, which are re-mobilized by PNPLA2. (D) Isomerization and cleavage of at-REs by RPE65. (E) Oxidation of 11-c-Rol to 11-c-Ral by 11-c-RDHs. (F) Shuttling of 11-c-Ral in the cytosolic space by RLBP1 (CRALBP) and handoff at the apical membrane to RBP3. (G) Transport of 11-c-Ral across the interphotoreceptor space by RBP3. (H)

An even more severe retinal-dystrophic phenotype is seen in human patients with pathogenic mutations in LRAT (Den Hollander et al., 2008). Following the production of at-REs by LRAT from at-Rol and suitable fatty-acyl donors in the sn-1 position of phosphatidylcholines, the resulting at-REs are stored in retinosomes, which bud off from the ER (Imanishi, Gerke, & Palczewski, 2004) (Fig. 2C). These RPE-localized retinosomes serve as a stable reservoir of retinoids that are poised to re-enter the visual cycle and replenish 11*cis*-Ral levels in photoreceptors following sustained bleaching with light. The enzyme patatin-like phospholipase domain containing 2 (PNPLA2) mobilizes and hydrolyzes at-REs from retinosomes, releasing at-Rol in the RPE for re-entry into the visual cycle (Hara et al., 2023). Genetic ablation of PNPLA2 in mice revealed an accumulation of lipid droplets in the RPE and impaired visual function, supporting PNPLA2's role as a critical regulator of retinyl ester levels in the RPE. As the visual cycle progresses, these at-REs are transferred to another ER-localized enzyme in the RPE, RPE-specific 65 kDa protein (RPE65). This enzyme is robustly expressed in the RPE and has a crucial retinoid-isomerase enzymatic activity; RPE65 simultaneously removes the fatty-acyl chain from the at-REs and isomerizes the retinol polyene chain from all-*trans* to 11-*cis* to generate 11*cis*-Rol (Hamel et al., 1993; Jin, Li, Moghrabi, Sun, & Travis, 2005; Kiser et al., 2014; Moiseyev, Chen, Takahashi, Wu, & Ma, 2005) (Fig. 2D). Initial characterization of RPE65-knockout mice was done before the enzymatic activity of RPE65 was validated with *in vitro* expression experiments. Phenotyping of these knockout mice revealed progressive changes to photoreceptor and RPE morphology, over-accumulation of at-REs, no 11*cis*-Ral present in the retina, and absence of scotopic (rod photoreceptor-based) ERG responses (Jin et al., 2005; Moiseyev et al., 2005; Redmond et al., 2005; Redmond et al., 1998). It was thought that RPE65 also serves as a key enzyme in an additional visual cycle pathway that is hypothesized to exist between cone photoreceptors and Müller glia as an alternative means of regenerating visual chromophore for cones (Wang & Kefalov, 2011). Further phenotyping of mouse cones from different inbred backgrounds coupled with the characterization of

Photoisomerization of 11-c-Ral as 11-c-retinylidene in RHO to At-retinylidene, then ejection as at-Ral to the inner leaflet of OS discs. (I) Flipping of at-Ret-PEs by ABCA4 from the inner leaflet to the outer leaflet of OS disc membranes. (J) Reduction of at-Ral to at-Rol by OS-localized at-RDHs. (K) Bistable photoisomerization of at-Ral and 11-c-Ral by RGR in the RPE. (L) Shuttling of at-Rol to the RPE across the interphotoreceptor space by RBP3.

mouse cones using single-cell transcriptomics revealed that RPE65 expression in cones is dependent on the background strain of the mice (for review, see (Kiser, 2022)). Transgenic mice with artificial, cone-localized expression of RPE65 under a GRK1 (rhodopsin kinase)(Palczewski & Benovic, 1991) promoter revealed no differences in cone ERG responses that would reflect a RPE65-dependent augmentation in cone-specific 11*cis*-Ral regeneration (Kolesnikov, Tang, & Kefalov, 2018). Generation and characterization of a RPE65-Cre-ER fluorescent reporter mouse revealed that no cones actively express RPE65 (Choi et al., 2021). As such, physiologically-relevant RPE65 expression in mouse cones remains dubious, as do other key steps of the purported alternative visual cycle in mammalian cones and Müller glia. Both *Lrat*-null and *Rpe65*-null mice have been proposed as models of a childhood blinding retinopathy known as Leber Congenital Amaurosis (LCA), stemming from pathogenic mutations in LRAT and RPE65. Due to its severity, LCA has been of great interest to the field of translational vision research and gene-augmentation/gene-editing therapeutic development (Fan, Rohrer, Frederick, Baehr, & Crouch, 2008; Maguire et al., 2009; Suh et al., 2021).

11*cis*-Rol generated by RPE65 is then oxidized by 11*cis*-RDHs, expressed in the RPE, yielding the key visual chromophore, 11*cis*-Ral (Fig. 2E). Alternatively, 11*cis*-Rol can be bound by retinal-binding protein 1 (RLBP1; a.k.a. CRALBP), which shuttles 11*cis*-Rol in the cytosol of Müller glia and RPE (for review, see Saari & Crabb, 2005) (Fig. 2F). *In vivo* studies on *Rlbp1*-null mice demonstrate that deletion of RLBP1 activity delays dark adaptation and 11*cis*-Ral regeneration, although dark-adapted photosensitivity remains comparable with that of wild-type mice (Saari et al., 2001). Later phenotyping of Müller glia- and RPE-specific *Rlbp1* knock-out mice indicated that RPE-derived RLBP1 facilitates the efficient delivery of 11*cis*-Ral to the photoreceptors (Bassetto et al., 2024). The role of Müller glia-derived RLBP1 remains unclear but is likely involved in the efficient shuttling of retinoids from the outer retina to the intrinsically photosensitive retinal ganglion cells (ipRGCs) of the inner retina, as they require 11*cis*-Ral for non-vision-forming responses to light (Do & Yau, 2010).

RPE-localized oxidation of 11*cis*-Rol to 11*cis*-Ral is primarily accomplished by three independent enzymes: RDH5, RDH10, and RDH11. RDH5 is abundantly expressed in the RPE, and is thought to catalyze the oxidation of 11*cis*-Rol in a NAD^+-dependent manner (Simon, Hellman, Wernstedt, & Eriksson, 1995). Performing HPLC-based retinoid analyses on eyes from mice with disruptions to the *Rdh5* gene revealed that they have a pronounced over-accumulation of 13-*cis*-RE species due to inefficient oxidation of 11*cis*-Rol to 11*cis*-Ral; instead, the 11*cis*-Rol isomerizes to 13-*cis*-Rol, and is looped back to LRAT, where it's esterified and subsequently accumulates in the RPE (Driessen et al., 2000; Jang et al., 2001). This accumulation of 13-*cis*-RE species suggests that they are not viable substrates for RPE65 and thus cannot re-enter the visual cycle. The *Rdh5*-null mice still exhibited normal dark-adaptation kinetics after low levels of bleaching with light and they displayed delayed dark-adaptation after intense bleaching with light, suggesting that there are other, uncharacterized RDHs that also oxidizes 11*cis*-Rol. Pathogenic mutations in RDH5 are present in certain patients with *fundus albipunctatus*, a condition that typically impairs night vision and photoreceptor dark adaptation. This pathology is occasionally accompanied by macular degeneration and cone dystrophy (Nakamura, Hotta, Tanikawa, Terasaki, & Miyake, 2000; Sergouniotis et al., 2011).

RDH11 is also expressed in the RPE, and was shown *in vitro* to have dual substrate specificity for 11*cis*-Rol and a*t*-Rol isoforms in a NADP$^+$-dependent manner, unlike RDH5 which is 11-*cis*-Rol-specific (Haeseleer et al., 2002). This activity was assessed *in vivo via* generation and characterization of *Rdh11*-knockout and *Rdh5/Rdh11* double-knockout mice (Kim et al., 2005). While *Rdh11* single knockout mice had no significant differences in baseline retinoid content or photopic (cone-derived) and scotopic (rod-derived) ERG responses, *Rdh5/Rdh11* double-knockout mice exhibited an even more pronounced accumulation of 11-*cis*-REs than *Rdh5* single-knockout mice, suggesting that RDH11 has a complimentary role to RDH5 in the oxidation of 11*cis*-Rol. A lack of 11-*cis*- and 13-*cis*-Ral accumulation in the *Rdh11* single-knockout mice, however, suggests that the absence of RDH11 activity is fully compensated by other RDHs (*i.e.*, RDH5, RDH10). As mentioned previously, 11*cis*-Rol oxidation is also carried out by RDH10, which is expressed in the RPE and Müller cells of the mouse retina (Wu et al., 2002; Wu et al., 2004). *In vitro* studies with recombinantly expressed human RDH10 in COS-1 cells demonstrated that RDH10 utilizes 11*cis*-Rol as a substrate (Farjo, Moiseyev, Takahashi, Crouch, & Ma, 2009). Conditional knock-out of *Rdh10* in the RPE of mice resulted in delayed regeneration of 11*cis*-Ral following bleaching with light, implying that RDH10 indeed participates in oxidation of 11*cis*-Rol in the RPE (Sahu et al., 2015).

After RDH5, RDH10, and RDH11 co-operatively synthesize 11*cis*-Ral in the RPE, 11*cis*-Ral is bound by RBP3 (Rbp3; a.k.a. Irbp), which shuttles 11*cis*-Ral from RPE membranes to photoreceptor outer segment membranes across the extracellular space (Saari, Bredberg, & Noy, 1994) (Fig. 2G). In contrast to RLBP1, which is produced by RPE and Müller glial cells and confined to the cytosol of those cells, RBP3 is produced by photoreceptors and secreted into the interphotoreceptor matrix, ferrying retinoids such as 11*cis*-Ral exclusively between the RPE and photoreceptors. Generation and study of *Rbp3*-null mice revealed early postnatal decreases in photoreceptor nuclei and compromised photoreceptor outer segment morphology (Liou et al., 1998). Later studies revealed that cones from RBP3-deficient mice have chromophore deficiencies under photopic lighting conditions, implicating RBP3 in regenerating cone chromophore in the alternative visual cycle, which is thought to be distributed between cones and Müller glia (Parker, Wang, Kefalov, & Crouch, 2011; Parker, Fan, Nickerson, Liou, & Crouch, 2009). RBP3 also participates in pre- and post-natal maturation of photoreceptor outer segments, as its expression increases as cones and rods differentiate from photoreceptor progenitor populations in the embryonic retina, preceding the earliest expression of opsin transcripts and maturation of outer segments (Liou, Wang, & Matragoon, 1994). Additionally, RBP3 is implicated in ocular development, as a different *Rbp3*-deficient mouse line manifests extreme myopia, which recapitulates the phenotype of humans with RBP3 deficiencies (Arno et al., 2015; Wisard et al., 2011).

Upon being shuttled or passively diffusing to photoreceptor outer segment membranes from the RPE, 11*cis*-Ral is incorporated into rhodopsin (Rho)(in the case of rod photoreceptors) or into M- and/or S-opsins (in the case of murine cone photoreceptors) (Orban, Jastrzebska, & Palczewski, 2014; Palczewski, 2006, 2012) (Fig. 2H). For the sake of clarity, we choose to focus on the canonical rod opsin, Rho, which was the first of the visual opsins to have its chromophore characterized as 11*cis*-Ral by George Wald, who went on to win the Nobel Prize in Physiology and Medicine for his discoveries on opsins and their chromophore (Wald, 1933) (for a historic timeline of retinoid-handling gene characterization, see Table 1). 11*cis*-Ral forms a protonated Schiff-base bond with a Lys residue in ground-state Rho, priming Rho to respond to an incoming photon of light (Bownds, 1967; Hong & Palczewski, 2023). When a photon of light contacts the 11-*cis*-retinylidene chromophore within Rho, the 11-*cis* bond photoisomerizes to all-*trans*, initiating a conformational change in the opsin protein. This conformational change in Rho recruits the heterotrimeric protein transducin, which initiates a transient G-protein-coupled phototransduction cascade that

Table 1 Proteins involved in retinoid biology.

Mouse protein ID:	Human ortholog gene ID (OMIM):	Full gene name (HGNC):	Structure:	Biological/ biochemical function in the eye:	Expression profile:	Cell/tissue localization of protein:	Associated human diseases (OMIM IDs):	Transgenic mice characterized:
Abca4	ABCA4	ATP-binding cassette, subfamily A, member 4	PDB ID: 7LKP Organism: Human Cryo-EM 3.3 Å Citation: (Liu, Lee, & Chen, 2021)	ATP-dependent transporter that functions as a flippase, transporting all-*trans* retinyl-PE from the inner leaflet of rod disc membranes to the outer leaflet (Molday, Garces, Scortecci, & Molday, 2022).	Highly enriched in photoreceptors, marginally expressed in RPE (Molday et al., 2022).	Localized to cell membranes, particularly photoreceptor outer segment membranes (Molday et al., 2022).	Gene ID: *601691 Disease IDs: #153800 (*Macular degeneration, age-related, 2*) #604116 (*cone-rod dystrophy 3*) #248200 (*fundus flavimaculatus; retinal dystrophy, early-onset severe; Stargardt disease 1*) #601718 (*Retinitis pigmentosa 19*)	*Abca4* knockout mice (Weng et al., 1999; Zhang et al., 2015).
Aldh1a1/ Aldh1a2/ Aldh1a3	ALDH1A1/ ALDH1A2/ ALDH1A3	Aldehyde dehydrogenase family 1, subfamilies A1/A2/A3 (syn. Raldh1, 2, 3)	Aldh1a1: PDB ID: 1BXS Organism: Sheep X-Ray 2.35 Å Citation: (Moore et al., 1998) Aldh1a2: PDB ID: Organism: Rat X-Ray 2.7 Å Citation: (Lamb & Newcomer,	Participates in the cytosolic biosynthesis of at-RA acid from at-Rol (Duester, 1998, 2008)	Spatially patterned in developing retina: -Aldh1a1 highly expressed in dorsal retina of developing mouse embryos (Haselbeck, Hoffmann, & Duester, 1999). -Aldh1a2- widely expressed in developing embryos; deletion is embryonically lethal	Cytosolic.	Aldh1a1: Gene ID: *100640 Disease IDs: none identified. Aldh1a2: Gene ID: *603687 Disease IDs: #620025 (*Diaphragmatic hernia 4, with cardiovascular defects*).	*Aldh1a1* knockout mouse (Fan et al., 2003). *Aldh1a2* knockout mouse (Mic et al., 2004). *Aldh1a3* knockout mouse

(*continued*)

Table 1 Proteins involved in retinoid biology. (cont'd)

Mouse protein ID:	Human ortholog gene ID (OMIM):	Full gene name (HGNC):	Structure:	Biological/ biochemical function in the eye:	Expression profile:	Cell/tissue localization of protein:	Associated human diseases (OMIM IDs):	Transgenic mice characterized:
			1999) ALDH1A3: PDB ID: 5FHZ Organism: Human X-Ray 2.90 Å Citation: (Moretti et al., 2016).		(Niederreither, Subbarayan, Dolle, & Chambon, 1999). No optic cup formation observed in embryos (Mic, Molotkov, Molotkova, & Duester, 2004). -Aldh1a3 highly expressed in ventral retina of developing mouse embryos. Defects in Aldh1a3 lead to severe oculonasal developmental defects (Dupe et al., 2003).		Aldh1a3: Gene ID: *600463 Disease IDs: #615113 (*Microphthalmia, isolated 8*).	(Dupe et al., 2003).
Crabp1	CRABP1	Cellular retinoic acid- binding protein 1	PDB ID: 1CBR Organism: Murine X-Ray 2.9 Å Citation: (Kleywegt et al., 1994).	Binds cytosolic retinoic acid and ferries it to cytosolic/ non-nuclear target proteins; also serve as a cytosolic R.A. sink (Napoli, 2017).	Expression in neural retina of developing mouse embryo (Perez-Castro, Toth- Rogler, Wei, & Nguyen-Huu, 1989). Expressed in adult tissues such as thyroid and spleen.	Cytosolic.	Gene ID: *180230 Disease IDs: None identified.	*Crabp1* knockout mice (Gorry et al., 1994).

Crabp2	CRABP2	Cellular retinoic acid-binding protein 2	PDB ID: 1OPA Organism: Rat X-Ray 1.90 Å Citation: (Winter, Bratt, & Banaszak, 1993).	Binds cytosolic retinoic acid and ferries it to nuclear targets (RARs/RXRs).	Expression in neural retina of developing mouse embryo (Perez-Castro et al., 1989). Also expressed in the skin of adult mice (Giguere, Lyn, Yip, Siu, & Amin, 1990). Restricted expression in adult tissues.	Cytosolic/nuclear.	Gene ID: *180231 Disease IDs: None identified.	Crabp2 knockout mice (Lampron et al., 1995).
Cyp26a1/ Cyp26c1	CYP26A1/ CYP26C1	Cytochrome P450 family 26 subfamily A member 1 Cytochrome P450 family 26 subfamily C member 1	Cyp26a1: Structure unsolved. Cyp26c1: Structure unsolved.	Degrades retinoic acid, controls embryonic/developmental patterning (Isoherranen & Zhong, 2019; Sakai et al., 2001; Thatcher & Isoherranen, 2009).	Widely expressed in developing embryos; RA-inducible gene (9-cis RA seems to induce Cyp26a1 expression more potently than all-trans-RA). Cyp26a1: degrades retinoic acid in a horizontal region that extends across developing retina; Cyp26c1 expression emerges at a much later stage of development (Sakai, Luo, McCaffery, Hamada, & Drager, 2004). *Cyp26b1 was found to be expressed in RGCs.	Endoplasmic reticulum-associated.	Cyp26a1: Gene ID: *602239 Disease IDs: none identified. Cyp26b1: Gene ID: *605207 Disease IDs: #614416 (Craniosynostosis with radiohumeral fusions and other skeletal and craniofacial anomalies). Cyp26c1: Gene ID: *608428 Disease IDs: #614974 (Focal facial dermal dysplasia 4).	Cyp26 knockout mice (Sakai et al., 2001).

(continued)

Table 1 Proteins involved in retinoid biology. (cont'd)

Mouse protein ID:	Human ortholog gene ID (OMIM):	Full gene name (HGNC):	Structure:	Biological/biochemical function in the eye:	Expression profile:	Cell/tissue localization of protein:	Associated human diseases (OMIM IDs):	Transgenic mice characterized:
Lrat	LRAT	Lecithin:retinol acyltransferase	Structure unsolved.	Catalyzes the acyl transfer of fatty acids in the sn−1 position of phosphatidylcholine to at-Rol (Sears & Palczewski, 2016).	Highly enriched in RPE and HSCs (Sears & Palczewski, 2016).	Anchored to endoplasmic reticulum membranes (Sears & Palczewski, 2016).	Gene ID: *604863 Disease IDs: #613341 (*Leber congenital amaurosis 14; retinal dystrophy, early-onset severe; retinitis pigmentosa, juvenile*) (den Hollander, Roepman, Koenekoop, & Cremers, 2008).	Lrat knockout mouse line (Batten et al., 2004).
Pnpla2	PNPLA2	Patatin-like phospholipase domain-containing protein 2	Structure unsolved.	Mobilization of fatty acids and retinoids from triglyceride and retinosome stores, respectively. Possesses retinyl ester hydrolase activity (Hara et al., 2023; Zimmermann et al., 2004).	Highly enriched in adipose tissue, expressed in RPE (Hara et al., 2023; Zimmermann et al., 2004).	Cytosolic, lipid droplet/retinosome-associated (Hara et al., 2023; Zimmermann et al., 2004).	Gene ID: *609059 Disease IDs: #610717 Neutral lipid storage disease with myopathy (Fischer et al., 2007).	Pnpla2 knockout mice (Bernardo-Colon et al., 2023).

Rara	RARA	Retinoic acid receptor, alpha	PDB ID: 1DKF Organism: Human X-Ray 2.50 Å Citation: (Bourguet et al., 2000).	Nuclear hormone receptor isoform that dimerizes with RXRs to drive retinoic acid-dependent gene expression at retinoic acid response elements (RAREs); binds both all-*trans* and 9*cis*-RA.	Expressed in a controlled, cell-type specific manner within the developing eye: High expression in the RPE and mesenchyme at early developmental timepoints (E10.5-E14.5). Expression becomes virtually ubiquitous in the eye at E14.5 onwards to adult. Highest expression levels at E14.5-P10 embryonic neural retina/INL/ONL/bipolar cell layer (Lufkin et al., 1993).	Nuclear.	Gene ID: *180240 Disease IDs: #612376 (Leukemia, acute promyelocytic).	*Rara* knockout mice (Lufkin et al., 1993).
Rarb	RARB	Retinoic acid receptor, beta	PDB ID: 1XAP Organism: Human X-Ray 2.10 Å Citation: (Germain et al., 2004).	Nuclear hormone receptor isoform that dimerizes with RXRs to drive retinoic acid-dependent gene expression at retinoic acid response elements (RAREs); binds both at-RA and 9*cis*-RA.	Expressed in a controlled, cell-type specific manner within the developing eye: Highest expression in the mesenchyme at early developmental timepoints (E10.5-E14.5). Transient expression in RPE, choroid, sclera, embryonic neural retina/INL/ONL/bipolar cell layer E14.5- P4 (Mori, Ghyselinck, Chambon, & Mark, 2001).	Nuclear.	Gene ID: *180220 Disease IDs: #615524 (Microphthalmia, syndromic 12).	*Rarb* knockout mice (Luo, Pasceri, Conlon, Rossant, & Giguere, 1995).

(continued)

Table 1 Proteins involved in retinoid biology.—cont'd

Mouse protein ID:	Human ortholog gene ID (OMIM):	Full gene name (HGNC):	Structure:	Biological/biochemical function in the eye:	Expression profile:	Cell/tissue localization of protein:	Associated human diseases (OMIM IDs):	Transgenic mice characterized:
Rarg	RARG	Retinoic acid receptor, gamma	PDB ID: 2LBD Organism: Human X-Ray 2.06 Å Citation: (Renaud et al., 1995).	Nuclear hormone receptor isoform that dimerizes with RXRs to drive retinoic acid-dependent gene expression at retinoic acid response elements (RAREs); binds both all-*trans*- and 9*cis*-RA.	Highest expression in the mesenchyme at early developmental timepoints (E10.5–E14.5). Sustained expression in cornea, conjunctiva, lids, choroid, and sclera at later developmental timepoints into adulthood (Mori et al., 2001).	Nuclear.	Gene ID: *180190 Disease IDs: None identified.	*Rarg* knockout mice (Subbarayan et al., 1997).
Rbp1	RBP1 (alias: CRBP1)	Retinol-binding protein 1 (alias: cellular retinol-binding protein 1)	PDB ID: 1CRB Organism: Rat X-Ray 2.10 Å Citation: (Cowan, Newcomer, & Jones, 1993)	Assists in intracellular vitamin A (at-Rol)/ retinoid transport (Ghyselinck et al., 1999; Napoli & Yoo, 2020).	Widely expressed (Ghyselinck et al., 1999; Napoli & Yoo, 2020).	Cytosolic (Ghyselinck et al., 1999; Napoli & Yoo, 2020).	Gene ID: *180260 Disease IDs: None identified.	*Rbp1* knockout mouse line (Napoli & Yoo, 2020).
Rbp3	RBP3 (alias: IRBP)	Retinol-binding protein 3 (alias interphotoreceptor-binding protein)	PDB ID: 1J7X Organism: Xenopus X-Ray 1.80 Å Citation: (Loew & Gonzalez-Fernandez, 2002).	Dual retinal/retinol-binding protein that shuttles at-Rol or 11-*cis*- retinal between photoreceptors and the RPE in the canonical visual cycle (Qtaishat, Wiggert, & Pepperberg, 2005; Zeng et al., 2020).	Expressed in photoreceptors and secreted into interphotoreceptor space (Bunt-Milam & Saari, 1983; Zeng et al., 2020).	Soluble, intercellular/cytosolic protein (Loew & Gonzalez-Fernandez, 2002; Zeng et al., 2020).	Gene ID: *180290 Disease IDs: #615233? (retinitis pigmentosa 66). Severe myopia (Arno et al., 2015).	*Rbp3* knockout mice (Liou et al., 1998).

Rbp4	RBP4	Retinol-binding protein 4	PDB ID: 1RBP Organism: Human X-Ray 2.0 Å (Cowan, Newcomer, & Jones, 1990)	Acts as a carrier of all-*trans* retinol in complex with transthyretin (Ttr); transports at-Rol from liver to peripheral tissues, including the eye (Steinhoff, Lass, & Schupp, 2021).	Highly enriched in liver (Steinhoff et al., 2021).	Secreted and circulates in serum (Steinhoff et al., 2021).	Gene ID: *180250 Disease IDs: #615147 (*Retinal dystrophy, iris coloboma, and comedogenic acne syndrome*) #616428 (*Microphthalmia, Isolated, with Coloboma 10*) (Cukras et al., 2012; Seeliger et al., 1999)	*Rbp4* knockout mouse (Quadro et al., 1999).
Rdh10	RDH10	Retinol dehydrogenase 10	Structure unsolved.	Retinol dehydrogenase that is the predominant synthesizer of the all-*trans*-retinal required for at-RA synthesis during embryonic development (Sandell et al., 2007; Wu et al., 2002).	Widespread distribution in developing embryos (Sandell et al., 2007).	ER-associated.	Gene ID: *607599 Disease IDs: None identified.	*Rdh10* catalytically inactive mutant mice (Sandell et al., 2007).

(continued)

Table 1 Proteins involved in retinoid biology. (cont'd)

Mouse protein ID:	Human ortholog gene ID (OMIM):	Full gene name (HGNC):	Structure:	Biological/biochemical function in the eye:	Expression profile:	Cell/tissue localization of protein:	Associated human diseases (OMIM IDs):	Transgenic mice characterized:
Rdh5	RDH5	Retinol dehydrogenase 5	Structure unsolved.	Short-chain dehydrogenase/reductase; predominant enzyme that catalyzes the oxidation of 11-*cis*-retinol to 11-*cis*-retinal in the visual cycle. Also hypothesized to catalyze production of 9-*cis*-retinal from 9-*cis*-retinol (Jang et al., 2001; Simon, Hellman, Wernstedt, & Eriksson, 1995).	Enriched in the RPE (Jang et al., 2001; Simon et al., 1995).	Associated with endoplasmic reticulum membranes. (Jang et al., 2001; Simon et al., 1995).	Gene ID: *601617 Disease IDs: #136880 (*Fundus albipunctatus; retinitis punctata albescens*).	*Rdh5* knockout mouse line (Driessen et al., 2000).
Rdh8	RDH8	Retinol dehydrogenase 8	Structure unsolved.	Short-chain dehydrogenase/reductase; predominant enzyme that catalyzes the reduction of all-*trans*-retinal to at-Rol in the visual cycle (Maeda, Golczak, Maeda, & Palczewski, 2009; Rattner, Smallwood, & Nathans, 2000; Saari, Garwin, Van Hooser, & Palczewski, 1998).	Enriched in rod photoreceptor (Maeda et al., 2009; Rattner et al., 2000; Saari et al., 1998).	Localized to outer segment disc membranes in rod photoreceptors (Maeda et al., 2009; Rattner et al., 2000; Saari et al., 1998).	Gene ID: *608575 Disease IDs: None identified.	*Rdh8* knockout mouse line (Maeda et al., 2005).

Rgr	RGR	Retinal G-protein coupled receptor	Structure unsolved.	Bleaching G-protein coupled receptor that photoisomerizes all-*trans*- to 11-*cis*-retinal upon stimulation with green light (Jiang, Pandey, & Fong, 1993; Morshedian et al., 2019; Palczewski & Kiser, 2020; Tworak et al., 2023).	Highly enriched in RPE and Müller glia (Morshedian et al., 2019; Palczewski & Kiser, 2020; Tworak et al., 2023).	Localized to cell membranes (Morshedian et al., 2019; Palczewski & Kiser, 2020; Pandey, Blanks, Spee, Jiang, & Fong, 1994; Tworak et al., 2023).	Gene ID: *600342 Disease IDs: #613769 (*Retinitis pigmentosa 44*)	*Rgr* knockout mice (Chen et al., 2001).
Rho	RHO	Rhodopsin	PDB ID: 1F88 Organism: Bovine X-Ray 2.80 Å Citation: (Palczewski et al., 2000)	G-protein coupled receptor responsible for the detection of light stimulus using an 11-*cis*- retinal chromophore; photoisomerization of the 11-*cis*-retinal chromophore to all-*trans*- retinal kicks off the phototransduction cascade (Ernst et al., 2014; Palczewski, 2006).	Highly enriched in rod photoreceptor (Ernst et al., 2014; Palczewski, 2006).	Localized to outer segment disc membranes (Ernst et al., 2014; Palczewski, 2006).	Gene ID: *180380 Disease IDs: #610445 (*Night blindness, congenital stationary, autosomal dominant 1*). #613731 (*Retinitis pigmentosa 4, autosomal dominant or recessive*) #136880 (*Retinitis punctata albescens*)	*Rho* knockout mice (Humphries et al., 1997).

(*continued*)

Table 1 Proteins involved in retinoid biology. (cont'd)

Mouse protein ID:	Human ortholog gene ID (OMIM):	Full gene name (HGNC):	Structure:	Biological/biochemical function in the eye:	Expression profile:	Cell/tissue localization of protein:	Associated human diseases (OMIM IDs):	Transgenic mice characterized:
Rlbp1	RLBP1 (alias: CRALBP)	Retinaldehyde-binding protein 1 (alias cellular retinaldehyde-binding protein)	PDB ID: 3HY5 Organism: Human X-Ray 3.04 Å Citation: (He, Lobsiger, & Stocker, 2009).	Retinal-binding protein that binds and shuttles 11-*cis*-retinal intracellularly within the RPE and Müller glia (Saari, Bredberg, & Noy, 1994; Saari et al., 2001; Xue et al., 2015).	Enriched in the RPE and Müller glia (Bunt-Milam & Saari, 1983; Xue et al., 2015).	Cytosolic protein (Liu, Jenwitheesuk, Teller, & Samudrala, 2005).	Gene ID: *180090 Disease IDs: #607475 (*Bothnia retinal dystrophy*) #136880 (*Fundus albipunctatus*) #607476 (*Newfoundland rod-cone dystrophy*) #136880 (*Fundus albipunctatus; retinitis punctata albescens*).	*Rlbp1* knockout mouse line (Saari et al., 2001).
Rpe65	RPE65	Retinoid isomerohydrolase RPE65	PDB ID: 3FSN Organism: Bovine X-Ray 2.14 Å Citation: (Kiser, Golczak, Lodowski, Chance, & Palczewski, 2009)	Catalyzes the coupled isomerization and cleavage of all-*trans* retinyl esters to produce 11-*cis*-retinol and free fatty acids (Moiseyev, Chen, Takahashi, Wu, & Ma, 2005).	Highly enriched in the RPE (Hamel et al., 1993).	Associated with endoplasmic reticulum membrane within the RPE (Hamel et al., 1993; Redmond et al., 1998).	Gene ID: *180069 Disease IDs: #204100 (*Leber congenital amaurosis 2*) *613794 (*Retinitis pigmentosa 20*) *618697 (*Retinitis pigmentosa 87 with choroidal involvement*).	*Rpe65* knockout mouse line (Redmond et al., 1998).

Rxra	RXRA	Retinoid X receptor, alpha	PDB ID: 1G5Y Organism: Human X-Ray 2.00 Å Citation: (Gampe et al., 2000).	Nuclear hormone receptor isoform that dimerizes with RARs to drive retinoic acid-dependent gene expression at retinoic acid response elements (RAREs). Preferential binding of 9*cis*- RA. Also binds DHA (Cao et al., 2020).	Highest expression in the RPE and mesenchyme at early developmental timepoints (E10.5-E14.5). Sustained expression in cornea, conjunctiva, lids, choroid, and sclera at later developmental timepoints into adulthood. Transient expression in neural retina that is eliminated by adulthood (Mori et al., 2001).	Nuclear.	Gene ID: *180245 Disease IDs: None identified.	*Rxra* knockouts are embryonically lethal (Sucov et al., 1994).
Rxrb	RXRB	Retinoid X receptor, beta	PDB ID: 1H9U Organism: Human X-Ray 2.70 Å Citation: (Love et al., 2002)	Nuclear hormone receptor isoform that dimerizes with RARs to drive retinoic acid-dependent gene expression at retinoic acid response elements (RAREs). Preferential binding of 9*cis*-RA (Yu et al., 1991).	Ubiquitous expression (Mori et al., 2001).	Nuclear.	Gene ID: *180246 Disease IDs: None identified.	*Rxrb* knockout mice (Kastner et al., 1996).

(continued)

Table 1 Proteins involved in retinoid biology. (cont'd)

Mouse protein ID:	Human ortholog gene ID (OMIM):	Full gene name (HGNC):	Structure:	Biological/biochemical function in the eye:	Expression profile:	Cell/tissue localization of protein:	Associated human diseases (OMIM IDs):	Transgenic mice characterized:
Rxrg	RXRG	Retinoid X receptor, gamma	PDB ID: 2GL8 Organism: Human X-Ray 2.40 Å Citation: (Min).	Nuclear hormone receptor isoform that dimerizes with RARs to drive at-RA–dependent gene expression at retinoic acid response elements (RAREs). Preferential binding of 9*cis*-RA.	Highest expression in the developing neural retina from E10.5 onwards. Transient expression in early lens tissue (Mori et al., 2001).	Nuclear.	Gene ID: *180247 Disease IDs: None identified.	*Rxrg* knockout mice (Krezel et al., 1998).
Stra6	STRA6	Stimulated by retinoic acid receptor 6 STRA6	PDB ID: 5SY1 Organism: Zebrafish Cryo-EM 3.90 Å (Chen et al., 2016)	Assists in at-Rol transport across plasma membranes in many cell types; essential for retinol uptake into the mature eye from serum (Chen & Heller, 1977; Heller, 1975; Kawaguchi et al., 2007).	Enriched in mature mouse brain, kidney, spleen, female genital tract; highly enriched in the RPE, choroid plexus, and Sertoli cells of the testis (Bouillet et al., 1997).	Localized to plasma membrane (Chen & Heller, 1977; Heller, 1975; Kawaguchi et al., 2007)	Gene ID: *610745 Disease IDs: #601186 (*Microphthalmia, isolated, with coloboma 8*; *Microphthalmia, syndromic 9*; alias: Matthew–Wood Syndrome)(Golzio et al., 2007).	*Stra6* knockout mice lines (Amengual et al., 2014; Ruiz et al., 2012).

leads to a hyperpolarization of the rod photoreceptors. This hyperpolarization leads to a relay of signals through other neuronal cell types in the retina; the signals are eventually integrated and transmitted to different regions of the brain responsible for processing visual stimuli (Fu & Yau, 2007). Cutting-edge native mass spectrometry conducted on bovine Rho and phototransduction proteins in native membranes has enabled profiling of native membrane compositions and their effects of on the rates of individual steps in the Rho phototransduction cascade (Chen et al., 2022).

The same general stepwise mechanism of opsin activation, isomerization of the 11-*cis*-retinylidene chromophore to all-*trans*-retinylidene, and initiation of the phototransduction cascade is conserved in cone photoreceptors, although there are key physiological differences between rod and cone responses to light stimuli (Fain & Sampath, 2021). Initial generation and characterization of *Rho*-null mice revealed a progressive and almost complete loss of photoreceptors in homozygous mice by three months of age (Humphries et al., 1997). Prior to the degeneration of the photoreceptor layer, there are disruptions to inner- and outer-segment morphology. ERGs of slightly younger, 8-week-old *Rho*-null animals revealed no scotopic (rod-derived) responses. Heterozygous mice exhibited less prominent photoreceptor loss, but still displayed disruptions to inner- and outer-segment morphology. A spectrum of mutations in the *RHO* gene have been identified in human patients with autosomal dominant RP. One of the most prominent mutations known to cause autosomal dominant RP is the Pro23His (P23H) Rho mutation, which has been introduced into a variety of animals for *in vivo* studies (Olsson et al., 1992; Ross et al., 2012; Sung et al., 1991). In-depth characterization of mice with fluorescently-tagged heterozygous and homozygous P23H-*Rho* mutant alleles revealed that the mutant RHO mislocalizes to rod inner segments, leading to an abnormal expansion of endoplasmic reticulum membranes and disruption of inner- and outer-segment morphology (Robichaux et al., 2022).

In normal, functioning rods following photoisomerization of the 11-*cis* bond, the all-*trans*-retinylidene adduct on the lysine residue of RHO is hydrolyzed and the *at*-Ral is released from RHO into the inner leaflet of the rod outer segment disc membranes. There, the *at*-Ral spontaneously forms all-*trans*-N-retinylidene-phosphatidylethanolamines with phosphatidylethanolamines that exist within the rod disc membranes. The rates of chromophore hydrolysis and release from bovine Rho, and subsequent formation of all-*trans*-N-retinylidene-phosphatidylethanolamine have been characterized in native membranes, using a combination of borohydride reduction, proteolysis, precipitation, and mass spectrometry (Hong et al., 2022). Efficient flipping of all-*trans*-N-retinylidene-phosphatidylethanolamines from the inner leaflet to the outer leaflet of disc membranes is essential for progression of the visual cycle and timely regeneration of 11*cis*-Ral (Anderson & Maude, 1970; Sparrow, Wu, Kim, & Zhou, 2010). This flippase activity is accomplished in an ATP-dependent manner by ATP-binding cassette transporter 4 (ABCA4), a large, 220-kDa protein localized to rod disc outer segment rims and the RPE (Illing, Molday, & Molday, 1997; Lenis et al., 2018; Molday, Zhong, & Quazi, 2009; Sun & Nathans, 1997) (Fig. 2I). Without functional ABCA4, *at*-Ral and all-*trans*-N-retinylidene-phosphatidylethanolamine accumulate in the inner leaflet of disc membranes and form toxic pyridinium bisretinoids including A2E, which progressively poison the RPE cells as they engulf and break down old outer segments, resulting in the death of both RPE and photoreceptor cells. This accumulation of toxic bisretinoid species manifests in a progressive retinal dystrophy in human patients known as Stargardt-1 disease (Tanna, Strauss, Fujinami, & Michaelides, 2017). *Abca4*-null mice display delayed dark adaptation after bleaching with light, a distinct accumulation of *at*-Ral and all-*trans*-N-retinylidene-phosphatidyletha-nolamine species in the retina, and a larger, progressive accumulation of bisretinoids such as A2E in the RPE of the *Abca4* mutants as a function of age. Researchers studying this initial *Abca4*-null mouse strain also noted dense melanosome or melanosome-phagosome fusion structures in the RPE and a thickening of Bruch's membrane. These cumulative changes recapitulate the A2E-rich lipofuscin accumulation seen in the RPEs of human patients with Stargardt-1 disease or age-related macular degeneration (AMD) (Eldred & Lasky, 1993). An additional *Abca4*-null mouse strain has been generated with a pair of proline-to-valine point mutations, mimicking mutations that were observed in human patients with a particularly severe Stargardt-1 phenotype (Zhang et al., 2015). Despite these two mutations causing ABCA4 misfolding and subsequent degradation in the photoreceptors, these mice exhibited few phenotypic differences compared to *Abca4*-null mice that completely lack expression of the ABCA4 protein. These data suggest that mouse photoreceptors and RPE are more resilient to misfolded ABCA4 production than their human counterparts. More extensive comparison of the two *Abca4*-mutant mouse strains revealed that the two stains exhibited differential degrees of responses to drugs targeting toxic ceramide accumulation and immune signaling through CCR5 (Engfer et al., 2023).

After ABCA4 flips all-*trans*-N-retinylidene-phosphatidylethanolamine from the inner leaflet of rod discs to the outer leaflet, *at*-Ral can be released from the phosphatidylethanolamine moieties by spontaneous hydrolysis and reduced to a*t*-Rol by RDHs present in the rod outer segments, primarily by the photoreceptor-specific RDH8 (Fig. 2J). Genetic deletion of RDH8 activity in mice had no apparent effect on photoreceptor and inner retinal morphology but resulted in an overaccumulation of a*t*-Ral in the retina following bleaching with light. Deletion of RDH8 activity also resulted in delayed rod dark adaptation, which was profiled with ERG (Maeda et al., 2005). This accumulation of a*t*-Ral led to slight increases in A2E levels when compared to age-matched wild-type controls. The maintenance of 11*cis*-Ral regeneration in the retinas of *Rdh8*-knockout mice, however, suggests that RDH8 is not essential for visual cycle function, and that reduction of a*t*-Ral can be carried out by other RDHs present in the photoreceptors, such as RDH12 (expressed in photoreceptor inner segments), which was demonstrated to protect mouse photoreceptors against light-induced degeneration (Maeda et al., 2006). Interestingly, a different, non-visual opsin called RGR-opsin (RGR) expressed in RPE and Müller glia is capable of directly regenerating 11*cis*-Ral from a*t*-Ral upon stimulation with green wavelengths of light (Fig. 2K). This alternative, photoisomerization-based mechanism of 11*cis*-Ral regeneration provides a means of rapidly clearing excess a*t*-Ral from RPE and Müller glia and rapidly producing visual chromophore for use by photoreceptors (Morshedian et al., 2019; Tworak et al., 2023).

Once a*t*-Ral has been reduced to a*t*-Rol, the alcohol then freely diffuses through the interphotoreceptor matrix or is actively shuttled back to the apical RPE surface by RBP3 (Fig. 2L). Following diffusion/ transport back into the RPE, a*t*-Rol then is re-esterified by LRAT (Fig. 2B) for stable storage in retinosomes (Fig. 2C) or begins another round of regenerating of the 11*cis*-Ral chromophore (Fig. 2D–H).

5. Conclusion

In summary, retinoids such as a*t*-Rol, 11*cis*-Ral, and a*t*-RA are essential for development and function of the retina of the eye. A host of different retinoid-handling proteins are responsible for salvaging retinoids from the diet, stably storing retinoids for controlled release in the bloodstream, light perception, and regulating a*t*-RA signaling during development, all of which operate in concert to maintain critical steady-state levels of retinoids in the gut, liver, bloodstream, and retina. The balance among these processes is critical in protecting the retina from morphological defects in the case of dysregulated *at*-RA production/signaling, and from vision loss and cell death in the case of dysregulated retinoid transport to the eye and retinoid flux through the visual cycle. Mutant mice with specific deletions in retinoid-handling genes have played an essential role in uncovering the biochemical basis of retinoid interconversion within different tissues. As many retinoid-related genes display a high degree of functional conservation between humans and mice, there is ample rationale for the continued use of modified mice to study human retinal degenerative diseases stemming from altered retinoid flux in the retina.

Acknowledgments

We would like to thank Drs. Philip Kiser, Johannes von Lintig, T. Michael Redmond, and Alexander R. Moise for helpful comments on this review. We also appreciate the insight on retinoid biochemistry from members of the Palczewski laboratory. This work was supported in part by grants from the National Institutes of Health, including R01EY009339 (K.P.), R01EY030873 (K.P.), and NIH training grant 5F31EY03402702 for Z.J.E. The authors acknowledge support to the Department of Ophthalmology Gavin Herbert Eye Institute at the University of California, Irvine from an unrestricted Research to Prevent Blindness award, from NIH core grant P30EY034070.

Author Contributions

Z.J.E. and K.P. contributed to the design and overall scope of the manuscript contents and figures.

Competing Interest Statement

KP is a consultant for Polgenix Inc., and serves on the Scientific Advisory Board at Hyperion Eye Ltd. All other authors have declared no conflict of interest exists.

References

Adams, M. K., Belyaeva, O. V., Wu, L., Chaple, I. F., Dunigan-Russell, K., Popov, K. M., & Kedishvili, N. Y. (2021). Characterization of subunit interactions in the hetero-oligomeric retinoid oxidoreductase complex. *The Biochemical Journal*, 478(19), 3597–3611. https://doi.org/10.1042/BCJ20210589.

Adams, M. K., Belyaeva, O. V., Wu, L., & Kedishvili, N. Y. (2014). The retinaldehyde reductase activity of DHRS3 is reciprocally activated by retinol dehydrogenase 10 to control retinoid homeostasis. *The Journal of Biological Chemistry*, 289(21), 14868–14880. https://doi.org/10.1074/jbc.M114.552257.

Alvarez, R., Vaz, B., Gronemeyer, H., & De Lera, A. R. (2014). Functions, therapeutic applications, and synthesis of retinoids and carotenoids. *Chemical Reviews*, 114(1), 1–125. https://doi.org/10.1021/cr400126u.

Amamoto, R., Wallick, G. K., & Cepko, C. L. (2022). Retinoic acid signaling mediates peripheral cone photoreceptor survival in a mouse model of retina degeneration. *Elife*, 11. https://doi.org/10.7554/eLife.76389.

Amengual, J., Zhang, N., Kemerer, M., Maeda, T., Palczewski, K., & Von Lintig, J. (2014). STRA6 is critical for cellular vitamin A uptake and homeostasis. *Human Molecular Genetics*, 23(20), 5402–5417. https://doi.org/10.1093/hmg/ddu258.

Anderson, R. E., & Maude, M. B. (1970). Phospholipids of bovine outer segments. *Biochemistry, 9*(18), 3624–3628. https://doi.org/10.1021/bi00820a019.

Andre, A., Ruivo, R., Gesto, M., Castro, L. F., & Santos, M. M. (2014). Retinoid metabolism in invertebrates: When evolution meets endocrine disruption. *General and Comparative Endocrinology, 208*, 134–145. https://doi.org/10.1016/j.ygcen.2014.08.005.

Ang, H. L., Deltour, L., Hayamizu, T. F., Zgombic-Knight, M., & Duester, G. (1996). Retinoic acid synthesis in mouse embryos during gastrulation and craniofacial development linked to class IV alcohol dehydrogenase gene expression. *The Journal of Biological Chemistry, 271*(16), 9526–9534. https://doi.org/10.1074/jbc.271.16.9526.

Arno, G., Hull, S., Robson, A. G., Holder, G. E., Cheetham, M. E., Webster, A. R., ... Moore, A. T. (2015). Lack of interphotoreceptor retinoid binding protein caused by homozygous mutation of RBP3 is associated with high myopia and retinal dystrophy. *Investigative Ophthalmology & Visual Science, 56*(4), 2358–2365. https://doi.org/10.1167/iovs.15-16520.

Balkan, W., Colbert, M., Bock, C., & Linney, E. (1992). Transgenic indicator mice for studying activated retinoic acid receptors during development. *Proceedings of the National Academy of Sciences of the United States of America, 89*(8), 3347–3351. https://doi.org/10.1073/pnas.89.8.3347.

Bassetto, M., Kolesnikov, A. V., Lewandowski, D., Kiser, J. Z., Halabi, M., Einstein, D. E., ... Kiser, P. D. (2024). Dominant role for pigment epithelial CRALBP in supplying visual chromophore to photoreceptors. *Cell Reports, 43*(5), 114143. https://doi.org/10.1016/j.celrep.2024.114143.

Bastien, J., & Rochette-Egly, C. (2004). Nuclear retinoid receptors and the transcription of retinoid-target genes. *Gene, 328*, 1–16. https://doi.org/10.1016/j.gene.2003.12.005.

Batten, M. L., Imanishi, Y., Maeda, T., Tu, D. C., Moise, A. R., Bronson, D., ... Palczewski, K. (2004). Lecithin-retinol acyltransferase is essential for accumulation of all-trans-retinyl esters in the eye and in the liver. *The Journal of Biological Chemistry, 279*(11), 10422–10432. https://doi.org/10.1074/jbc.M312410200.

Bavik, C., Henry, S. H., Zhang, Y., Mitts, K., McGinn, T., Budzynski, E., ... Kubota, R. (2015). Visual cycle modulation as an approach toward preservation of retinal integrity. *PLoS One, 10*(5), e0124940. https://doi.org/10.1371/journal.pone.0124940.

Bernardo-Colon, A., Dong, L., Abu-Asab, M., Brush, R. S., Agbaga, M. P., & Becerra, S. P. (2023). Ablation of pigment epithelium-derived factor receptor (PEDF-R/Pnpla2) causes photoreceptor degeneration. *Journal of Lipid Research, 64*(5), 100358. https://doi.org/10.1016/j.jlr.2023.100358.

Billings, S. E., Pierzchalski, K., Butler Tjaden, N. E., Pang, X. Y., Trainor, P. A., Kane, M. A., & Moise, A. R. (2013). The retinaldehyde reductase DHRS3 is essential for preventing the formation of excess retinoic acid during embryonic development. *The FASEB Journal, 27*(12), 4877–4889. https://doi.org/10.1096/fj.13-227967.

Blaner, W. S., Dixon, J. L., Moriwaki, H., Martino, R. A., Stein, O., Stein, Y., & Goodman, D. S. (1987). Studies on the in vivo transfer of retinoids from parenchymal to stellate cells in rat liver. *European Journal of Biochemistry/FEBS, 164*(2), 301–307. https://doi.org/10.1111/j.1432-1033.1987.tb11058.x.

Blaner, W. S., Hendriks, H. F., Brouwer, A., de Leeuw, A. M., Knook, D. L., & Goodman, D. S. (1985). Retinoids, retinoid-binding proteins, and retinyl palmitate hydrolase distributions in different types of rat liver cells. *Journal of Lipid Research, 26*(10), 1241–1251.

Blaner, W. S., Li, Y., Brun, P. J., Yuen, J. J., Lee, S. A., & Clugston, R. D. (2016). Vitamin A absorption, storage and mobilization. *Sub-Cellular Biochemistry, 81*, 95–125. https://doi.org/10.1007/978-94-024-0945-1_4.

Blomhoff, R., Helgerud, P., Rasmussen, M., Berg, T., & Norum, K. R. (1982). In vivo uptake of chylomicron [3H]retinyl ester by rat liver: Evidence for retinol transfer from parenchymal to nonparenchymal cells. *Proceedings of the National Academy of Sciences of the United States of America, 79*(23), 7326–7330. https://doi.org/10.1073/pnas.79.23.7326.

Boleda, M. D., Saubi, N., Farres, J., & Pares, X. (1993). Physiological substrates for rat alcohol dehydrogenase classes: Aldehydes of lipid peroxidation, omega-hydroxyfatty acids, and retinoids. *Archives of Biochemistry and Biophysics, 307*(1), 85–90. https://doi.org/10.1006/abbi.1993.1564.

Borel, P., Lietz, G., Goncalves, A., Szabo de Edelenyi, F., Lecompte, S., Curtis, P., ... Reboul, E. (2013). CD36 and SR-BI are involved in cellular uptake of provitamin A carotenoids by Caco-2 and HEK cells, and some of their genetic variants are associated with plasma concentrations of these micronutrients in humans. *The Journal of Nutrition, 143*(4), 448–456. https://doi.org/10.3945/jn.112.172734.

Bouillet, P., Oulad-Abdelghani, M., Vicaire, S., Garnier, J. M., Schuhbaur, B., Dolle, P., & Chambon, P. (1995). Efficient cloning of cDNAs of retinoic acid-responsive genes in P19 embryonal carcinoma cells and characterization of a novel mouse gene, Stra1 (mouse LERK-2/Eplg2). *Developmental Biology, 170*(2), 420–433. https://doi.org/10.1006/dbio.1995.1226.

Bouillet, P., Sapin, V., Chazaud, C., Messaddeq, N., Decimo, D., Dolle, P., & Chambon, P. (1997). Developmental expression pattern of Stra6, a retinoic acid-responsive gene encoding a new type of membrane protein. *Mechanisms of Development, 63*(2), 173–186. https://doi.org/10.1016/s0925-4773(97)00039-7.

Bourguet, W., Vivat, V., Wurtz, J. M., Chambon, P., Gronemeyer, H., & Moras, D. (2000). Crystal structure of a heterodimeric complex of RAR and RXR ligand-binding domains. *Molecular Cell, 5*(2), 289–298. https://doi.org/10.1016/s1097-2765(00)80424-4.

Bownds, D. (1967). Site of attachment of retinal in rhodopsin. *Nature, 216*(5121), 1178–1181. https://doi.org/10.1038/2161178a0.

Brown, D. M., Yu, J., Kumar, P., Paulus, Q. M., Kowalski, M. A., Patel, J. M., ... Pardue, M. T. (2023). Exogenous all-trans retinoic acid induces myopia and alters scleral biomechanics in mice. *Investigative Ophthalmology & Visual Science, 64*(5), 22. https://doi.org/10.1167/iovs.64.5.22.

Bunt-Milam, A. H., & Saari, J. C. (1983). Immunocytochemical localization of two retinoid-binding proteins in vertebrate retina. *The Journal of Cell Biology, 97*(3), 703–712. https://doi.org/10.1083/jcb.97.3.703.

Cao, H., Li, M. Y., Li, G., Li, S. J., Wen, B., Lu, Y., & Yu, X. (2020). Retinoid X receptor alpha regulates DHA-dependent spinogenesis and functional synapse formation in vivo. *Cell Reports, 31*(7), 107649. https://doi.org/10.1016/j.celrep.2020.107649.

Carazo, A., Macakova, K., Matousova, K., Krcmova, L. K., Protti, M., & Mladenka, P. (2021). Vitamin A update: Forms, sources, kinetics, detection, function, deficiency, therapeutic use and toxicity. *Nutrients, 13*(5), https://doi.org/10.3390/nu13051703.

Cerignoli, F., Guo, X., Cardinali, B., Rinaldi, C., Casaletto, J., Frati, L., ... Giannini, G. (2002). retSDR1, a short-chain retinol dehydrogenase/reductase, is retinoic acid-inducible and frequently deleted in human neuroblastoma cell lines. *Cancer Research, 62*(4), 1196–1204.

Chang, B., Hawes, N. L., Hurd, R. E., Davisson, M. T., Nusinowitz, S., & Heckenlively, J. R. (2002). Retinal degeneration mutants in the mouse. *Vision Research, 42*(4), 517–525. https://doi.org/10.1016/s0042-6989(01)00146-8.

Chen, C. C., & Heller, J. (1977). Uptake of retinol and retinoic acid from serum retinol-binding protein by retinal pigment epithelial cells. *The Journal of Biological Chemistry, 252*(15), 5216–5221.

Chen, P., Hao, W., Rife, L., Wang, X. P., Shen, D., Chen, J., ... Fong, H. K. (2001). A photic visual cycle of rhodopsin regeneration is dependent on Rgr. *Nature Genetics, 28*(3), 256–260. https://doi.org/10.1038/90089.

Chen, S., Getter, T., Salom, D., Wu, D., Quetschlich, D., Chorev, D. S., ... Robinson, C. V. (2022). Capturing a rhodopsin receptor signalling cascade across a native membrane. *Nature, 604*(7905), 384–390. https://doi.org/10.1038/s41586-022-04547-x.

Chen, Y., Clarke, O. B., Kim, J., Stowe, S., Kim, Y. K., Assur, Z., ... Mancia, F. (2016). Structure of the STRA6 receptor for retinol uptake. *Science (New York, N. Y.), 353*(6302), https://doi.org/10.1126/science.aad8266.

Chithalen, J. V., Luu, L., Petkovich, M., & Jones, G. (2002). HPLC-MS/MS analysis of the products generated from all-trans-retinoic acid using recombinant human CYP26A. *Journal of Lipid Research, 43*(7), 1133–1142. https://doi.org/10.1194/jlr.m100343-jlr200.

Choi, E. H., Suh, S., Einstein, D. E., Leinonen, H., Dong, Z., Rao, S. R., ... Kiser, P. D. (2021). An inducible Cre mouse for studying roles of the RPE in retinal physiology and disease. *JCI Insight, 6*(9), https://doi.org/10.1172/jci.insight.146604.

Chou, C. M., Nelson, C., Tarle, S. A., Pribila, J. T., Bardakjian, T., Woods, S., ... Glaser, T. (2015). Biochemical basis for dominant inheritance, variable penetrance, and maternal effects in RBP4 congenital eye disease. *Cell, 161*(3), 634–646. https://doi.org/10.1016/j.cell.2015.03.006.

Clagett-Dame, M., & Knutson, D. (2011). Vitamin A in reproduction and development. *Nutrients, 3*(4), 385–428. https://doi.org/10.3390/nu3040385.

Cohlan, S. Q. (1954). Congenital anomalies in the rat produced by excessive intake of vitamin A during pregnancy. *Pediatrics, 13*(6), 556–567.

Cooper, A. D. (1997). Hepatic uptake of chylomicron remnants. *Journal of Lipid Research, 38*(11), 2173–2192.

Cowan, S. W., Newcomer, M. E., & Jones, T. A. (1990). Crystallographic refinement of human serum retinol binding protein at 2A resolution. *Proteins, 8*(1), 44–61. https://doi.org/10.1002/prot.340080108.

Cowan, S. W., Newcomer, M. E., & Jones, T. A. (1993). Crystallographic studies on a family of cellular lipophilic transport proteins. Refinement of P2 myelin protein and the structure determination and refinement of cellular retinol-binding protein in complex with all-trans-retinol. *Journal of Molecular Biology, 230*(4), 1225–1246. https://doi.org/10.1006/jmbi.1993.1238.

Cukras, C., Gaasterland, T., Lee, P., Gudiseva, H. V., Chavali, V. R., Pullakhandam, R., ... Ayyagari, R. (2012). Exome analysis identified a novel mutation in the RBP4 gene in a consanguineous pedigree with retinal dystrophy and developmental abnormalities. *PLoS One, 7*(11), e50205. https://doi.org/10.1371/journal.pone.0050205.

Cvekl, A., & Wang, W. L. (2009). Retinoic acid signaling in mammalian eye development. *Experimental Eye Research, 89*(3), 280–291. https://doi.org/10.1016/j.exer.2009.04.012.

Deltour, L., Foglio, M. H., & Duester, G. (1999a). Impaired retinol utilization in Adh4 alcohol dehydrogenase mutant mice. *Developmental Genetics, 25*(1), 1–10. https://doi.org/10.1002/(SICI)1520-6408(1999)25:1 < 1::AID-DVG1 > 3.0.CO;2-W.

Deltour, L., Foglio, M. H., & Duester, G. (1999b). Metabolic deficiencies in alcohol dehydrogenase Adh1, Adh3, and Adh4 null mutant mice. Overlapping roles of Adh1 and Adh4 in ethanol clearance and metabolism of retinol to retinoic acid. *The Journal of Biological Chemistry, 274*(24), 16796–16801. https://doi.org/10.1074/jbc.274.24.16796.

Den Hollander, A. I., Roepman, R., Koenekoop, R. K., & Cremers, F. P. (2008). Leber congenital amaurosis: Genes, proteins and disease mechanisms. *Progress in Retinal and Eye Research, 27*(4), 391–419. https://doi.org/10.1016/j.preteyeres.2008.05.003.

Do, M. T., & Yau, K. W. (2010). Intrinsically photosensitive retinal ganglion cells. *Physiological Reviews, 90*(4), 1547–1581. https://doi.org/10.1152/physrev.00013.2010.

Donovan, M., Olofsson, B., Gustafson, A. L., Dencker, L., & Eriksson, U. (1995). The cellular retinoic acid binding proteins. *The Journal of Steroid Biochemistry and Molecular Biology, 53*(1–6), 459–465. https://doi.org/10.1016/0960-0760(95)00092-e.

Dowling, J. E., & Wald, G. (1960). The biological function of vitamin a acid. *Proceedings of the National Academy of Sciences of the United States of America, 46*(5), 587–608. https://doi.org/10.1073/pnas.46.5.587.

Driessen, C. A., Winkens, H. J., Hoffmann, K., Kuhlmann, L. D., Janssen, B. P., Van Vugt, A. H., ... Janssen, J. J. (2000). Disruption of the 11-cis-retinol dehydrogenase gene leads to accumulation of cis-retinols and cis-retinyl esters. *Molecular and Cellular Biology, 20*(12), 4275–4287. https://doi.org/10.1128/MCB.20.12.4275-4287.2000.

Duester, G. (1998). Alcohol dehydrogenase as a critical mediator of retinoic acid synthesis from vitamin A in the mouse embryo. *The Journal of Nutrition, 128*(2 Suppl), 459S–462S. https://doi.org/10.1093/jn/128.2.459S.

Duester, G. (2000). Families of retinoid dehydrogenases regulating vitamin A function: Production of visual pigment and retinoic acid. *European Journal of Biochemistry / FEBS, 267*(14), 4315–4324. https://doi.org/10.1046/j.1432-1327.2000.01497.x.

Duester, G. (2008). Retinoic acid synthesis and signaling during early organogenesis. *Cell, 134*(6), 921–931. https://doi.org/10.1016/j.cell.2008.09.002.

Dupe, V., Matt, N., Garnier, J. M., Chambon, P., Mark, M., & Ghyselinck, N. B. (2003). A newborn lethal defect due to inactivation of retinaldehyde dehydrogenase type 3 is prevented by maternal retinoic acid treatment. *Proceedings of the National Academy of Sciences of the United States of America, 100*(24), 14036–14041. https://doi.org/10.1073/pnas.2336223100.

During, A., & Harrison, E. H. (2007). Mechanisms of provitamin A (carotenoid) and vitamin A (retinol) transport into and out of intestinal Caco-2 cells. *Journal of Lipid Research, 48*(10), 2283–2294. https://doi.org/10.1194/jlr.M700263-JLR200.

Eldred, G. E., & Lasky, M. R. (1993). Retinal age pigments generated by self-assembling lysosomotropic detergents. *Nature, 361*(6414), 724–726. https://doi.org/10.1038/361724a0.

Engfer, Z. J., Lewandowski, D., Dong, Z., Palczewska, G., Zhang, J., Kordecka, K., ... Palczewski, K. (2023). Distinct mouse models of Stargardt disease display differences in pharmacological targeting of ceramides and inflammatory responses. *Proceedings of the National Academy of Sciences of the United States of America, 120*(50), e2314698120. https://doi.org/10.1073/pnas.2314698120.

Ernst, O. P., Lodowski, D. T., Elstner, M., Hegemann, P., Brown, L. S., & Kandori, H. (2014). Microbial and animal rhodopsins: Structures, functions, and molecular mechanisms. *Chemical Reviews, 114*(1), 126–163. https://doi.org/10.1021/cr4003769.

Fain, G. L., & Sampath, A. P. (2021). Light responses of mammalian cones. *Pflugers Archiv: European Journal of Physiology, 473*(9), 1555–1568. https://doi.org/10.1007/s00424-021-02551-0.

Fan, J., Rohrer, B., Frederick, J. M., Baehr, W., & Crouch, R. K. (2008). Rpe65-/- and Lrat-/- mice: Comparable models of leber congenital amaurosis. *Investigative Ophthalmology & Visual Science, 49*(6), 2384–2389. https://doi.org/10.1167/iovs.08-1727.

Fan, X., Molotkov, A., Manabe, S., Donmoyer, C. M., Deltour, L., Foglio, M. H., ... Duester, G. (2003). Targeted disruption of Aldh1a1 (Raldh1) provides evidence for a complex mechanism of retinoic acid synthesis in the developing retina. *Molecular and Cellular Biology, 23*(13), 4637–4648. https://doi.org/10.1128/MCB.23.13.4637-4648.2003.

Farjo, K. M., Moiseyev, G., Nikolaeva, O., Sandell, L. L., Trainor, P. A., & Ma, J. X. (2011). RDH10 is the primary enzyme responsible for the first step of embryonic Vitamin A metabolism and retinoic acid synthesis. *Developmental Biology, 357*(2), 347–355. https://doi.org/10.1016/j.ydbio.2011.07.011.

Farjo, K. M., Moiseyev, G., Takahashi, Y., Crouch, R. K., & Ma, J. X. (2009). The 11-cis-retinol dehydrogenase activity of RDH10 and its interaction with visual cycle proteins. *Investigative Ophthalmology & Visual Science, 50*(11), 5089–5097. https://doi.org/10.1167/iovs.09-3797.

Fischer, J., Lefevre, C., Morava, E., Mussini, J. M., Laforet, P., Negre-Salvayre, A., ... Salvayre, R. (2007). The gene encoding adipose triglyceride lipase (PNPLA2) is mutated in neutral lipid storage disease with myopathy. *Nature Genetics, 39*(1), 28–30. https://doi.org/10.1038/ng1951.

Fritsch, P. O. (1992). Retinoids in psoriasis and disorders of keratinization. *Journal of the American Academy of Dermatology, 27*(6 Pt 2), S8–S14. https://doi.org/10.1016/s0190-9622(08)80253-8.

Fu, Y., & Yau, K. W. (2007). Phototransduction in mouse rods and cones. *Pflugers Archiv: European Journal of Physiology, 454*(5), 805–819. https://doi.org/10.1007/s00424-006-0194-y.

Gampe, R. T., Jr., Montana, V. G., Lambert, M. H., Wisely, G. B., Milburn, M. V., & Xu, H. E. (2000). Structural basis for autorepression of retinoid X receptor by tetramer formation and the AF-2 helix. *Genes & Development, 14*(17), 2229–2241. https://doi.org/10.1101/gad.802300.

Germain, P., Kammerer, S., Perez, E., Peluso-Iltis, C., Tortolani, D., Zusi, F. C., ... Gronemeyer, H. (2004). Rational design of RAR-selective ligands revealed by RARbeta crystal stucture. *EMBO Reports, 5*(9), 877–882. https://doi.org/10.1038/sj.embor.7400235.

Ghyselinck, N. B., Bavik, C., Sapin, V., Mark, M., Bonnier, D., Hindelang, C., ... Chambon, P. (1999). Cellular retinol-binding protein I is essential for vitamin A homeostasis. *The EMBO Journal, 18*(18), 4903–4914. https://doi.org/10.1093/emboj/18.18.4903.

Ghyselinck, N. B., & Duester, G. (2019). Retinoic acid signaling pathways. *Development (Cambridge, England), 146*(13), https://doi.org/10.1242/dev.167502.

Ghyselinck, N. B., Dupe, V., Dierich, A., Messaddeq, N., Garnier, J. M., Rochette-Egly, C., ... Mark, M. (1997). Role of the retinoic acid receptor beta (RARbeta) during mouse development. *The International Journal of Developmental Biology, 41*(3), 425–447.

Giguere, V., Lyn, S., Yip, P., Siu, C. H., & Amin, S. (1990). Molecular cloning of cDNA encoding a second cellular retinoic acid-binding protein. *Proceedings of the National Academy of Sciences of the United States of America, 87*(16), 6233–6237. https://doi.org/10.1073/pnas.87.16.6233.

Golzio, C., Martinovic-Bouriel, J., Thomas, S., Mougou-Zrelli, S., Grattagliano-Bessieres, B., Bonniere, M., ... Etchevers, H. C. (2007). Matthew-Wood syndrome is caused by truncating mutations in the retinol-binding protein receptor gene STRA6. *American Journal of Human Genetics, 80*(6), 1179–1187. https://doi.org/10.1086/518177.

Gorry, P., Lufkin, T., Dierich, A., Rochette-Egly, C., Decimo, D., Dolle, P., ... Chambon, P. (1994). The cellular retinoic acid binding protein I is dispensable. *Proceedings of the National Academy of Sciences of the United States of America, 91*(19), 9032–9036. https://doi.org/10.1073/pnas.91.19.9032.

Grumet, L., Taschler, U., & Lass, A. (2016). Hepatic retinyl ester hydrolases and the mobilization of retinyl ester stores. *Nutrients, 9*(1), https://doi.org/10.3390/nu9010013.

Haeseleer, F., Huang, J., Lebioda, L., Saari, J. C., & Palczewski, K. (1998). Molecular characterization of a novel short-chain dehydrogenase/reductase that reduces all-trans-retinal. *The Journal of Biological Chemistry, 273*(34), 21790–21799. https://doi.org/10.1074/jbc.273.34.21790.

Haeseleer, F., Jang, G. F., Imanishi, Y., Driessen, C., Matsumura, M., Nelson, P. S., & Palczewski, K. (2002). Dual-substrate specificity short chain retinol dehydrogenases from the vertebrate retina. *The Journal of Biological Chemistry, 277*(47), 45537–45546. https://doi.org/10.1074/jbc.M208882200.

Hall, J. A., Grainger, J. R., Spencer, S. P., & Belkaid, Y. (2011). The role of retinoic acid in tolerance and immunity. *Immunity, 35*(1), 13–22. https://doi.org/10.1016/j.immuni.2011.07.002.

Hamel, C. P., Tsilou, E., Harris, E., Pfeffer, B. A., Hooks, J. J., Detrick, B., & Redmond, T. M. (1993). A developmentally regulated microsomal protein specific for the pigment epithelium of the vertebrate retina. *Journal of Neuroscience Research, 34*(4), 414–425. https://doi.org/10.1002/jnr.490340406.

Hara, M., Wu, W., Malechka, V. V., Takahashi, Y., Ma, J. X., & Moiseyev, G. (2023). PNPLA2 mobilizes retinyl esters from retinosomes and promotes the generation of 11-cis-retinal in the visual cycle. *Cell Reports, 42*(2), 112091. https://doi.org/10.1016/j.celrep.2023.112091.

Hasegawa, T., McLeod, D. S., Bhutto, I. A., Prow, T., Merges, C. A., Grebe, R., & Lutty, G. A. (2007). The embryonic human choriocapillaris develops by hemo-vasculogenesis. *Developmental Dynamics: An Official Publication of the American Association of Anatomists, 236*(8), 2089–2100. https://doi.org/10.1002/dvdy.21231.

Haselbeck, R. J., Hoffmann, I., & Duester, G. (1999). Distinct functions for Aldh1 and Raldh2 in the control of ligand production for embryonic retinoid signaling pathways. *Developmental Genetics, 25*(4), 353–364. https://doi.org/10.1002/(SICI)1520-6408(1999)25:4 < 353::AID-DVG9 > 3.0.CO;2-G.

He, X., Lobsiger, J., & Stocker, A. (2009). Bothnia dystrophy is caused by domino-like rearrangements in cellular retinaldehyde-binding protein mutant R234W. *Proceedings of the National Academy of Sciences of the United States of America, 106*(44), 18545–18550. https://doi.org/10.1073/pnas.0907454106.

Heller, J. (1975). Interactions of plasma retinol-binding protein with its receptor. Specific binding of bovine and human retinol-binding protein to pigment epithelium cells from bovine eyes. *The Journal of Biological Chemistry, 250*(10), 3613–3619.

Hellgren, M., Stromberg, P., Gallego, O., Martras, S., Farres, J., Persson, B., ... Hoog, J. O. (2007). Alcohol dehydrogenase 2 is a major hepatic enzyme for human retinol metabolism. *Cellular and Molecular Life Sciences: CMLS, 64*(4), 498–505. https://doi.org/10.1007/s00018-007-6449-8.

Herr, F. M., & Ong, D. E. (1992). Differential interaction of lecithin-retinol acyltransferase with cellular retinol binding proteins. *Biochemistry, 31*(29), 6748–6755. https://doi.org/10.1021/bi00144a014.

Heyman, R. A., Mangelsdorf, D. J., Dyck, J. A., Stein, R. B., Eichele, G., Evans, R. M., & Thaller, C. (1992). 9-cis retinoic acid is a high affinity ligand for the retinoid X receptor. *Cell, 68*(2), 397–406. https://doi.org/10.1016/0092-8674(92)90479-v.

Hong, J. D., & Palczewski, K. (2023). A short story on how chromophore is hydrolyzed from rhodopsin for recycling. *Bioessays: News and Reviews in Molecular, Cellular and Developmental Biology, 45*(9), e2300068. https://doi.org/10.1002/bies.202300068.

Hong, J. D., Salom, D., Kochman, M. A., Kubas, A., Kiser, P. D., & Palczewski, K. (2022). Chromophore hydrolysis and release from photoactivated rhodopsin in native membranes. *Proceedings of the National Academy of Sciences of the United States of America, 119*(45), e2213911119. https://doi.org/10.1073/pnas.2213911119.

Humphries, M. M., Rancourt, D., Farrar, G. J., Kenna, P., Hazel, M., Bush, R. A., ... Humphries, P. (1997). Retinopathy induced in mice by targeted disruption of the rhodopsin gene. *Nature Genetics, 15*(2), 216–219. https://doi.org/10.1038/ng0297-216.

Hurley, J. B. (2021). Retina metabolism and metabolism in the pigmented epithelium: A busy intersection. *Annual Review of Vision Science, 7*, 665–692. https://doi.org/10.1146/annurev-vision-100419-115156.

Illing, M., Molday, L. L., & Molday, R. S. (1997). The 220-kDa rim protein of retinal rod outer segments is a member of the ABC transporter superfamily. *The Journal of Biological Chemistry, 272*(15), 10303–10310. https://doi.org/10.1074/jbc.272.15.10303.

Imanishi, Y., Gerke, V., & Palczewski, K. (2004). Retinosomes: New insights into intracellular managing of hydrophobic substances in lipid bodies. *The Journal of Cell Biology, 166*(4), 447–453. https://doi.org/10.1083/jcb.200405110.

Isoherranen, N., & Zhong, G. (2019). Biochemical and physiological importance of the CYP26 retinoic acid hydroxylases. *Pharmacology & Therapeutics, 204*, 107400. https://doi.org/10.1016/j.pharmthera.2019.107400.

Jang, G. F., Van Hooser, J. P., Kuksa, V., McBee, J. K., He, Y. G., Janssen, J. J., ... Palczewski, K. (2001). Characterization of a dehydrogenase activity responsible for oxidation of 11-cis-retinol in the retinal pigment epithelium of mice with a disrupted RDH5 gene. A model for the human hereditary disease fundus albipunctatus. *The Journal of Biological Chemistry, 276*(35), 32456–32465. https://doi.org/10.1074/jbc.M104949200.

Jiang, M., Pandey, S., & Fong, H. K. (1993). An opsin homologue in the retina and pigment epithelium. *Investigative Ophthalmology & Visual Science, 34*(13), 3669–3678.
Jin, M., Li, S., Moghrabi, W. N., Sun, H., & Travis, G. H. (2005). Rpe65 is the retinoid isomerase in bovine retinal pigment epithelium. *Cell, 122*(3), 449–459. https://doi.org/10.1016/j.cell.2005.06.042.
Kane, M. A. (2012). Analysis, occurrence, and function of 9-cis-retinoic acid. *Biochimica et Biophysica Acta, 1821*(1), 10–20. https://doi.org/10.1016/j.bbalip.2011.09.012.
Kastner, P., Grondona, J. M., Mark, M., Gansmuller, A., LeMeur, M., Decimo, D., ... Chambon, P. (1994). Genetic analysis of RXR alpha developmental function: Convergence of RXR and RAR signaling pathways in heart and eye morphogenesis. *Cell, 78*(6), 987–1003. https://doi.org/10.1016/0092-8674(94)90274-7.
Kastner, P., Mark, M., Leid, M., Gansmuller, A., Chin, W., Grondona, J. M., ... Chambon, P. (1996). Abnormal spermatogenesis in RXR beta mutant mice. *Genes & Development, 10*(1), 80–92. https://doi.org/10.1101/gad.10.1.80.
Kawaguchi, R., Yu, J., Honda, J., Hu, J., Whitelegge, J., Ping, P., ... Sun, H. (2007). A membrane receptor for retinol binding protein mediates cellular uptake of vitamin A. *Science (New York, N. Y.), 315*(5813), 820–825. https://doi.org/10.1126/science.1136244.
Kim, T. S., Maeda, A., Maeda, T., Heinlein, C., Kedishvili, N., Palczewski, K., & Nelson, P. S. (2005). Delayed dark adaptation in 11-cis-retinol dehydrogenase-deficient mice: A role of RDH11 in visual processes in vivo. *The Journal of Biological Chemistry, 280*(10), 8694–8704. https://doi.org/10.1074/jbc.M413172200.
Kiser, P. D. (2022). Retinal pigment epithelium 65 kDa protein (RPE65): An update. *Progress in Retinal and Eye Research, 88*, 101013. https://doi.org/10.1016/j.preteyeres.2021.101013.
Kiser, P. D., Golczak, M., Lodowski, D. T., Chance, M. R., & Palczewski, K. (2009). Crystal structure of native RPE65, the retinoid isomerase of the visual cycle. *Proceedings of the National Academy of Sciences of the United States of America, 106*(41), 17325–17330. https://doi.org/10.1073/pnas.0906600106.
Kiser, P. D., Golczak, M., & Palczewski, K. (2014). Chemistry of the retinoid (visual) cycle. *Chemical Reviews, 114*(1), 194–232. https://doi.org/10.1021/cr400107q.
Kiser, P. D., & Palczewski, K. (2016). Retinoids and retinal diseases. *Annual Review of Vision Science, 2*, 197–234. https://doi.org/10.1146/annurev-vision-111815-114407.
Kiser, P. D., & Palczewski, K. (2021). Pathways and disease-causing alterations in visual chromophore production for vertebrate vision. *The Journal of Biological Chemistry, 296*, 100072. https://doi.org/10.1074/jbc.REV120.014405.
Kleywegt, G. J., Bergfors, T., Senn, H., Le Motte, P., Gsell, B., Shudo, K., & Jones, T. A. (1994). Crystal structures of cellular retinoic acid binding proteins I and II in complex with all-trans-retinoic acid and a synthetic retinoid. *Structure (London, England: 1993), 2*(12), 1241–1258. https://doi.org/10.1016/s0969-2126(94)00125-1.
Kolesnikov, A. V., Tang, P. H., & Kefalov, V. J. (2018). Examining the role of cone-expressed RPE65 in mouse cone function. *Scientific Reports, 8*(1), 14201. https://doi.org/10.1038/s41598-018-32667-w.
Krezel, W., Ghyselinck, N., Samad, T. A., Dupe, V., Kastner, P., Borrelli, E., & Chambon, P. (1998). Impaired locomotion and dopamine signaling in retinoid receptor mutant mice. *Science (New York, N. Y.), 279*(5352), 863–867. https://doi.org/10.1126/science.279.5352.863.
Krois, C. R., Vuckovic, M. G., Huang, P., Zaversnik, C., Liu, C. S., Gibson, C. E., ... Napoli, J. L. (2019). RDH1 suppresses adiposity by promoting brown adipose adaptation to fasting and re-feeding. *Cellular and Molecular Life Sciences: CMLS, 76*(12), 2425–2447. https://doi.org/10.1007/s00018-019-03046-z.
Kumar, S., Sandell, L. L., Trainor, P. A., Koentgen, F., & Duester, G. (2012). Alcohol and aldehyde dehydrogenases: retinoid metabolic effects in mouse knockout models. *Biochimica et Biophysica Acta, 1821*(1), 198–205. https://doi.org/10.1016/j.bbalip.2011.04.004.

Lamb, A. L., & Newcomer, M. E. (1999). The structure of retinal dehydrogenase type II at 2.7 Å resolution: Implications for retinal specificity. *Biochemistry, 38*(19), 6003–6011. https://doi.org/10.1021/bi9900471.

Lampron, C., Rochette-Egly, C., Gorry, P., Dolle, P., Mark, M., Lufkin, T., ... Chambon, P. (1995). Mice deficient in cellular retinoic acid binding protein II (CRABPII) or in both CRABPI and CRABPII are essentially normal. *Development (Cambridge, England), 121*(2), 539–548. https://doi.org/10.1242/dev.121.2.539.

Laursen, K. B., Kashyap, V., Scandura, J., & Gudas, L. J. (2015). An alternative retinoic acid-responsive Stra6 promoter regulated in response to retinol deficiency. *The Journal of Biological Chemistry, 290*(7), 4356–4366. https://doi.org/10.1074/jbc.M114.613968.

Lefebvre, P., Benomar, Y., & Staels, B. (2010). Retinoid X receptors: Common heterodimerization partners with distinct functions. *Trends in Endocrinology and Metabolism: TEM, 21*(11), 676–683. https://doi.org/10.1016/j.tem.2010.06.009.

Lejoyeux, R., Benillouche, J., Ong, J., Errera, M. H., Rossi, E. A., Singh, S. R., ... Chhablani, J. (2022). Choriocapillaris: Fundamentals and advancements. *Progress in Retinal and Eye Research, 87*, 100997. https://doi.org/10.1016/j.preteyeres.2021.100997.

Lenis, T. L., Hu, J., Ng, S. Y., Jiang, Z., Sarfare, S., Lloyd, M. B., ... Radu, R. A. (2018). Expression of ABCA4 in the retinal pigment epithelium and its implications for Stargardt macular degeneration. *Proceedings of the National Academy of Sciences of the United States of America, 115*(47), E11120–E11127. https://doi.org/10.1073/pnas.1802519115.

Li, H., Wagner, E., McCaffery, P., Smith, D., Andreadis, A., & Drager, U. C. (2000). A retinoic acid synthesizing enzyme in ventral retina and telencephalon of the embryonic mouse. *Mechanisms of Development, 95*(1-2), 283–289. https://doi.org/10.1016/s0925-4773(00)00352-x.

Liou, G. I., Fei, Y., Peachey, N. S., Matragoon, S., Wei, S., Blaner, W. S., ... Ripps, H. (1998). Early onset photoreceptor abnormalities induced by targeted disruption of the interphotoreceptor retinoid-binding protein gene. *The Journal of Neuroscience, 18*(12), 4511–4520. https://doi.org/10.1523/JNEUROSCI.18-12-04511.1998.

Liou, G. I., Wang, M., & Matragoon, S. (1994). Timing of interphotoreceptor retinoid-binding protein (IRBP) gene expression and hypomethylation in developing mouse retina. *Developmental Biology, 161*(2), 345–356. https://doi.org/10.1006/dbio.1994.1036.

Liu, F., Lee, J., & Chen, J. (2021). Molecular structures of the eukaryotic retinal importer ABCA4. *Elife, 10*. https://doi.org/10.7554/eLife.63524.

Liu, T., Jenwitheesuk, E., Teller, D. C., & Samudrala, R. (2005). Structural insights into the cellular retinaldehyde-binding protein (CRALBP). *Proteins, 61*(2), 412–422. https://doi.org/10.1002/prot.20621.

Lobo, G. P., Hessel, S., Eichinger, A., Noy, N., Moise, A. R., Wyss, A., ... von Lintig, J. (2010). ISX is a retinoic acid-sensitive gatekeeper that controls intestinal beta,beta-carotene absorption and vitamin A production. *The FASEB Journal, 24*(6), 1656–1666. https://doi.org/10.1096/fj.09-150995.

Loew, A., & Gonzalez-Fernandez, F. (2002). Crystal structure of the functional unit of interphotoreceptor retinoid binding protein. *Structure (London, England: 1993), 10*(1), 43–49. https://doi.org/10.1016/s0969-2126(01)00698-0.

Lohnes, D., Kastner, P., Dierich, A., Mark, M., LeMeur, M., & Chambon, P. (1993). Function of retinoic acid receptor gamma in the mouse. *Cell, 73*(4), 643–658. https://doi.org/10.1016/0092-8674(93)90246-m.

Love, J. D., Gooch, J. T., Benko, S., Li, C., Nagy, L., Chatterjee, V. K., ... Schwabe, J. W. (2002). The structural basis for the specificity of retinoid-X receptor-selective agonists: new insights into the role of helix H12. *The Journal of Biological Chemistry, 277*(13), 11385–11391. https://doi.org/10.1074/jbc.M110869200.

Lufkin, T., Lohnes, D., Mark, M., Dierich, A., Gorry, P., Gaub, M. P., ... Chambon, P. (1993). High postnatal lethality and testis degeneration in retinoic acid receptor alpha mutant mice. *Proceedings of the National Academy of Sciences of the United States of America, 90*(15), 7225–7229. https://doi.org/10.1073/pnas.90.15.7225.

Luo, J., Pasceri, P., Conlon, R. A., Rossant, J., & Giguere, V. (1995). Mice lacking all isoforms of retinoic acid receptor beta develop normally and are susceptible to the teratogenic effects of retinoic acid. *Mechanisms of Development, 53*(1), 61–71. https://doi.org/10.1016/0925-4773(95)00424-6.

Maeda, A., Golczak, M., Maeda, T., & Palczewski, K. (2009). Limited roles of Rdh8, Rdh12, and Abca4 in all-trans-retinal clearance in mouse retina. *Investigative Ophthalmology & Visual Science, 50*(11), 5435–5443. https://doi.org/10.1167/iovs.09-3944.

Maeda, A., Maeda, T., Imanishi, Y., Kuksa, V., Alekseev, A., Bronson, J. D., ... Palczewski, K. (2005). Role of photoreceptor-specific retinol dehydrogenase in the retinoid cycle in vivo. *The Journal of Biological Chemistry, 280*(19), 18822–18832. https://doi.org/10.1074/jbc.M501757200.

Maeda, A., Maeda, T., Imanishi, Y., Sun, W., Jastrzebska, B., Hatala, D. A., ... Palczewski, K. (2006). Retinol dehydrogenase (RDH12) protects photoreceptors from light-induced degeneration in mice. *The Journal of Biological Chemistry, 281*(49), 37697–37704. https://doi.org/10.1074/jbc.M608375200.

Maguire, A. M., High, K. A., Auricchio, A., Wright, J. F., Pierce, E. A., Testa, F., ... Bennett, J. (2009). Age-dependent effects of RPE65 gene therapy for Leber's congenital amaurosis: A phase 1 dose-escalation trial. *Lancet, 374*(9701), 1597–1605. https://doi.org/10.1016/S0140-6736(09)61836-5.

Mark, M., Ghyselinck, N. B., & Chambon, P. (2006). Function of retinoid nuclear receptors: lessons from genetic and pharmacological dissections of the retinoic acid signaling pathway during mouse embryogenesis. *Annual Review of Pharmacology and Toxicology, 46*, 451–480. https://doi.org/10.1146/annurev.pharmtox.46.120604.141156.

Mark, M., Ghyselinck, N. B., & Chambon, P. (2009). Function of retinoic acid receptors during embryonic development. *Nuclear Receptor Signaling, 7*, e002. https://doi.org/10.1621/nrs.07002.

Matt, N., Dupe, V., Garnier, J. M., Dennefeld, C., Chambon, P., Mark, M., & Ghyselinck, N. B. (2005). Retinoic acid-dependent eye morphogenesis is orchestrated by neural crest cells. *Development (Cambridge, England), 132*(21), 4789–4800. https://doi.org/10.1242/dev.02031.

McBee, J. K., Palczewski, K., Baehr, W., & Pepperberg, D. R. (2001). Confronting complexity: The interlink of phototransduction and retinoid metabolism in the vertebrate retina. *Progress in Retinal and Eye Research, 20*(4), 469–529. https://doi.org/10.1016/s1350-9462(01)00002-7.

McCarthy, P. T., & Cerecedo, L. R. (1952). Vitamin A deficiency in the mouse. *The Journal of Nutrition, 46*(3), 361–376. https://doi.org/10.1093/jn/46.3.361.

Mic, F. A., Molotkov, A., Fan, X., Cuenca, A. E., & Duester, G. (2000). RALDH3, a retinaldehyde dehydrogenase that generates retinoic acid, is expressed in the ventral retina, otic vesicle and olfactory pit during mouse development. *Mechanisms of Development, 97*(1-2), 227–230. https://doi.org/10.1016/s0925-4773(00)00434-2.

Mic, F. A., Molotkov, A., Molotkova, N., & Duester, G. (2004). Raldh2 expression in optic vesicle generates a retinoic acid signal needed for invagination of retina during optic cup formation. *Developmental Dynamics: An Official Publication of the American Association of Anatomists, 231*(2), 270–277. https://doi.org/10.1002/dvdy.20128.

Min, J. R., Schuetz, A., Loppnau, P., Weigelt, J., Sundstrom, M., Arrowsmith, C. H., Edwards, A. M., Bochkarev, A., & Plotnikov, A. N. Structural Genomics Consortium (SGC).

Moiseyev, G., Chen, Y., Takahashi, Y., Wu, B. X., & Ma, J. X. (2005). RPE65 is the isomerohydrolase in the retinoid visual cycle. *Proceedings of the National Academy of Sciences of the United States of America, 102*(35), 12413–12418. https://doi.org/10.1073/pnas.0503460102.

Molday, R. S., Garces, F. A., Scortecci, J. F., & Molday, L. L. (2022). Structure and function of ABCA4 and its role in the visual cycle and Stargardt macular degeneration. *Progress in Retinal and Eye Research, 89*, 101036. https://doi.org/10.1016/j.preteyeres.2021.101036.

Molday, R. S., Zhong, M., & Quazi, F. (2009). The role of the photoreceptor ABC transporter ABCA4 in lipid transport and Stargardt macular degeneration. *Biochimica et Biophysica Acta, 1791*(7), 573–583. https://doi.org/10.1016/j.bbalip.2009.02.004.

Molotkov, A., Deltour, L., Foglio, M. H., Cuenca, A. E., & Duester, G. (2002). Distinct retinoid metabolic functions for alcohol dehydrogenase genes Adh1 and Adh4 in protection against vitamin A toxicity or deficiency revealed in double null mutant mice. *The Journal of Biological Chemistry, 277*(16), 13804–13811. https://doi.org/10.1074/jbc.M112039200.

Molotkov, A., Fan, X., Deltour, L., Foglio, M. H., Martras, S., Farres, J., ... Duester, G. (2002). Stimulation of retinoic acid production and growth by ubiquitously expressed alcohol dehydrogenase Adh3. *Proceedings of the National Academy of Sciences of the United States of America, 99*(8), 5337–5342. https://doi.org/10.1073/pnas.082093299.

Molotkov, A., Ghyselinck, N. B., Chambon, P., & Duester, G. (2004). Opposing actions of cellular retinol-binding protein and alcohol dehydrogenase control the balance between retinol storage and degradation. *The Biochemical Journal, 383*(Pt 2), 295–302. https://doi.org/10.1042/BJ20040621.

Molotkov, A., Molotkova, N., & Duester, G. (2006). Retinoic acid guides eye morphogenetic movements via paracrine signaling but is unnecessary for retinal dorsoventral patterning. *Development (Cambridge, England), 133*(10), 1901–1910. https://doi.org/10.1242/dev.02328.

Monaco, H. L. (2000). The transthyretin-retinol-binding protein complex. *Biochimica et Biophysica Acta, 1482*(1-2), 65–72. https://doi.org/10.1016/s0167-4838(00)00140-0.

Moore, S. A., Baker, H. M., Blythe, T. J., Kitson, K. E., Kitson, T. M., & Baker, E. N. (1998). Sheep liver cytosolic aldehyde dehydrogenase: The structure reveals the basis for the retinal specificity of class 1 aldehyde dehydrogenases. *Structure (London, England: 1993), 6*(12), 1541–1551. https://doi.org/10.1016/s0969-2126(98)00152-x.

Moretti, A., Li, J., Donini, S., Sobol, R. W., Rizzi, M., & Garavaglia, S. (2016). Crystal structure of human aldehyde dehydrogenase 1A3 complexed with NAD(+) and retinoic acid. *Scientific Reports, 6*, 35710. https://doi.org/10.1038/srep35710.

Mori, M., Ghyselinck, N. B., Chambon, P., & Mark, M. (2001). Systematic immunolocalization of retinoid receptors in developing and adult mouse eyes. *Investigative Ophthalmology & Visual Science, 42*(6), 1312–1318.

Morse, P. H. (1983). Ocular anatomy, embryology, and teratology. *JAMA: The Journal of the American Medical Association, 249*(20), 2830–2831. https://doi.org/10.1001/jama.1983.03330440066046.

Morshedian, A., Kaylor, J. J., Ng, S. Y., Tsan, A., Frederiksen, R., Xu, T., ... Travis, G. H. (2019). Light-driven regeneration of cone visual pigments through a mechanism involving RGR opsin in muller glial cells. *Neuron, 102*(6), 1172–1183.e1175. https://doi.org/10.1016/j.neuron.2019.04.004.

Muto, Y., Smith, J. E., Milch, P. O., & Goodman, D. S. (1972). Regulation of retinol-binding protein metabolism by vitamin A status in the rat. *The Journal of Biological Chemistry, 247*(8), 2542–2550.

Nakamura, M., Hotta, Y., Tanikawa, A., Terasaki, H., & Miyake, Y. (2000). A high association with cone dystrophy in Fundus albipunctatus caused by mutations of the RDH5 gene. *Investigative Ophthalmology & Visual Science, 41*(12), 3925–3932.

Napoli, J. L. (2012). Physiological insights into all-trans-retinoic acid biosynthesis. *Biochimica et Biophysica Acta, 1821*(1), 152–167. https://doi.org/10.1016/j.bbalip.2011.05.004.

Napoli, J. L. (2017). Cellular retinoid binding-proteins, CRBP, CRABP, FABP5: Effects on retinoid metabolism, function and related diseases. *Pharmacology & Therapeutics, 173*, 19–33. https://doi.org/10.1016/j.pharmthera.2017.01.004.

Napoli, J. L., & Yoo, H. S. (2020). Retinoid metabolism and functions mediated by retinoid binding-proteins. *Methods Enzymol, 637*, 55–75. https://doi.org/10.1016/bs.mie.2020.02.004.

Nickla, D. L., & Wallman, J. (2010). The multifunctional choroid. *Progress in Retinal and Eye Research, 29*(2), 144–168. https://doi.org/10.1016/j.preteyeres.2009.12.002.

Niederreither, K., Fraulob, V., Garnier, J. M., Chambon, P., & Dolle, P. (2002). Differential expression of retinoic acid-synthesizing (RALDH) enzymes during fetal development and organ differentiation in the mouse. *Mechanisms of Development, 110*(1-2), 165–171. https://doi.org/10.1016/s0925-4773(01)00561-5.

Niederreither, K., McCaffery, P., Drager, U. C., Chambon, P., & Dolle, P. (1997). Restricted expression and retinoic acid-induced downregulation of the retinaldehyde dehydrogenase type 2 (RALDH-2) gene during mouse development. *Mechanisms of Development, 62*(1), 67–78. https://doi.org/10.1016/s0925-4773(96)00653-3.

Niederreither, K., Subbarayan, V., Dolle, P., & Chambon, P. (1999). Embryonic retinoic acid synthesis is essential for early mouse post-implantation development. *Nature Genetics, 21*(4), 444–448. https://doi.org/10.1038/7788.

Niederreither, K., Vermot, J., Fraulob, V., Chambon, P., & Dolle, P. (2002). Retinaldehyde dehydrogenase 2 (RALDH2)- independent patterns of retinoic acid synthesis in the mouse embryo. *Proceedings of the National Academy of Sciences of the United States of America, 99*(25), 16111–16116. https://doi.org/10.1073/pnas.252626599.

O'Byrne, S. M., Wongsiriroj, N., Libien, J., Vogel, S., Goldberg, I. J., Baehr, W., ... Blaner, W. S. (2005). Retinoid absorption and storage is impaired in mice lacking lecithin:retinol acyltransferase (LRAT). *The Journal of Biological Chemistry, 280*(42), 35647–35657. https://doi.org/10.1074/jbc.M507924200.

Olsson, J. E., Gordon, J. W., Pawlyk, B. S., Roof, D., Hayes, A., Molday, R. S., ... Dryja, T. P. (1992). Transgenic mice with a rhodopsin mutation (Pro23His): A mouse model of autosomal dominant retinitis pigmentosa. *Neuron, 9*(5), 815–830. https://doi.org/10.1016/0896-6273(92)90236-7.

Ong, D. E. (1993). Retinoid metabolism during intestinal absorption. *The Journal of Nutrition, 123*(2 Suppl)), 351–355. https://doi.org/10.1093/jn/123.suppl_2.351.

Ong, D. E., & Page, D. L. (1987). Cellular retinol-binding protein (type two) is abundant in human small intestine. *Journal of Lipid Research, 28*(6), 739–745.

Orban, T., Jastrzebska, B., & Palczewski, K. (2014). Structural approaches to understanding retinal proteins needed for vision. *Current Opinion in Cell Biology, 27*, 32–43. https://doi.org/10.1016/j.ceb.2013.11.001.

Palczewska, G., Dong, Z., Golczak, M., Hunter, J. J., Williams, D. R., Alexander, N. S., & Palczewski, K. (2014). Noninvasive two-photon microscopy imaging of mouse retina and retinal pigment epithelium through the pupil of the eye. *Nature Medicine, 20*(7), 785–789. https://doi.org/10.1038/nm.3590.

Palczewska, G., Maeda, T., Imanishi, Y., Sun, W., Chen, Y., Williams, D. R., ... Palczewski, K. (2010). Noninvasive multiphoton fluorescence microscopy resolves retinol and retinal condensation products in mouse eyes. *Nature Medicine, 16*(12), 1444–1449. https://doi.org/10.1038/nm.2260.

Palczewski, K. (2006). G protein-coupled receptor rhodopsin. *Annual Review of Biochemistry, 75*, 743–767. https://doi.org/10.1146/annurev.biochem.75.103004.142743.

Palczewski, K. (2010). Retinoids for treatment of retinal diseases. *Trends in Pharmacological Sciences, 31*(6), 284–295. https://doi.org/10.1016/j.tips.2010.03.001.

Palczewski, K. (2012). Chemistry and biology of vision. *The Journal of Biological Chemistry, 287*(3), 1612–1619. https://doi.org/10.1074/jbc.R111.301150.

Palczewski, K., & Benovic, J. L. (1991). G-protein-coupled receptor kinases. *Trends in Biochemical Sciences, 16*(10), 387–391. https://doi.org/10.1016/0968-0004(91)90157-q.

Palczewski, K., & Kiser, P. D. (2020). Shedding new light on the generation of the visual chromophore. *Proceedings of the National Academy of Sciences of the United States of America, 117*(33), 19629–19638. https://doi.org/10.1073/pnas.2008211117.

Palczewski, K., Kumasaka, T., Hori, T., Behnke, C. A., Motoshima, H., Fox, B. A., ... Miyano, M. (2000). Crystal structure of rhodopsin: A G protein-coupled receptor. *Science (New York, N. Y.), 289*(5480), 739–745. https://doi.org/10.1126/science.289.5480.739.

Pandey, S., Blanks, J. C., Spee, C., Jiang, M., & Fong, H. K. (1994). Cytoplasmic retinal localization of an evolutionary homolog of the visual pigments. *Experimental Eye Research, 58*(5), 605–613. https://doi.org/10.1006/exer.1994.1055.

Pares, X., Farres, J., Kedishvili, N., & Duester, G. (2008). Medium- and short-chain dehydrogenase/reductase gene and protein families: Medium-chain and short-chain dehydrogenases/reductases in retinoid metabolism. *Cellular and Molecular Life Sciences: CMLS, 65*(24), 3936–3949. https://doi.org/10.1007/s00018-008-8591-3.

Parker, R., Wang, J. S., Kefalov, V. J., & Crouch, R. K. (2011). Interphotoreceptor retinoid-binding protein as the physiologically relevant carrier of 11-cis-retinol in the cone visual cycle. *The Journal of Neuroscience, 31*(12), 4714–4719. https://doi.org/10.1523/JNEUROSCI.3722-10.2011.

Parker, R. O., Fan, J., Nickerson, J. M., Liou, G. I., & Crouch, R. K. (2009). Normal cone function requires the interphotoreceptor retinoid binding protein. *The Journal of Neuroscience, 29*(14), 4616–4621. https://doi.org/10.1523/JNEUROSCI.0063-09.2009.

Pasutto, F., Sticht, H., Hammersen, G., Gillessen-Kaesbach, G., Fitzpatrick, D. R., Nurnberg, G., ... Rauch, A. (2007). Mutations in STRA6 cause a broad spectrum of malformations including anophthalmia, congenital heart defects, diaphragmatic hernia, alveolar capillary dysplasia, lung hypoplasia, and mental retardation. *American Journal of Human Genetics, 80*(3), 550–560. https://doi.org/10.1086/512203.

Perez-Castro, A. V., Toth-Rogler, L. E., Wei, L. N., & Nguyen-Huu, M. C. (1989). Spatial and temporal pattern of expression of the cellular retinoic acid-binding protein and the cellular retinol-binding protein during mouse embryogenesis. *Proceedings of the National Academy of Sciences of the United States of America, 86*(22), 8813–8817. https://doi.org/10.1073/pnas.86.22.8813.

Plaisancie, J., Martinovic, J., Chesneau, B., Whalen, S., Rodriguez, D., Audebert-Bellanger, S., ... Chassaing, N. (2023). Clinical, genetic and biochemical signatures of RBP4-related ocular malformations. *Journal of Medical Genetics, 61*(1), 84–92. https://doi.org/10.1136/jmg-2023-109331.

Qtaishat, N. M., Wiggert, B., & Pepperberg, D. R. (2005). Interphotoreceptor retinoid-binding protein (IRBP) promotes the release of all-trans retinol from the isolated retina following rhodopsin bleaching illumination. *Experimental Eye Research, 81*(4), 455–463. https://doi.org/10.1016/j.exer.2005.03.005.

Quadro, L., Blaner, W. S., Salchow, D. J., Vogel, S., Piantedosi, R., Gouras, P., ... Gottesman, M. E. (1999). Impaired retinal function and vitamin A availability in mice lacking retinol-binding protein. *The EMBO Journal, 18*(17), 4633–4644. https://doi.org/10.1093/emboj/18.17.4633.

Quadro, L., Hamberger, L., Gottesman, M. E., Wang, F., Colantuoni, V., Blaner, W. S., & Mendelsohn, C. L. (2005). Pathways of vitamin A delivery to the embryo: Insights from a new tunable model of embryonic vitamin A deficiency. *Endocrinology, 146*(10), 4479–4490. https://doi.org/10.1210/en.2005-0158.

Ramkumar, S., Moon, J., Golczak, M., & Von Lintig, J. (2021). LRAT coordinates the negative-feedback regulation of intestinal retinoid biosynthesis from beta-carotene. *Journal of Lipid Research, 62*, 100055. https://doi.org/10.1016/j.jlr.2021.100055.

Rattner, A., Smallwood, P. M., & Nathans, J. (2000). Identification and characterization of all-trans-retinol dehydrogenase from photoreceptor outer segments, the visual cycle enzyme that reduces all-trans-retinal to all-trans-retinol. *The Journal of Biological Chemistry, 275*(15), 11034–11043. https://doi.org/10.1074/jbc.275.15.11034.

Redmond, T. M., Poliakov, E., Yu, S., Tsai, J. Y., Lu, Z., & Gentleman, S. (2005). Mutation of key residues of RPE65 abolishes its enzymatic role as isomerohydrolase in the visual cycle. *Proceedings of the National Academy of Sciences of the United States of America, 102*(38), 13658–13663. https://doi.org/10.1073/pnas.0504167102.

Redmond, T. M., Yu, S., Lee, E., Bok, D., Hamasaki, D., Chen, N., ... Pfeifer, K. (1998). Rpe65 is necessary for production of 11-cis-vitamin A in the retinal visual cycle. *Nature Genetics, 20*(4), 344–351. https://doi.org/10.1038/3813.

Renaud, J. P., Rochel, N., Ruff, M., Vivat, V., Chambon, P., Gronemeyer, H., & Moras, D. (1995). Crystal structure of the RAR-gamma ligand-binding domain bound to all-trans retinoic acid. *Nature, 378*(6558), 681–689. https://doi.org/10.1038/378681a0.

Robichaux, M. A., Nguyen, V., Chan, F., Kailasam, L., He, F., Wilson, J. H., & Wensel, T. G. (2022). Subcellular localization of mutant P23H rhodopsin in an RFP fusion knock-in mouse model of retinitis pigmentosa. *Disease Models & Mechanism, 15*(5), https://doi.org/10.1242/dmm.049336.

Ronne, H., Ocklind, C., Wiman, K., Rask, L., Obrink, B., & Peterson, P. A. (1983). Ligand-dependent regulation of intracellular protein transport: Effect of vitamin a on the secretion of the retinol-binding protein. *The Journal of Cell Biology, 96*(3), 907–910. https://doi.org/10.1083/jcb.96.3.907.

Ross, J. W., Fernandez de Castro, J. P., Zhao, J., Samuel, M., Walters, E., Rios, C., ... Kaplan, H. J. (2012). Generation of an inbred miniature pig model of retinitis pigmentosa. *Investigative Ophthalmology & Visual Science, 53*(1), 501–507. https://doi.org/10.1167/iovs.11-8784.

Rossant, J., Zirngibl, R., Cado, D., Shago, M., & Giguere, V. (1991). Expression of a retinoic acid response element-hsplacZ transgene defines specific domains of transcriptional activity during mouse embryogenesis. *Genes & Development, 5*(8), 1333–1344. https://doi.org/10.1101/gad.5.8.1333.

Ruiz, A., Ghyselinck, N. B., Mata, N., Nusinowitz, S., Lloyd, M., Dennefeld, C., ... Bok, D. (2007). Somatic ablation of the Lrat gene in the mouse retinal pigment epithelium drastically reduces its retinoid storage. *Investigative Ophthalmology & Visual Science, 48*(12), 5377–5387. https://doi.org/10.1167/iovs.07-0673.

Ruiz, A., Mark, M., Jacobs, H., Klopfenstein, M., Hu, J., Lloyd, M., ... Bok, D. (2012). Retinoid content, visual responses, and ocular morphology are compromised in the retinas of mice lacking the retinol-binding protein receptor, STRA6. *Investigative Ophthalmology & Visual Science, 53*(6), 3027–3039. https://doi.org/10.1167/iovs.11-8476.

Saari, J. C., Bredberg, D. L., & Noy, N. (1994). Control of substrate flow at a branch in the visual cycle. *Biochemistry, 33*(10), 3106–3112. https://doi.org/10.1021/bi00176a045.

Saari, J. C., & Crabb, J. W. (2005). Focus on molecules: Cellular retinaldehyde-binding protein (CRALBP). *Experimental Eye Research, 81*(3), 245–246. https://doi.org/10.1016/j.exer.2005.06.015.

Saari, J. C., Garwin, G. G., Van Hooser, J. P., & Palczewski, K. (1998). Reduction of all-trans-retinal limits regeneration of visual pigment in mice. *Vision Research, 38*(10), 1325–1333. https://doi.org/10.1016/s0042-6989(97)00198-3.

Saari, J. C., Nawrot, M., Garwin, G. G., Kennedy, M. J., Hurley, J. B., Ghyselinck, N. B., & Chambon, P. (2002). Analysis of the visual cycle in cellular retinol-binding protein type I (CRBPI) knockout mice. *Investigative Ophthalmology & Visual Science, 43*(6), 1730–1735.

Saari, J. C., Nawrot, M., Kennedy, B. N., Garwin, G. G., Hurley, J. B., Huang, J., ... Crabb, J. W. (2001). Visual cycle impairment in cellular retinaldehyde binding protein (CRALBP) knockout mice results in delayed dark adaptation. *Neuron, 29*(3), 739–748. https://doi.org/10.1016/s0896-6273(01)00248-3.

Sahu, B., Sun, W., Perusek, L., Parmar, V., Le, Y. Z., Griswold, M. D., ... Maeda, A. (2015). Conditional ablation of retinol dehydrogenase 10 in the retinal pigmented epithelium causes delayed dark adaption in mice. *The Journal of Biological Chemistry, 290*(45), 27239–27247. https://doi.org/10.1074/jbc.M115.682096.

Sakai, Y., Luo, T., McCaffery, P., Hamada, H., & Drager, U. C. (2004). CYP26A1 and CYP26C1 cooperate in degrading retinoic acid within the equatorial retina during later eye development. *Developmental Biology, 276*(1), 143–157. https://doi.org/10.1016/j.ydbio.2004.08.032.

Sakai, Y., Meno, C., Fujii, H., Nishino, J., Shiratori, H., Saijoh, Y., ... Hamada, H. (2001). The retinoic acid-inactivating enzyme CYP26 is essential for establishing an uneven distribution of retinoic acid along the anterio-posterior axis within the mouse embryo. *Genes & Development, 15*(2), 213–225. https://doi.org/10.1101/gad.851501.

Sandell, L. L., Sanderson, B. W., Moiseyev, G., Johnson, T., Mushegian, A., Young, K., ... Trainor, P. A. (2007). RDH10 is essential for synthesis of embryonic retinoic acid and is required for limb, craniofacial, and organ development. *Genes & Development, 21*(9), 1113–1124. https://doi.org/10.1101/gad.1533407.

Schreiber, R., Taschler, U., Preiss-Landl, K., Wongsiriroj, N., Zimmermann, R., & Lass, A. (2012). Retinyl ester hydrolases and their roles in vitamin A homeostasis. *Biochimica et Biophysica Acta, 1821*(1), 113–123. https://doi.org/10.1016/j.bbalip.2011.05.001.

Sears, A. E., Bernstein, P. S., Cideciyan, A. V., Hoyng, C., Charbel Issa, P., Palczewski, K., ... Scholl, H. P. N. (2017). Towards treatment of stargardt disease: Workshop organized and sponsored by the foundation fighting blindness. *Translational Vision Science & Technology, 6*(5), 6. https://doi.org/10.1167/tvst.6.5.6.

Sears, A. E., & Palczewski, K. (2016). Lecithin:Retinol acyltransferase: A key enzyme involved in the retinoid (visual) cycle. *Biochemistry, 55*(22), 3082–3091. https://doi.org/10.1021/acs.biochem.6b00319.

Seeliger, M. W., Biesalski, H. K., Wissinger, B., Gollnick, H., Gielen, S., Frank, J., ... Zrenner, E. (1999). Phenotype in retinol deficiency due to a hereditary defect in retinol binding protein synthesis. *Investigative Ophthalmology & Visual Science, 40*(1), 3–11.

Sergouniotis, P. I., Sohn, E. H., Li, Z., McBain, V. A., Wright, G. A., Moore, A. T., ... Webster, A. R. (2011). Phenotypic variability in RDH5 retinopathy (Fundus Albipunctatus). *Ophthalmology, 118*(8), 1661–1670. https://doi.org/10.1016/j.ophtha.2010.12.031.

Sessler, R. J., & Noy, N. (2005). A ligand-activated nuclear localization signal in cellular retinoic acid binding protein-II. *Molecular Cell, 18*(3), 343–353. https://doi.org/10.1016/j.molcel.2005.03.026.

Shen, J., Shi, D., Suzuki, T., Xia, Z., Zhang, H., Araki, K., ... Li, Z. (2016). Severe ocular phenotypes in Rbp4-deficient mice in the C57BL/6 genetic background. *Laboratory Investigation; A Journal of Technical Methods and Pathology, 96*(6), 680–691. https://doi.org/10.1038/labinvest.2016.39.

Shete, V., & Quadro, L. (2013). Mammalian metabolism of beta-carotene: Gaps in knowledge. *Nutrients, 5*(12), 4849–4868. https://doi.org/10.3390/nu5124849.

Shih, M. Y., Kane, M. A., Zhou, P., Yen, C. L., Streeper, R. S., Napoli, J. L., & Farese, R. V., Jr. (2009). Retinol esterification by DGAT1 is essential for retinoid homeostasis in murine skin. *The Journal of Biological Chemistry, 284*(7), 4292–4299. https://doi.org/10.1074/jbc.M807503200.

Simon, A., Hellman, U., Wernstedt, C., & Eriksson, U. (1995). The retinal pigment epithelial-specific 11-cis retinol dehydrogenase belongs to the family of short chain alcohol dehydrogenases. *The Journal of Biological Chemistry, 270*(3), 1107–1112.

Sommer, A. (2008). Vitamin a deficiency and clinical disease: an historical overview. *The Journal of Nutrition, 138*(10), 1835–1839. https://doi.org/10.1093/jn/138.10.1835.

Sparrow, J. R., Wu, Y., Kim, C. Y., & Zhou, J. (2010). Phospholipid meets all-trans-retinal: The making of RPE bisretinoids. *Journal of Lipid Research, 51*(2), 247–261. https://doi.org/10.1194/jlr.R000687.

Srour, M., Chitayat, D., Caron, V., Chassaing, N., Bitoun, P., Patry, L., ... Michaud, J. L. (2013). Recessive and dominant mutations in retinoic acid receptor beta in cases with microphthalmia and diaphragmatic hernia. *American Journal of Human Genetics, 93*(4), 765–772. https://doi.org/10.1016/j.ajhg.2013.08.014.

Steinhoff, J. S., Lass, A., & Schupp, M. (2021). Biological functions of RBP4 and its relevance for human diseases. *Frontiers in Physiology, 12*, 659977. https://doi.org/10.3389/fphys.2021.659977.

Stevens, G. A., Bennett, J. E., Hennocq, Q., Lu, Y., De-Regil, L. M., Rogers, L., ... Ezzati, M. (2015). Trends and mortality effects of vitamin A deficiency in children in 138 low-income and middle-income countries between 1991 and 2013: A pooled analysis of population-based surveys. *The Lancet Global Health, 3*(9), e528–e536. https://doi.org/10.1016/S2214-109X(15)00039-X.

Subbarayan, V., Kastner, P., Mark, M., Dierich, A., Gorry, P., & Chambon, P. (1997). Limited specificity and large overlap of the functions of the mouse RAR gamma 1 and RAR gamma 2 isoforms. *Mechanisms of Development, 66*(1–2), 131–142. https://doi.org/10.1016/s0925-4773(97)00098-1.

Sucov, H. M., Dyson, E., Gumeringer, C. L., Price, J., Chien, K. R., & Evans, R. M. (1994). RXR alpha mutant mice establish a genetic basis for vitamin A signaling in heart morphogenesis. *Genes & Development, 8*(9), 1007–1018. https://doi.org/10.1101/gad.8.9.1007.

Suh, S., Choi, E. H., Leinonen, H., Foik, A. T., Newby, G. A., Yeh, W. H., ... Palczewski, K. (2021). Restoration of visual function in adult mice with an inherited retinal disease via adenine base editing. *Nature Biomedical Engineering, 5*(2), 169–178. https://doi.org/10.1038/s41551-020-00632-6.

Sun, H., & Nathans, J. (1997). Stargardt's ABCR is localized to the disc membrane of retinal rod outer segments. *Nature Genetics, 17*(1), 15–16. https://doi.org/10.1038/ng0997-15.

Sung, C. H., Davenport, C. M., Hennessey, J. C., Maumenee, I. H., Jacobson, S. G., Heckenlively, J. R., ... Nathans, J. (1991). Rhodopsin mutations in autosomal dominant retinitis pigmentosa. *Proceedings of the National Academy of Sciences of the United States of America, 88*(15), 6481–6485. https://doi.org/10.1073/pnas.88.15.6481.

Taneja, R., Bouillet, P., Boylan, J. F., Gaub, M. P., Roy, B., Gudas, L. J., & Chambon, P. (1995). Reexpression of retinoic acid receptor (RAR) gamma or overexpression of RAR alpha or RAR beta in RAR gamma-null F9 cells reveals a partial functional redundancy between the three RAR types. *Proceedings of the National Academy of Sciences of the United States of America, 92*(17), 7854–7858. https://doi.org/10.1073/pnas.92.17.7854.

Tang, X. H., & Gudas, L. J. (2011). Retinoids, retinoic acid receptors, and cancer. *Annual Review of Pathology, 6*, 345–364. https://doi.org/10.1146/annurev-pathol-011110-130303.

Tanna, P., Strauss, R. W., Fujinami, K., & Michaelides, M. (2017). Stargardt disease: Clinical features, molecular genetics, animal models and therapeutic options. *The British Journal of Ophthalmology, 101*(1), 25–30. https://doi.org/10.1136/bjophthalmol-2016-308823.

Taschler, U., Schreiber, R., Chitraju, C., Grabner, G. F., Romauch, M., Wolinski, H., ... Zimmermann, R. (2015). Adipose triglyceride lipase is involved in the mobilization of triglyceride and retinoid stores of hepatic stellate cells. *Biochimica et Biophysica Acta, 1851*(7), 937–945. https://doi.org/10.1016/j.bbalip.2015.02.017.

Telias, M., Denlinger, B., Helft, Z., Thornton, C., Beckwith-Cohen, B., & Kramer, R. H. (2019). Retinoic acid induces hyperactivity, and blocking its receptor unmasks light responses and augments vision in retinal degeneration. *Neuron, 102*(3), 574–586.e575. https://doi.org/10.1016/j.neuron.2019.02.015.

Thatcher, J. E., & Isoherranen, N. (2009). The role of CYP26 enzymes in retinoic acid clearance. *Expert Opinion on Drug Metabolism & Toxicology, 5*(8), 875–886. https://doi.org/10.1517/17425250903032681.

Thompson, S. J., Sargsyan, A., Lee, S. A., Yuen, J. J., Cai, J., Smalling, R., ... Graham, T. E. (2017). Hepatocytes are the principal source of circulating RBP4 in mice. *Diabetes, 66*(1), 58–63. https://doi.org/10.2337/db16-0286.

Tso, P., & Balint, J. A. (1986). Formation and transport of chylomicrons by enterocytes to the lymphatics. *The American Journal of Physiology, 250*(6 Pt 1), G715–G726. https://doi.org/10.1152/ajpgi.1986.250.6.G715.

Tworak, A., Kolesnikov, A. V., Hong, J. D., Choi, E. H., Luu, J. C., Palczewska, G., ... Palczewski, K. (2023). Rapid RGR-dependent visual pigment recycling is mediated by the RPE and specialized Muller glia. *Cell Reports, 42*(8), 112982. https://doi.org/10.1016/j.celrep.2023.112982.

Vivat-Hannah, V., Bourguet, W., Gottardis, M., & Gronemeyer, H. (2003). Separation of retinoid X receptor homo- and heterodimerization functions. *Molecular and Cellular Biology, 23*(21), 7678–7688. https://doi.org/10.1128/MCB.23.21.7678-7688.2003.

von Lintig, J. (2012). Provitamin A metabolism and functions in mammalian biology. *The American Journal of Clinical Nutrition, 96*(5), 1234S–1244S. https://doi.org/10.3945/ajcn.112.034629.

von Lintig, J., Moon, J., Lee, J., & Ramkumar, S. (2020). Carotenoid metabolism at the intestinal barrier. *Biochimica Et Biophysica Acta-Molecular and Cell Biology of Lipids, 1865*(11), ARTN158580. https://doi.org/10.1016/j.bbalip.2019.158580.

Wald, G. (1933). Vitamin A in the retina. *Nature, 132*(3330), 316–317. https://doi.org/10.1038/132316a0.

Wang, J. S., & Kefalov, V. J. (2011). The cone-specific visual cycle. *Progress in Retinal and Eye Research, 30*(2), 115–128. https://doi.org/10.1016/j.preteyeres.2010.11.001.

Wang, Z. G., Delva, L., Gaboli, M., Rivi, R., Giorgio, M., Cordon-Cardo, C., ... Pandolfi, P. P. (1998). Role of PML in cell growth and the retinoic acid pathway. *Science (New York, N. Y.), 279*(5356), 1547–1551. https://doi.org/10.1126/science.279.5356.1547.

Wei, L. N. (2016). Cellular retinoic acid binding proteins: Genomic and non-genomic functions and their regulation. *Sub-Cellular Biochemistry, 81*, 163–178. https://doi.org/10.1007/978-94-024-0945-1_6.

Weng, J., Mata, N. L., Azarian, S. M., Tzekov, R. T., Birch, D. G., & Travis, G. H. (1999). Insights into the function of Rim protein in photoreceptors and etiology of Stargardt's disease from the phenotype in abcr knockout mice. *Cell, 98*(1), 13–23. https://doi.org/10.1016/S0092-8674(00)80602-9.

Werder, M., Han, C. H., Wehrli, E., Bimmler, D., Schulthess, G., & Hauser, H. (2001). Role of scavenger receptors SR-BI and CD36 in selective sterol uptake in the small intestine. *Biochemistry, 40*(38), 11643–11650. https://doi.org/10.1021/bi0109820.

White, J. C., Highland, M., Kaiser, M., & Clagett-Dame, M. (2000). Vitamin A deficiency results in the dose-dependent acquisition of anterior character and shortening of the caudal hindbrain of the rat embryo. *Developmental Biology, 220*(2), 263–284. https://doi.org/10.1006/dbio.2000.9635.

Widjaja-Adhi, M. A. K., Palczewski, G., Dale, K., Knauss, E. A., Kelly, M. E., Golczak, M., ... Von Lintig, J. (2017). Transcription factor ISX mediates the cross talk between diet and immunity. *Proceedings of the National Academy of Sciences of the United States of America, 114*(43), 11530–11535. https://doi.org/10.1073/pnas.1714963114.

Winter, N. S., Bratt, J. M., & Banaszak, L. J. (1993). Crystal structures of holo and apo-cellular retinol-binding protein II. *Journal of Molecular Biology, 230*(4), 1247–1259. https://doi.org/10.1006/jmbi.1993.1239.

Wisard, J., Faulkner, A., Chrenek, M. A., Waxweiler, T., Waxweiler, W., Donmoyer, C., ... Nickerson, J. M. (2011). Exaggerated eye growth in IRBP-deficient mice in early development. *Investigative Ophthalmology & Visual Science, 52*(8), 5804–5811. https://doi.org/10.1167/iovs.10-7129.

Won, J. Y., Nam, E. C., Yoo, S. J., Kwon, H. J., Um, S. J., Han, H. S., ... Kim, S. Y. (2004). The effect of cellular retinoic acid binding protein-I expression on the CYP26-mediated catabolism of all-trans retinoic acid and cell proliferation in head and neck squamous cell carcinoma. *Metabolism: Clinical and Experimental, 53*(8), 1007–1012. https://doi.org/10.1016/j.metabol.2003.12.015.

Wongsiriroj, N., Piantedosi, R., Palczewski, K., Goldberg, I. J., Johnston, T. P., Li, E., & Blaner, W. S. (2008). The molecular basis of retinoid absorption: A genetic dissection. *The Journal of Biological Chemistry, 283*(20), 13510–13519. https://doi.org/10.1074/jbc.M800777200.

Wu, B. X., Chen, Y., Chen, Y., Fan, J., Rohrer, B., Crouch, R. K., & Ma, J. X. (2002). Cloning and characterization of a novel all-trans retinol short-chain dehydrogenase/reductase from the RPE. *Investigative Ophthalmology & Visual Science, 43*(11), 3365–3372.

Wu, B. X., Moiseyev, G., Chen, Y., Rohrer, B., Crouch, R. K., & Ma, J. X. (2004). Identification of RDH10, an all-trans retinol dehydrogenase, in retinal muller cells. *Investigative Ophthalmology & Visual Science, 45*(11), 3857–3862. https://doi.org/10.1167/iovs.03-1302.

Wu, J. I., Lin, Y. P., Tseng, C. W., Chen, H. J., & Wang, L. H. (2019). Crabp2 promotes metastasis of lung cancer cells via HuR and integrin beta1/FAK/ERK signaling. *Scientific Reports, 9*(1), 845. https://doi.org/10.1038/s41598-018-37443-4.

Xue, Y., Shen, S. Q., Jui, J., Rupp, A. C., Byrne, L. C., Hattar, S., ... Kefalov, V. J. (2015). CRALBP supports the mammalian retinal visual cycle and cone vision. *Journal of Clinical Investigation, 125*(2), 727–738. https://doi.org/10.1172/JCI79651.

Yang, Z. N., Davis, G. J., Hurley, T. D., Stone, C. L., Li, T. K., & Bosron, W. F. (1994). Catalytic efficiency of human alcohol dehydrogenases for retinol oxidation and retinal reduction. *Alcoholism, Clinical and Experimental Research, 18*(3), 587–591. https://doi.org/10.1111/j.1530-0277.1994.tb00914.x.

Yen, C. L., Monetti, M., Burri, B. J., & Farese, R. V., Jr. (2005). The triacylglycerol synthesis enzyme DGAT1 also catalyzes the synthesis of diacylglycerols, waxes, and retinyl esters. *Journal of Lipid Research, 46*(7), 1502–1511. https://doi.org/10.1194/jlr.M500036-JLR200.

Yu, V. C., Delsert, C., Andersen, B., Holloway, J. M., Devary, O. V., Naar, A. M., ... Rosenfeld, M. G. (1991). RXR beta: A coregulator that enhances binding of retinoic acid, thyroid hormone, and vitamin D receptors to their cognate response elements. *Cell, 67*(6), 1251–1266. https://doi.org/10.1016/0092-8674(91)90301-e.

Zeng, S., Zhang, T., Madigan, M. C., Fernando, N., Aggio-Bruce, R., Zhou, F., ... Zhu, L. (2020). Interphotoreceptor retinoid-binding protein (IRBP) in retinal health and disease. *Frontiers in Cellular Neuroscience, 14*, 577935. https://doi.org/10.3389/fncel.2020.577935.

Zhang, J., Dong, Z., Mundla, S. R., Hu, X. E., Seibel, W., Papoian, R., ... Golczak, M. (2015). Expansion of first-in-class drug candidates that sequester toxic all-trans-retinal and prevent light-induced retinal degeneration. *Molecular Pharmacology, 87*(3), 477–491. https://doi.org/10.1124/mol.114.096560.

Zhang, M., Hu, P., Krois, C. R., Kane, M. A., & Napoli, J. L. (2007). Altered vitamin A homeostasis and increased size and adiposity in the rdh1-null mouse. *The FASEB Journal, 21*(11), 2886–2896. https://doi.org/10.1096/fj.06-7964com.

Zhang, N., Tsybovsky, Y., Kolesnikov, A. V., Rozanowska, M., Swider, M., Schwartz, S. B., ... Palczewski, K. (2015). Protein misfolding and the pathogenesis of ABCA4-associated retinal degenerations. *Human Molecular Genetics, 24*(11), 3220–3237. https://doi.org/10.1093/hmg/ddv073.

Zhong, M., Kawaguchi, R., Costabile, B., Tang, Y., Hu, J., Cheng, G., ... Sun, H. (2020). Regulatory mechanism for the transmembrane receptor that mediates bidirectional vitamin A transport. *Proceedings of the National Academy of Sciences of the United States of America, 117*(18), 9857–9864. https://doi.org/10.1073/pnas.1918540117.

Zimmermann, R., Strauss, J. G., Haemmerle, G., Schoiswohl, G., Birner-Gruenberger, R., Riederer, M., ... Zechner, R. (2004). Fat mobilization in adipose tissue is promoted by adipose triglyceride lipase. *Science (New York, N. Y.), 306*(5700), 1383–1386. https://doi.org/10.1126/science.1100747.

Zolfaghari, R., Chen, Q., & Ross, A. C. (2012). DHRS3, a retinal reductase, is differentially regulated by retinoic acid and lipopolysaccharide-induced inflammation in THP-1 cells and rat liver. *American Journal of Physiology. Gastrointestinal and Liver Physiology, 303*(5), G578–G588. https://doi.org/10.1152/ajpgi.00234.2012.

CHAPTER TEN

Retinoid signaling in pancreas development, islet function, and disease

Manuj Bandral[a], Lori Sussel[b], and David S. Lorberbaum[a,*]
[a]University of Michigan, Department of Pharmacology, Caswell Diabetes Institute, Ann Arbor, MI, United States
[b]University of Colorado Denver Anschutz Medical Campus, Barbara Davis Center for Diabetes, Aurora, CO, United States
*Corresponding author. e-mail address: dslorber@med.umich.edu

Contents

1. Introduction	297
2. The ATRA signaling pathway	298
3. Pancreas development	300
4. hPSC differentiations and therapeutic interventions	306
5. Maintaining islet function	308
6. ATRA signaling in disease	311
7. Conclusion	313
Acknowledgments	313
References	313

Abstract

All-*trans* retinoic acid (ATRA) signaling is essential in numerous different biological contexts. This review highlights the diverse roles of ATRA during development, function, and diseases of the pancreas. ATRA is essential to specify pancreatic progenitors from gut tube endoderm, endocrine and exocrine differentiation, and adult islet function. ATRA concentration must be carefully regulated during the derivation of islet-like cells from human pluripotent stem cells (hPSCs) to optimize the expression of key pancreatic transcription factors while mitigating adverse and unwanted cell-types in these cultures. The ATRA pathway is integral to the pancreas and here we will present selected studies from decades of research that has laid the essential groundwork for ongoing projects dedicated to unraveling the complexities of ATRA signaling in the pancreas.

1. Introduction

Signaling pathways are used repeatedly throughout development to specify a plethora of different cell types, aid in the maturation of these cells and tissues, and maintain their diverse functions in adults. Remarkably, there are only a handful of highly conserved signaling pathways that are used and reused in all of these spatial and temporal contexts including (but not limited to) WNT, Sonic Hedgehog (Shh), JAK/STAT, Fibroblast Growth Factor (FGF), Notch, and ATRA signaling. Each pathway regulates such a variety of functions that depend upon cellular context and its interactions with other signaling pathways that a literature review will never be able to cover all aspects and full importance of each pathway or how they work together. A prime example is the ATRA pathway that is essential for many developmental signaling cascades that regulate specification of the eye, spinal cord, brain, limb, liver, genitourinary tract, heart, and pancreas. Nearly all major organs are shaped by ATRA signaling during development and rely on ATRA to maintain their respective functions. While substantial effort has been dedicated to defining how ATRA signaling functions in these distinct contexts and has been reviewed extensively (Cunningham & Duester, 2015; Ghyselinck & Duester, 2019; Rhinn & Dolle, 2012; Theodosiou, Laudet, & Schubert, 2010), here we focus specifically on the ATRA signaling pathway and its roles in pancreas development, islet function, and disease. Considerable research using a variety of model systems from zebrafish to human cells has convincingly demonstrated that ATRA signaling is essential for specifying the pancreas from the gut tube endoderm in vertebrates (reviewed in Duester, 2008; Niederreither & Dolle, 2008). Here we emphasize the role of ATRA signaling during pancreatic endocrine progenitor specification, but also address several other roles in different pancreas tissue types and how dysregulation leads to disease and holds key therapeutic potential for the treatment of diabetes.

2. The ATRA signaling pathway

ATRA is a processed form of vitamin A (a. k. a. retinol), a nutrient that is an essential part of the mammalian diet. Consumption of foods that are high in retinol including liver, fish, sweet potatoes, and leafy greens, in addition to many foods fortified with retinol ensure that ample amounts are consumed (Berner, Keast, Bailey, & Dwyer, 2014). Once ingested, more than 75 % of retinol can be absorbed during digestion and may be used to activate the signaling cascade that is initiated with the conversion of retinol to retinaldehyde by retinol-dehydrogenase, or RDH10 (Fig. 1). This process is also reversible to ensure appropriate levels of retinaldehyde are maintained using enzymes such as

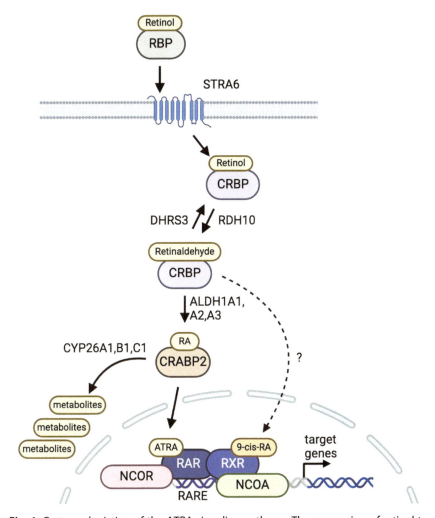

Fig. 1 Cartoon depiction of the ATRA signaling pathway. The conversion of retinol to ATRA requires several enzymes represented here. Upon entering the nucleus it will regulate target gene expression via RAR:RXR dimers along with nuclear co-repressors and co-activators (NCOR and NCOA, respectively). While there are context-dependent variants of most ATRA components, we have highlighted the versions most frequently seen in pancreatic ATRA signaling. *This figure was made with Biorender.*

dehydrogenase reductase, or DHRS3 (Billings et al., 2013; Feng, Hernandez, Waxman, Yelon, & Moens, 2010). Upon retinaldehyde generation, the conversion to ATRA is facilitated by context specific retinaldehyde dehydrogenases ALDH1A1, ALDH1A2, or ALDH1A3 (Cho et al., 2021). Unlike the conversion from retinol to retinaldehyde, conversion from retinaldehyde to ATRA are maintained, excess ATRA can be rapidly degraded by the P450 hydroxylase family of enzymes (including CYP26A1, CYP26B1 and CYP26C1) (Pennimpede et al., 2010). ATRA is then shuttled into the nucleus with the help of cellular retinoic acid-binding protein 2 (CRABP2), where it serves as the ligand for the transcriptional machinery including ATRA receptors (RARA, RARB, and RARG). These RARs form dimers with retinoid X receptors (RXRA, RXRA, and RXRA) that bind to DNA via retinoic acid response elements, or RAREs (Germain et al., 2006). Once bound to chromatin, they interact with a variety of nuclear co-activators and co-repressors to regulate target gene transcription. The targets of ATRA signaling vary in different contexts, but frequently the CYP26 enzymes are targets to help autoregulate and fine tune the pathway in real time (Craig & Moon, 2011).

3. Pancreas development

The pancreas is a glandular organ with a variety of functions including regulation of blood-glucose homeostasis and secretion of enzymes to aid in digestion. These two disparate processes are controlled by endocrine and exocrine compartments, respectively. The endocrine cells are known as islets of Langerhans and contain hormone producing cells, the most abundant of which are insulin producing β cells (Jørgensen et al., 2006; Mastracci & Sussel, 2012). Working in concert with the β cells are the glucagon-positive α cells, somatostatin-positive δ, and pancreatic polypeptide-positive cells. During development, ghrelin-positive ε cells and gastrin-positive cells are also present, although these are mostly lost during islet maturation after birth (Arnes, Hill, Gross, Magnuson, & Sussel, 2012; Suissa et al., 2013). The exocrine compartment is made up of acinar cells that secrete digestive enzymes including both trypsin and chymotrypsin (to break down proteins), amylase (to break down carbohydrates), and lipase (to break down fats). These secreted enzymes are released in response to food ingestion via the pancreatic ducts into the duodenum where they interact with bile from the gall bladder to extract the key nutrients required to survive. A healthy adult pancreas is composed of more than 95% exocrine and less than 5% endocrine cells. Despite this representative imbalance, both cell types are indispensable for the health of the individual.

All cells of the pancreas, including endocrine, acinar, and ductal cells, share a common progenitor that first arises as a single PDX1 positive bud on the dorsal side of the foregut endoderm around embryonic day (E)8.25 in mice and day 23 in humans (~Carnegie Stage, CS11) (Table 1) (Jennings, Berry, Strutt, Gerrard, & Hanley, 2015). While PDX1 is one of the earliest transcription factors (TFs) expressed in PPs, it is not the first because in a mouse model deleted for PDX1, a small pancreatic bud can still be identified, although it does not develop into a fully formed pancreas (Jonsson, Carlsson, Edlund, & Edlund, 1994; Rosanas-Urgell, Marfany, & Garcia-Fernandez, 2005). Another early factor, PTF1A, is also essential for pancreas development, although double knockout (KO) of *Ptf1a* with *Pdx1* still allows pancreatic bud formation suggesting there are even earlier TFs that have yet to be identified regulating the very earliest induction of pancreagenesis (Kumar, Jordan, Melton, & Grapin-Botton, 2003). The dorsal pancreas receives important signals from the dorsal aorta and notochord including FGFs, activin, and ATRA signaling (Kim, Hebrok, & Melton, 1997). Importantly, ATRA signaling induces *Pdx1* in the foregut endoderm via conserved molecular mechanisms in humans, chicken, mice, and fish (Kumar et al., 2003; Pérez, Benoit, & Gudas, 2013).

Within 24–48 h (E9-E10.5 in mice and CS12–13, days 28–32 in humans, Table 1), a second pancreatic bud will arise on the ventral side of the foregut endoderm and develops next to the liver primordia. This ventral bud receives signals from the cardiac mesoderm and septum transversum mesenchyme, predominantly involving FGFs and activins. There are mixed reports regarding whether ATRA signaling is also required for ventral pancreas development. Two studies suggested that deletion of *Aldh1a2* preferentially impacted dorsal compared to ventral pancreas development (Martín et al., 2005; Molotkov, Molotkova, & Duester, 2005). However, another group suggested that there was an important role for ATRA in the regulation of both ventral and dorsal pancreatic lobes in studies that used a dominant-negative RARA molecule (called dnRARA-403) to examine loss of ATRA function in vivo (Öström et al., 2008). This allele was generated by removing the activation domain of human RARA at position 403 that was shown to transcriptionally inhibit ATRA signaling in vitro, and when this dnRARA-403 fragment was injected as a randomly integrated transgene in mice, led to severe malformation of the palate and early postnatal lethality (Damm, Heyman, Umesono, & Evans, 1993). Due to redundancy built into the ATRA pathway, this tool is widely used to impair ATRA signaling in vivo in mice. It is also important to note that in addition to disrupting ATRA signaling, the dnRARA-403 allele could

Table 1 Summary of the role of ATRA signaling during pancreas development.

Pancreas development	Mouse stage	Human stage	Summarized roles of ATRA signaling	References
Dorsal bud formation, PDX1[+]	E8.25–E8.5	~CS11, day 23	ATRA regulates PDX1 expression and induces pancreatic bud formation. $Rdh10-/-$ and $Aldh1a2-/-$ mice had dorsal pancreas agenesis by E9.5	Martín et al. (2005), Molotkov et al. (2005)
Ventral bud formation, PDX1[+]	E9.0–E10.5	~CS13–14, days 28–32	Some Debate: $Rdh10-/-$ had dorsal agenesis and ventral hypoplasia. dnRARa403 expressing mice had complete agenesis when examined at birth	Arregi et al. (2016), Öström et al. (2008)
Expand, branch, begin exo-and endocrine specification	E10.5–E12.5	~CS14–16, days 33–38	Promotes endocrine specification: Adding ATRA to mouse explants increased Neurog3 + Cells	Öström et al. (2008)
D/V fusion endocrine progenitors are NEUROG3[+]	E12.5–E14.5	~CS16–19 days 38–47	Some Debate: ATRA has promoted and inhibited beta cell formation in human stem cell derived islets. In mice, ATRA generally promotes beta cell development, but seems to inhibit beta cell formation in zebrafish	Lorberbaum, Kishore et al. (2020), Öström et al. (2008), Huang et al. (2014), Vinckier et al. (2020)

A selection of the major phases of pancreas development are summarized in each row with corresponding timepoints in mice(Embryonic Day, E) and humans (Carnegie Stage, CS). Ongoing debates exist around dorsal/ventral specification and endocrine progenitor differentiation. This table was made with Biorender. This table was made with Biorender.

also inadvertently disrupt other signaling cascades. For example, The Evans group, who also created the model, later demonstrated the nuclear corepressor SMRT, renamed nuclear corepressor 2 (NCOR2), is an important regulator of both retinoid and thyroid-hormone receptors and could be impacted by dnRARA-403 action (Chen & Evans, 1995). Further, under conditions in which RAR function is impaired, RXR interactions with other pathways could also be disrupted leading to functional defects that are not normally regulated by ATRA signaling (Mangelsdorf et al., 1995). These potential caveats should be considered when interpreting results using this, or any dominant negative allele due to potential off target effects. To target pancreas development, a 4.5 kb fragment containing the *Pdx1* gene promoter was used to drive expression of dnRARA-403 that was randomly integrated into the genome to create transgenic mice (Apelqvist, Ahlgren, & Edlund, 1997; Ostrőm et al., 2008). This experiment yielded complete pancreas agenesis by birth, suggesting that ATRA signaling impacts both dorsal and ventral development. These results were supported by additional studies in mice carrying a deletion of *Rdh10* that caused dorsal pancreas agenesis and hypoplastic ventral pancreas with impaired growth and branching (Arregi et al., 2016). The discrepant results could be due to the inhibition of different pathway components in dnRARA-403-expressing mice and/or the existence of pathway redundancies that can at least partially compensate in some experimental contexts in *Rdh10*-null mice. Despite the differences, these studies collectively indicate a substantial role for ATRA signaling during PP development, likely in both the dorsal and ventral regions of the pancreas. Additional studies will be important to determine the extent to which ATRA signaling is required for ventral pancreas specification, as this could potentially highlight non-ATRA regulation of *Pdx1* expression in the ventral pancreas. Addressing these outstanding questions will likely provide additional novel mechanisms by which a pancreas is generated.

During the subsequent stages of development – at approximately E10.5-E12.5 in mice and about 33–38 days in humans (Table 1) – each region of the pancreas expands, branches, and begins apparently identical differentiation programs to form the endo- and exocrine tissues. In mice, glucagon positive cells arise first, followed by insulin positive cells in what is referred to as the primary transition (Gu, Dubauskaite, & Melton, 2002). The fate of these early hormone-positive cells is not completely known, but most likely they do not contribute to the final endocrine cells of the islet. Fusion of the dorsal and ventral pancreas regions occurs between E12.5-E14.5 in mice and 38–47 days in humans by morphological movements facilitated by rotations around the gut tube endoderm. Both regions of the pancreas give rise to all cell-types of the pancreas but remain in somewhat morphologically distinct regions: the head, which is derived from the ventral pancreas, will be directly attached to the duodenum, while the tail, which is derived from the dorsal pancreas will be attached to the stomach and spleen. During the developmental stages when these two buds fuse, the secondary transition begins when multipotency of the progenitor population is lost and the endocrine and exocrine compartments are specified. One of the major events required for this transition is the transient expression of the transcription factor NEUROG3, which is expressed around E12.5 during mouse development. Within about two days, insulin positive β cells begin to form and expand 4–10-fold more than other endocrine cell-types until birth (Herrera, 2002). Studies from Ostrom et al. demonstrated a role for ATRA signaling to promote endocrine progenitor specification and differentiation of β cells (Öström et al., 2008). Upon identifying that the key ATRA pathway component ALDH1A is expressed in the epithelium of both mouse and human pancreata near endocrine progenitors, they used an *ex vivo* model in which E10.5 mouse pancreas was isolated and grown in culture with or without ATRA. This model allowed them to circumvent the pancreas agenesis phenotype associated with early disruption of ATRA signaling. In this study, the authors demonstrated a significant increase in NEUROG3-positive cells, indicating that ATRA signaling likely contributes to endocrine progenitor specification. Interestingly, there was also a significant (>3 fold) increase in insulin positive cells with no significant change in glucagon-positive α cell numbers, suggesting an additional role for ATRA signaling in regulating endocrine progenitor specification. Since only insulin- and glucagon-positive cells were examined in these explants, it is still possible that the other endocrine cell-types may also be impacted, something that future studies have now addressed and are discussed below.

A handful of subsequent studies have addressed the role of ATRA in NEUROG3-positive endocrine progenitors using a variety of model systems. Huang et al. created a zebrafish dnRARA-403 construct [*Tg(ubb:loxP-eCFP-loxP-dnRAR-GFP)^{jh39}*],combined with human stem cell derived pancreatic β like cells to examine this question (Huang et al., 2014). After confirming a role for ATRA signaling in specifying PPs from nascent endoderm, like previous work (Stafford & Prince, 2002; Stafford, Hornbruch, Mueller, & Prince, 2004), they identified a second role for ATRA signaling that impairs endocrine formation. They confirmed these results using chemical inhibition of ALDH1A to inhibit the ATRA pathway via the drug 4-diethylaminobenzaldehyde (DEAB), which showed that blocking ATRA signaling increased the number of islet cells as defined by insulin, glucagon and somatostatin staining. Individual endocrine cell-types were not examined in this zebrafish model, but the authors performed similar experiments in human stem cell-derived islet differentiations. The latter requires high levels of ATRA signaling during pancreas progenitor specification, with a reduced requirement over time (Ramzy et al., 2022; Rezania et al., 2012; Rezania et al., 2014). Interestingly, the authors showed that within stage 4 of the differentiation program when endocrine progenitors are beginning to be specified, an endogenous source of ALDH1a is induced and can produce ATRA. This would suggest that, even though ATRA is not added to the media at this stage, any retinol supplied in the media would still be capable of activating the ATRA pathway. Therefore, the authors used the chemical inhibitor DEAB to block ATRA signaling by inhibiting ALDH1A during this time-window and found an increase in insulin, glucagon and somatostatin expression in addition to a significant increase in insulin-positive cells and a trending increase in glucagon-positive cells upon ATRA inhibition, largely agreeing with their in vivo zebrafish experiments. These results also agree with work done by the Sander group in which ATRA was continuously provided to hPSC-derived β cell past PP specification stage, leading to reductions in endocrine cells (Vinckier et al., 2020). Interestingly, these results are in contrast to the observations from the Edlund group in which they identified ATRA as a signal that promotes the formation of β cells during similar stages of pancreas development (Öström et al., 2008). It is possible that there are different mechanisms at play between the different models used, including species differences, the use of different dnRARA-403-expressing alleles, and the different chemicals used to activate or inhibit ATRA signaling. Additional studies will be important to untangle these differences and will likely uncover additional interacting partners of ATRA signaling important for this stage of pancreas specification.

To further examine the role of ATRA during endocrine progenitor specification, the Sussel group made use of the murine dnRARA-403 that had been integrated into the *Rosa26* locus downstream of a loxP-stop-loxP cassette and hence could be activated upon expression of CRE recombinase ($Gt(ROSA)26Sor^{tm1(RARA*)Soc}$ transgene) (Rosselot et al., 2010). After confirming that ATRA inhibition via this dnRARA-403 with a tamoxifen inducible *Pdx1:CreER* driver [*Tg(Pdx1-cre/Esr1*)#Dam* transgene] had a significantly smaller pancreas with fewer endocrine cells as compared to controls, which agreed with previous studies (Öström et al., 2008), they used a *Neurog3:Cre* driver (*Tg(Neurog3-cre)C1Able* transgene) to inhibit ATRA signaling in endocrine progenitors. Impairment of ATRA signaling specifically in the endocrine progenitor population caused a significant reduction in insulin transcripts and insulin-positive cells by E16.5 that persisted through postnatal day (P)2 where increased blood glucose was also observed (Lorberbaum, Kishore et al., 2020). These mice continued to be glucose intolerant, as adults, suggesting the existence of a functional endocrine defect, although they never became overtly diabetic. Interestingly, during endocrine specification in fetal mice, there was also an increase in somatostatin expression and other δ cell transcripts not accompanied by an increase in δ cell number, highlighting a potentially novel role for ATRA signaling in regulating δ cell functions in addition to a role in β cell specification (DiGruccio et al., 2016).

The potential novel role of ATRA in promoting the β cell lineage in the mice prompted Lorberbaum and colleagues to determine whether there was an analogous role for ATRA in a human model of β-like cell differentiation from human embryonic stem cells (Lorberbaum, Kishore et al., 2020). Using a panRAR chemical inhibitor to block signaling at the transcriptional level starting on day 11 of the differentiation, corresponding to endocrine progenitor specification as NEUROG3 is turning on (Tiyaboonchai et al., 2017), they discovered that instead of increasing insulin-positive, glucagon-positive, and/or somatostatin-positive expression as demonstrated previously (Huang et al., 2014), ATRA signaling inhibition led to a significant reduction in insulin expression with a trending loss of β cells without significantly impacting the other cell-types. There are many differences in experimental design that could have led to these apparently conflicting results, including the embryonic stem cell (ESC) lines, the differentiation protocols, and the growth factors and signaling molecules that were used. Furthermore, the ALDH1A inhibitor acts upstream of RAR and does not inhibit all aldehyde dehydrogenases, whereas the experiments that inhibit the RAR/RXR complex should avoid potential compensatory events. Regardless of the conflicting results, it seems likely that ATRA signaling is impacting the specification of islet cells during endocrine progenitor specification.

4. hPSC differentiations and therapeutic interventions

As previously discussed, hPSC differentiation protocols are increasingly used to better understand pancreas development and serve as a potential therapeutic for treating people with diabetes by the production of islet-like clusters from an undifferentiated embryonic state. Broadly, these protocols aim to replicate aspects of in vivo pancreatic development by guiding uncommitted ESC through successive pancreatic differentiation stages. In order to generate a β- or islet-like cell, hPSCs are directed to complete developmental stages corresponding to in vivo pancreatic embryogenesis that has mostly been studied in rodent models of development (Lorberbaum, Docherty, & Sussel, 2020), resulting in hPSC-derived cells characterized by the several stages including definitive endoderm; primitive gut tube; pancreatic progenitors (PP), endocrine progenitors, and finally stem cell-derived islets (Pagliuca et al., 2014; Rezania et al., 2014; Balboa et al., 2022) (Fig. 2).

Using these *in-vitro* models that mimic in vivo development, stage-specific cocktails of growth factors and compounds are employed to provide the precise molecular cues and modulation of signaling pathways to drive differentiation towards the pancreatic lineage. One of the critical molecular cues in the differentiation of hPSC-derived islet cells is ATRA. Similar to what was shown in animal models, it is clear that ATRA facilitates the differentiation of PP cells from endoderm (Lorberbaum, Kishore et al., 2020; Martín et al., 2005; Öström et al., 2008), and elevated ATRA levels, in conjunction with other signaling factors, are utilized to enhance the expression of pancreatic transcription factors such as PDX1 and NKX6.1, two critical markers of PPs (Jiang et al., 2007). However, increased levels of ATRA also led to the upregulation of retinoid receptors in the pancreatic exocrine cells (Kadison et al., 2001). Consequently, in most in vitro differentiation protocols, RA concentration is initially high to induce PDX1 expression but is gradually reduced to maintain PDX1 levels

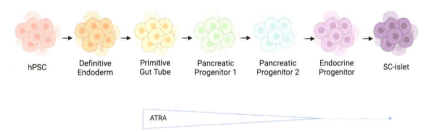

Fig. 2 Summary of the addition of retinoic acid signaling to hPSC derived islet differentiations. ATRA or retinol is added to the cultures in high concentrations starting at the primitive gut tube stage of development and reduced as stem cell derived-islets are formed. *This figure was made with Biorender.*

and promote or sustain NKX6.1 expression. After PPs have been specified, ATRA is administered at much lower concentrations, alongside additional factors, to facilitate the formation of endocrine progenitor cells. A recent report by the Gavalas group demonstrated that autocrine ATRA signaling likely occurs in PP cells allowing them to differentiate, even when ATRA is not present in the media (Jarc et al., 2024), which is in agreement with the previous studies (Huang et al., 2014). Interestingly, when retinol was removed from their cultures, the PP differentiations were stabilized. Furthermore, simultaneous inhibition of PDGF signaling allowed these cells to be reproducibly expanded, representing a potentially crucial advancement in the generation of hPSC-derived islet cells that could save time, effort, and funds when relying on hPSC derived islet-like cells for experiments and as a possible therapy for patients with diabetes.

In summary, ATRA signaling is a required component during differentiation of hPSCs into islet-like clusters as it regulates gene expression, promotes pancreatic lineage specification, influences developmental pathways, and enhances endocrine function. A better understanding of ATRA signaling and its function in these essential processes can lead to more effective and reliable methods for generating islet-like clusters, including functional β cells, for research and therapeutic purposes.

5. Maintaining islet function

Retinoid signaling is not only essential for healthy pancreas development but is also active in adult islets. For example, a well characterized ATRA-dependent reporter line, called RARE-lacZ (*Tg(RARE-Hspa1b/lacZ)12Jrt* transgene) (Rossant, Zirngibl, Cado, Shago, & Giguere, 1991), definitively marks retinoid-responsive cells in adult islets using β-galactosidase staining (Colvin et al., 2011). Knowing that retinoid signaling occurs in healthy islets, here we have highlighted a handful of important contributions spanning several decades that use chemical and genetic strategies to clarify the role of retinoid signaling in adult islet function.

One example of a study that identified a role for retinoid signaling in adult islets showed how different concentrations of 13-*cis* retinoic acid (13-*cis* RA) impacted islet cell adhesion dynamics and glucose-stimulated insulin secretion (GSIS) in rat models of β cell function (Chertow et al., 1983). Subjecting isolated rat islets to high levels (100 μM) of 13-*cis* RA significantly stimulated insulin secretion at different glucose concentrations

(9.7, 12.5, 16.7 and 27.7 mM). These results indicate that islet function can be altered by retinoid availability, and are especially interesting because they were obtained prior to identification of ATRA and 9-*cis* retinoic acid (9-*cis* RA), as the most transcriptionally active retinoid species (Gudas, 2012). Although ATRA and 9-*cis* RA have since become prevalent in contemporary research, 13-*cis* RA was still able to stimulate these islets, albeit to a lower extent than the other versions, but still highlights the conserved mechanisms by which retinoid signaling impacts insulin secretion. Decades later, 9-*cis* RA was identified within the pancreas, introducing the concept of autocrine retinoid signaling maintaining islet function (Kane et al., 2010). Here, the Napoli group used liquid chromatography to define the presence of 9-*cis* RA in the pancreas and showed reduced GSIS in mice and a rat β cell line. Along with these functional readouts, they also identified down regulated β cell factors like *Gck*, *Pdx1*, and *Hnf4a*. While it is not reported if these changes in gene expression are a direct result of retinoid manipulation in their model systems, there are clear changes to the β cell in both molecular makeup and function regarding insulin secretion and regulation of blood glucose homeostasis.

Recent work has shown that 9-*cis* RA has an expanded role in more organs than previously thought (Yoo, Moss, Cockrum, Woo, & Napoli, 2023) and several groups have sought to identify the key enzymes by which 9-*cis* RA is generated (Labrecque, Dumas, Lacroix, & Bhat, 1995; Romert, Tuvendal, Simon, Dencker, & Eriksson, 1998). In particular, an enzyme that was originally characterized for how it processes 9-*cis* RA is ALDH8A1 (Lin & Napoli, 2000), however, later studies have suggested that it should be reclassified as part of the kynurenine pathway (Davis, Yang, Wherritt, & Liu, 2018), and expression of this enzyme is undetectable in pancreatic islets (Hrovatin et al., 2023; Tabula Muris et al., 2018). Only a few studies have examined a mechanistic role for 9-*cis* RA signaling in the pancreas. Kane and colleagues demonstrated a role for 9-*cis* RA in maintaining islet function and attempted to identify the regulatory pathway functioning upstream of 9-*cis* RA by examining *Rbp1* knockout mice (*Rbp1^{tm1Ipc}*). They hypothesized that losing *Rbp1* would impact retinoid homeostasis in the pancreas based on the role of CRBP1 (the protein encoded by *Rbp1*) as a chaperone of retinoid homeostasis (Fig. 1). As predicted, they found substantial disruption of energy metabolism and glucose homeostasis in these mice along with increased levels of 9-*cis* RA, indicating that CRBP1 is an important mediator or 9-*cis* RA processing in the pancreas (Kane et al., 2011). While there remain several unaddressed questions about the full complement of enzymes involved in processing 9-*cis* RA, there is growing evidence for a role for 9-*cis* RA in pancreatic regulation of energy homeostasis.

In addition to studies using chemical reagents to elicit changes in islet function, including those examining retinoid produced within the pancreas itself, several groups have also used genetic models to modulate this pathway in adults. For instance, Miyazaki et al. utilized a robust genetic strategy involving an inducible dnRXRB, which inhibits RXR transcriptional activity (Miyazaki et al., 2010). By expressing dnRXRB using a Tet-On system under control of the human insulin gene promoter, they were able to fine tune expression in the adult β cell but not impact other islet/pancreas cell-types. This also effectively avoids altering developmental processes that require retinoid signaling and focuses more specifically on the role of adult islets. Upon examination of these islets, they found reduced insulin secretion in response to elevated glucose levels, suggesting that the role of RXR, and therefore retinoid signaling, in the islet is to negatively regulate GSIS from β cells (Umemiya et al., 1997). There was also speculation of interactions with PPARG in this case, as RXR is involved with multiple signaling pathways (Brun et al., 2015; Brun, Wongsiriroj, & Blaner, 2016). In another study using the dnRARA-403 as in developmental contexts (Lorberbaum, Kishore et al., 2020; Oström et al., 2008; Rosselot et al., 2010), the Blaner group also effectively inhibited retinoid signaling in the adult islets. Similar to Miyazaki et al., the RARa dominant negative (RARdn) bypassed early developmental roles of retinoid-dependent signaling as it was controlled by the *Tg(Pdx1-cre/Esr1*) #Dam* transgene (Brun et al., 2015). To accomplish this, the authors completed intraperitoneal injections of tamoxifen at days 63, 65, and 67 and waited 4–5 weeks to complete GSIS, gene expression, and blood test analyses. Interestingly, they found reduced insulin levels in dnRARA-403-expressing mice as compared to control islets in response to different glucose concentrations (2.8, 11.2 and 16.8 mM) and KCl (20 mM). To confirm these genetic results, they also used LE540, a panRAR-antagonist, to pharmacologically block retinoid signaling in isolated wild type islets. GSIS on these samples demonstrated decreased insulin secretion. Mice that had defects in insulin secretion due to the presence of the dnRARA-403 were euglycemic, however in either the fed state or upon glucose challenge, the dnRARA-403-expressing mice underperformed compared to controls, again indicating a role for retinoid signaling in the β cell. Histological analyses of their pancreata showed loss of insulin-positive staining compared to control islets, despite seeing no change in the mRNA levels of *Ins1* or *Ins2,* suggesting there may be a post-transcriptional effect on insulin expression. Finally, Brun and colleagues also found reduced levels of *Isl1*, an important islet gene, although other β cell genes like *Pdx1, Nkx2.2, Nkx6.1, Pax4, Pax6, Neurod1, Neurog3, Mafa,* and *Mafb* were not reduced. Many retinoid pathway components were significantly increased, together suggesting there was possible upregulation of retinoid processing in the islet to compensate for the effects of dnRARA-403 and maintain expression of some key islet transcription factors (Brun et al., 2015; Brun et al., 2016).

In summary, findings from various studies using a wide range of pharmacological and genetic approaches in multiple different models underscore the vital roles of retinoid-associated proteins and signaling in supporting the overall endocrine functions of pancreatic islets, including glucose-stimulated insulin secretion and the maintenance of β cell mass. Importantly, in the studies mentioned here, the islet defects were examined in adults bypassing the known developmental roles of ATRA signaling during pancreas development.

6. ATRA signaling in disease

While most of the focus of this review has been on the role of ATRA signaling during development and maintaining islet function, it is perhaps not surprising that this pathway has also been implicated in pancreatic disease. There is a wealth of promising literature and several reviews demonstrating the importance of ATRA signaling in pancreatic cancer, including pancreatic ductal adenocarcinoma (PDAC), a disease of the exocrine, glandular pancreas (Mere Del Aguila, Tang, & Gudas, 2022; Sun, Zheng, Gao, Brigstock, & Gao, 2024). Understanding the exact molecular mechanisms by which PDAC develops is an active area of study. It seems likely that this cancer develops from either pancreatic ducts or acinar cells via pancreatic intraepithelial neoplasias (PanINs), which are susceptible to KRAS mutations. Together, this leads to a depletion of ATRA availability and signaling in the pancreas – supplementation with ATRA has been demonstrated to inhibit tumor progression in mouse models (Arima et al., 2019). Interestingly, pancreatic acinar cells can undergo a process of de- or transdifferentiation in which, due to injury and/or disease, these cells become more progenitor-like and begin re-expressing developmental signatures (Storz, 2017). When thinking about PDAC in this overly simplified manner, it is perhaps not surprising that ATRA has a role in this disease (Colvin et al., 2011) as there is substantial reactivation of developmental signaling pathways, such as ATRA, when cells transition from a healthy to cancerous state.

In addition to its role in PDAC, there is also considerable evidence that ATRA signaling can contribute to metabolic disease related to blood glucose dysregulation, including diabetes. In studies from the late 1980's, Basu and colleagues demonstrated that patients with type 1 diabetes (T1D) had lower levels of circulating retinol (Basu, Tze, & Leichter, 1989), and although there was debate over these findings in the following decades, it seems likely that there is at least some association between circulating retinol and diabetes status, in which retinol is decreased in patients T1D and potentially increased with type 2 diabetes (T2D) (Lu et al., 2022). Interestingly, a handful of case studies also indicate that ATRA pathway components, like RBP4 (Fig. 1), could serve as a risk factor for gestational diabetes (Chen et al., 2017; Zhaoxia, Mengkai, Qin, & Danqing, 2014). These studies, however, are limited in scope both in numbers of patients examined and definition of the mechanism by which the changes in retinol or ATRA signaling components were altered and how they might directly impact disease. Still, these observations led to more targeted studies in animal models demonstrating a protective function of ATRA in models of pancreas injury that induces diabetes conditions in rodents (Chertow, Webb, Leidy, & Cordle, 1989; Eltony, Elmottaleb, Gomaa, Anwar, & El-Metwally, 2016). Along these lines, a more recent study made use of nanoparticles containing ATRA and other signaling agonists that were provided to mice as an oral supplement that were able to significantly reduce the severity and incidence of diabetes induction via streptozotocin (Koprivica et al., 2019). Koprivica et al. found that this inhibition was likely due to altering the immune response, another means by which ATRA could be used to modulate disease progression.

Research focused more specifically on islet biology should also be considered in the context of metabolic disease. For instance, when Brun and colleagues inhibited ATRA signaling using the dnRARA-403 in adult islets they found a loss of β cell mass and insulin content as well as disrupted function as measured via glucose stimulated insulin secretion in isolated islets and glucose tolerance tests in vivo (Brun et al., 2015; Brun et al., 2016). Similar results were reported using dietary retinol depletion in rodent models, further demonstrating that ATRA signaling and retinol are essential for maintaining blood glucose homeostasis (Trasino, Benoit, & Gudas, 2015).

This work from the Gudas group also demonstrated that retinol-deficient murine islets resemble islets affected by advanced T2D and went on to reintroduce retinol into the diets of these sick mice leading to a restoration of both pancreas function by restoration of euglycemia, islet number, and structure. These studies collectively advocate for more studies, like those performed by Koprivica and colleagues, that explore the use of retinol and its derivates in the treatment of diabetes. Defining these mechanisms by which ATRA acts in cases of disease could provide critical improvements to disease treatment.

7. Conclusion

While a few discrepancies remain amongst the studies discussed here, it is clear there is an essential requirement for ATRA signaling at many different states of pancreas development, for maintaining adult islet function, and in pancreas disease. There are numerous groups currently exploring ATRA function in the pancreas to continuously identify new roles for ATRA signaling in the pancreas, and in other contexts. New experimental tools and techniques along with greater access to patient samples will likely clarify these issues and better define the mechanistic roles of ATRA signaling in the future.

Acknowledgments

We would like to thank members of the Lorberbaum Lab, especially Andrea Laurin, for their helpful reading and discussion of this manuscript. This work is supported by the National Institutes of Health R00DK128537 (to DSL), P30DK020572 (to DSL), R01DK082590 (to LS), and R01DK118155 (to LS).

References

Apelqvist, A., Ahlgren, U., & Edlund, H. (1997). Sonic hedgehog directs specialised mesoderm differentiation in the intestine and pancreas. *Current Biology: CB, 7*, 801–804.

Arima, K., Ohmuraya, M., Miyake, K., Koiwa, M., Uchihara, T., Izumi, D., ... Ishimoto, T. (2019). Inhibition of 15-PGDH causes Kras-driven tumor expansion through prostaglandin E2-ALDH1 signaling in the pancreas. *Oncogene, 38*, 1211–1224.

Arnes, L., Hill, J. T., Gross, S., Magnuson, M. A., & Sussel, L. (2012). Ghrelin expression in the mouse pancreas defines a unique multipotent progenitor population. *PLoS One, 7*, e52026.

Arregi, I., Climent, M., Iliev, D., Strasser, J., Gouignard, N., Johansson, J. K., ... Pera, E. M. (2016). Retinol dehydrogenase-10 regulates pancreas organogenesis and endocrine cell differentiation via paracrine retinoic acid signaling. *Endocrinology, 157*, 4615–4631.

Balboa, D., Barsby, T., Lithovius, V., Saarimaki-Vire, J., Omar-Hmeadi, M., Dyachok, O., ... Otonkoski, T. (2022). Functional, metabolic and transcriptional maturation of human pancreatic islets derived from stem cells. *Nature Biotechnology, 40*, 1042–1055.

Basu, T. K., Tze, W. J., & Leichter, J. (1989). Serum vitamin A and retinol-binding protein in patients with insulin-dependent diabetes mellitus. *The American Journal of Clinical Nutrition, 50*, 329–331.

Berner, L. A., Keast, D. R., Bailey, R. L., & Dwyer, J. T. (2014). Fortified foods are major contributors to nutrient intakes in diets of US children and adolescents. *Journal of the Academy of Nutrition and Dietetics, 114*, 1009–1022 e1008.

Billings, S. E., Pierzchalski, K., Butler Tjaden, N. E., Pang, X. Y., Trainor, P. A., Kane, M. A., & Moise, A. R. (2013). The retinaldehyde reductase DHRS3 is essential for preventing the formation of excess retinoic acid during embryonic development. *The FASEB Journal, 27*, 4877–4889.

Brun, P.-J., Grijalva, A., Rausch, R., Watson, E., Yuen, J. J., Das, B. C., ... Blaner, W. S. (2015). Retinoic acid receptor signaling is required to maintain glucose-stimulated insulin secretion and β-cell mass. *FASEB Journal: Official Publication of the Federation of American Societies for Experimental Biology, 29*, 671–683.

Brun, P. J., Wongsiriroj, N., & Blaner, W. S. (2016). Retinoids in the pancreas. *Hepatobiliary Surgery and Nutrition, 5*, 1–14.

Chen, J. D., & Evans, R. M. (1995). A transcriptional co-repressor that interacts with nuclear hormone receptors. *Nature, 377*, 454–457.

Chen, Y., Lv, P., Du, M., Liang, Z., Zhou, M., & Chen, D. (2017). Increased retinol-free RBP4 contributes to insulin resistance in gestational diabetes mellitus. *Archives of Gynecology and Obstetrics, 296*, 53–61.

Chertow, B. S., Baranetsky, N. G., Sivitz, W. I., Meda, P., Webb, M. D., & Shih, J. C. (1983). Cellular mechanisms of insulin release. Effects of retinoids on rat islet cell-to-cell adhesion, reaggregation, and insulin release. *Diabetes, 32*, 568–574.

Chertow, B. S., Webb, M. D., Leidy, J. W., Jr., & Cordle, M. B. (1989). Protective effects of retinyl palmitate on streptozotocin- and alloxan-induced beta cell toxicity and diabetes in the rat. *Research Communications in Chemical Pathology and Pharmacology, 63*, 27–44.

Cho, K., Lee, S. M., Heo, J., Kwon, Y. M., Chung, D., Yu, W. J., ... Kim, Y. (2021). Retinaldehyde dehydrogenase inhibition-related adverse outcome pathway: Potential risk of retinoic acid synthesis inhibition during embryogenesis. *Toxins (Basel), 13*.

Colvin, E. K., Susanto, J. M., Kench, J. G., Ong, V. N., Mawson, A., Pinese, M., ... Biankin, A. V. (2011). Retinoid signaling in pancreatic cancer, injury and regeneration. *PLoS One, 6*, e29075.

Craig, P. M., & Moon, T. W. (2011). Fasted zebrafish mimic genetic and physiological responses in mammals: A model for obesity and diabetes? *Zebrafish, 8*, 109–117.

Cunningham, T. J., & Duester, G. (2015). Mechanisms of retinoic acid signalling and its roles in organ and limb development. *Nature Reviews. Molecular Cell Biology, 16*, 110–123.

Damm, K., Heyman, R. A., Umesono, K., & Evans, R. M. (1993). Functional inhibition of retinoic acid response by dominant negative retinoic acid receptor mutants. *Proceedings of the National Academy of Sciences of the United States of America, 90*, 2989–2993.

Davis, I., Yang, Y., Wherritt, D., & Liu, A. (2018). Reassignment of the human aldehyde dehydrogenase ALDH8A1 (ALDH12) to the kynurenine pathway in tryptophan catabolism. *The Journal of Biological Chemistry, 293*, 9594–9603.

DiGruccio, M. R., Mawla, A. M., Donaldson, C. J., Noguchi, G. M., Vaughan, J., Cowing-Zitron, C., ... Huising, M. O. (2016). Comprehensive alpha, beta and delta cell transcriptomes reveal that ghrelin selectively activates delta cells and promotes somatostatin release from pancreatic islets. *Molecular Metabolism, 5*, 449–458.

Duester, G. (2008). Retinoic acid synthesis and signaling during early organogenesis. *Cell, 134*, 921–931.

Eltony, S. A., Elmottaleb, N. A., Gomaa, A. M., Anwar, M. M., & El-Metwally, T. H. (2016). Effect of all-trans retinoic acid on the pancreas of streptozotocin-induced diabetic rat. *Anatomical Record (Hoboken, N. J.: 2007), 299*, 334–351.

Feng, L., Hernandez, R. E., Waxman, J. S., Yelon, D., & Moens, C. B. (2010). Dhrs3a regulates retinoic acid biosynthesis through a feedback inhibition mechanism. *Developmental Biology, 338*, 1–14.

Germain, P., Chambon, P., Eichele, G., Evans, R. M., Lazar, M. A., Leid, M., ... Gronemeyer, H. (2006). International Union of Pharmacology. LX. Retinoic acid receptors. *Pharmacological Reviews, 58*, 712–725.

Ghyselinck, N. B., & Duester, G. (2019). Retinoic acid signaling pathways. *Development (Cambridge, England), 146*.

Gu, G., Dubauskaite, J., & Melton, D. A. (2002). Direct evidence for the pancreatic lineage: NGN3+ cells are islet progenitors and are distinct from duct progenitors. *Development (Cambridge, England), 129*, 2447–2457.

Gudas, L. J. (2012). Emerging roles for retinoids in regeneration and differentiation in normal and disease states. *Biochimica et Biophysica Acta, 1821*, 213–221.

Herrera, P. L. (2002). Defining the cell lineages of the islets of Langerhans using transgenic mice. *The International Journal of Developmental Biology, 46*, 97–103.

Hrovatin, K., Bastidas-Ponce, A., Bakhti, M., Zappia, L., Buttner, M., Salinno, C., ... Theis, F. J. (2023). Delineating mouse beta-cell identity during lifetime and in diabetes with a single cell atlas. *Nature Metabolism, 5*, 1615–1637.

Huang, W., Wang, G., Delaspre, F., Vitery Mdel, C., Beer, R. L., & Parsons, M. J. (2014). Retinoic acid plays an evolutionarily conserved and biphasic role in pancreas development. *Developmental Biology, 394*, 83–93.

Jarc, L., Bandral, M., Zanfrini, E., Lesche, M., Kufrin, V., Sendra, R., ... Gavalas, A. (2024). Regulation of multiple signaling pathways promotes the consistent expansion of human pancreatic progenitors in defined conditions. *Elife, 12*.

Jennings, R. E., Berry, A. A., Strutt, J. P., Gerrard, D. T., & Hanley, N. A. (2015). Human pancreas development. *Development (Cambridge, England), 142*, 3126–3137.

Jiang, W., Shi, Y., Zhao, D., Chen, S., Yong, J., Zhang, J., ... Deng, H. (2007). In vitro derivation of functional insulin-producing cells from human embryonic stem cells. *Cell Research, 17*, 333–344.

Jonsson, J., Carlsson, L., Edlund, T., & Edlund, H. (1994). Insulin-promoter-factor 1 is required for pancreas development in mice. *Nature, 371*, 606–609.

Jørgensen, M. C., Ahnfelt-Rønne, J., Hald, J., Madsen, O. D., Serup, P., & Heckshøer-Sørensen, J. (2006). An illustrated review of early pancreas development in the mouse. *Endocrine Reviews*.

Kadison, A., Kim, J., Maldonado, T., Crisera, C., Prasadan, K., Manna, P., ... Gittes, G. (2001). Retinoid signaling directs secondary lineage selection in pancreatic organogenesis. *Journal of Pediatric Surgery, 36*, 1150–1156.

Kane, M. A., Folias, A. E., Pingitore, A., Perri, M., Krois, C. R., Ryu, J. Y., ... Napoli, J. L. (2011). CrbpI modulates glucose homeostasis and pancreas 9-cis-retinoic acid concentrations. *Molecular and Cellular Biology, 31*, 3277–3285.

Kane, M. A., Folias, A. E., Pingitore, A., Perri, M., Obrochta, K. M., Krois, C. R., ... Napoli, J. L. (2010). Identification of 9-cis-retinoic acid as a pancreas-specific autacoid that attenuates glucose-stimulated insulin secretion. *Proceedings of the National Academy of Sciences of the United States of America, 107*, 21884–21889.

Kim, S. K., Hebrok, M., & Melton, D. A. (1997). Notochord to endoderm signaling is required for pancreas development. *Development (Cambridge, England), 124*, 4243–4252.

Koprivica, I., Gajic, D., Saksida, T., Cavalli, E., Auci, D., Despotovic, S., ... Stojanovic, I. (2019). Orally delivered all-trans-retinoic acid- and transforming growth factor-beta-loaded microparticles ameliorate type 1 diabetes in mice. *European Journal of Pharmacology, 864*, 172721.

Kumar, M., Jordan, N., Melton, D., & Grapin-Botton, A. (2003). Signals from lateral plate mesoderm instruct endoderm toward a pancreatic fate. *Developmental Biology, 259*, 109–122.

Labrecque, J., Dumas, F., Lacroix, A., & Bhat, P. V. (1995). A novel isoenzyme of aldehyde dehydrogenase specifically involved in the biosynthesis of 9-cis and all-trans retinoic acid. *The Biochemical Journal, 305*(Pt 2), 681–684.

Lin, M., & Napoli, J. L. (2000). cDNA cloning and expression of a human aldehyde dehydrogenase (ALDH) active with 9-cis-retinal and identification of a rat ortholog, ALDH12. *The Journal of Biological Chemistry, 275*, 40106–40112.

Lorberbaum, D. S., Docherty, F. M., & Sussel, L. (2020). Animal models of pancreas development, developmental disorders, and disease. *Advances in Experimental Medicine and Biology, 1236*, 65–85.

Lorberbaum, D. S., Kishore, S., Rosselot, C., Sarbaugh, D., Brooks, E. P., Aragon, E., ... Sussel, L. (2020). Retinoic acid signaling within pancreatic endocrine progenitors regulates mouse and human beta cell specification. *Development, 147*.

Lu, J., Wang, D., Ma, B., Gai, X., Kang, X., Wang, J., & Xiong, K. (2022). Blood retinol and retinol-binding protein concentrations are associated with diabetes: A systematic review and meta-analysis of observational studies. *European Journal of Nutrition, 61*, 3315–3326.

Mangelsdorf, D. J., Thummel, C., Beato, M., Herrlich, P., Schutz, G., Umesono, K., ... Evans, R. M. (1995). The nuclear receptor superfamily: The second decade. *Cell, 83*, 835–839.

Martín, M., Gallego-Llamas, J., Ribes, V., Kedinger, M., Niederreither, K., Chambon, P., ... Gradwohl, G. (2005). Dorsal pancreas agenesis in retinoic acid-deficient Raldh2 mutant mice. *Developmental Biology, 284*, 399–411.

Mastracci, T. L., & Sussel, L. (2012). The endocrine pancreas: insights into development, differentiation, and diabetes. *Wiley Interdisciplinary Reviews: Developmental Biology, 1*, 609–628.

Mere Del Aguila, E., Tang, X. H., & Gudas, L. J. (2022). Pancreatic ductal adenocarcinoma: New insights into the actions of vitamin A. *Oncology Research and Treatment, 45*, 291–298.

Miyazaki, S., Taniguchi, H., Moritoh, Y., Tashiro, F., Yamamoto, T., Yamato, E., ... Miyazaki, J. (2010). Nuclear hormone retinoid X receptor (RXR) negatively regulates the glucose-stimulated insulin secretion of pancreatic ss-cells. *Diabetes, 59*, 2854–2861.

Molotkov, A., Molotkova, N., & Duester, G. (2005). Retinoic acid generated by Raldh2 in mesoderm is required for mouse dorsal endodermal pancreas development. *Developmental Dynamics, 232*, 950–957.

Niederreither, K., & Dolle, P. (2008). Retinoic acid in development: Towards an integrated view. *Nature Reviews. Genetics, 9*, 541–553.

Oström, M., Loffler, K. A., Edfalk, S., Selander, L., Dahl, U., Ricordi, C., ... Edlund, H. (2008). Retinoic acid promotes the generation of pancreatic endocrine progenitor cells and their further differentiation into beta-cells. *PLoS One, 3*, e2841.

Pagliuca, F. W., Millman, J. R., Gurtler, M., Segel, M., Van Dervort, A., Ryu, J. H., ... Melton, D. A. (2014). Generation of functional human pancreatic beta cells in vitro. *Cell, 159*, 428–439.

Pennimpede, T., Cameron, D. A., MacLean, G. A., Li, H., Abu-Abed, S., & Petkovich, M. (2010). The role of CYP26 enzymes in defining appropriate retinoic acid exposure during embryogenesis. *Birth Defects Research. Part A, Clinical and Molecular Teratology, 88*, 883–894.

Pérez, R. J., Benoit, Y. D., & Gudas, L. J. (2013). Deletion of retinoic acid receptor β (RARβ) impairs pancreatic endocrine differentiation. *Experimental Cell Research, 319*, 2196–2204.

Ramzy, A., Belmonte, P. J., Braam, M. J. S., Ida, S., Wilts, E. M., Levings, M. K., ... Kieffer, T. J. (2022). A century long journey from the discovery of insulin to the implantation of stem cell derived islets. *Endocrine Reviews*.

Rezania, A., Bruin, J. E., Arora, P., Rubin, A., Batushansky, I., Asadi, A., ... Kieffer, T. J. (2014). Reversal of diabetes with insulin-producing cells derived in vitro from human pluripotent stem cells. *Nature Biotechnology, 32*, 1121–1133.

Rezania, A., Bruin, J. E., Riedel, M. J., Mojibian, M., Asadi, A., Xu, J., ... Kieffer, T. J. (2012). Maturation of human embryonic stem cell-derived pancreatic progenitors into functional islets capable of treating pre-existing diabetes in mice. *Diabetes, 61*, 2016–2029.

Rhinn, M., & Dolle, P. (2012). Retinoic acid signalling during development. *Development (Cambridge, England), 139*, 843–858.

Romert, A., Tuvendal, P., Simon, A., Dencker, L., & Eriksson, U. (1998). The identification of a 9-cis retinol dehydrogenase in the mouse embryo reveals a pathway for synthesis of 9-cis retinoic acid. *Proceedings of the National Academy of Sciences of the United States of America, 95*, 4404–4409.

Rosanas-Urgell, A., Marfany, G., & Garcia-Fernandez, J. (2005). Pdx1-related homeodomain transcription factors are distinctly expressed in mouse adult pancreatic islets. *Molecular and Cellular Endocrinology, 237*, 59–66.

Rossant, J., Zirngibl, R., Cado, D., Shago, M., & Giguere, V. (1991). Expression of a retinoic acid response element-hsplacZ transgene defines specific domains of transcriptional activity during mouse embryogenesis. *Genes & Development, 5*, 1333–1344.

Rosselot, C., Spraggon, L., Chia, I., Batourina, E., Riccio, P., Lu, B., ... Mendelsohn, C. (2010). Non-cell-autonomous retinoid signaling is crucial for renal development. *Development (Cambridge, England), 137*, 283–292.

Stafford, D., Hornbruch, A., Mueller, P. R., & Prince, V. E. (2004). A conserved role for retinoid signaling in vertebrate pancreas development. *Development Genes and Evolution, 214*, 432–441.

Stafford, D., & Prince, V. E. (2002). Retinoic acid signaling is required for a critical early step in zebrafish pancreatic development. *Current Biology: CB, 12*, 1215–1220.

Storz, P. (2017). Acinar cell plasticity and development of pancreatic ductal adenocarcinoma. *Nature Reviews Gastroenterology & Hepatology, 14*, 296–304.

Suissa, Y., Magenheim, J., Stolovich-Rain, M., Hija, A., Collombat, P., Mansouri, A., ... Glaser, B. (2013). Gastrin: A distinct fate of neurogenin3 positive progenitor cells in the embryonic pancreas. *PLoS One, 8*, e70397.

Sun, L., Zheng, M., Gao, Y., Brigstock, D. R., & Gao, R. (2024). Retinoic acid signaling pathway in pancreatic stellate cells: Insight into the anti-fibrotic effect and mechanism. *European Journal of Pharmacology, 967*, 176374.

Tabula Muris Consortium, Overall coordination, Logistical coordination, Organ collection and processing, Library preparation and sequencing, Computational data analysis, Cell type annotation, Writing group, Supplemental text writing group & Principal investigators. (2018). Single-cell transcriptomics of 20 mouse organs creates a Tabula Muris. *Nature, 562*, 367–372.

Theodosiou, M., Laudet, V., & Schubert, M. (2010). From carrot to clinic: An overview of the retinoic acid signaling pathway. *Cellular and Molecular Life Sciences: CMLS, 67*, 1423–1445.

Tiyaboonchai, A., Cardenas-Diaz, F. L., Ying, L., Maguire, J. A., Sim, X., Jobaliya, C., ... Gadue, P. (2017). GATA6 plays an important role in the induction of human definitive endoderm, development of the pancreas, and functionality of pancreatic β cells. *Stem Cell Reports*.

Trasino, S. E., Benoit, Y. D., & Gudas, L. J. (2015). Vitamin A deficiency causes hyperglycemia and loss of pancreatic beta-cell mass. *The Journal of Biological Chemistry, 290*, 1456–1473.

Umemiya, H., Fukasawa, H., Ebisawa, M., Eyrolles, L., Kawachi, E., Eisenmann, G., ... Kagechika, H. (1997). Regulation of retinoidal actions by diazepinylbenzoic acids. Retinoid synergists which activate the RXR-RAR heterodimers. *Journal of Medicinal Chemistry, 40*, 4222–4234.

Vinckier, N. K., Patel, N. A., Geusz, R. J., Wang, A., Wang, J., Matta, I., ... Sander, M. (2020). LSD1-mediated enhancer silencing attenuates retinoic acid signalling during pancreatic endocrine cell development. *Nature Communications, 11*, 2082.

Yoo, H. S., Moss, K. O., Cockrum, M. A., Woo, W., & Napoli, J. L. (2023). Energy status regulates levels of the RAR/RXR ligand 9-cis-retinoic acid in mammalian tissues: Glucose reduces its synthesis in beta-cells. *The Journal of Biological Chemistry, 299*, 105255.

Zhaoxia, L., Mengkai, D., Qin, F., & Danqing, C. (2014). Significance of RBP4 in patients with gestational diabetes mellitus: A case-control study of Han Chinese women. *Gynecological Endocrinology: The Official Journal of the International Society of Gynecological Endocrinology, 30*, 161–164.

CHAPTER ELEVEN

Vitamin A supply in the eye and establishment of the visual cycle

Sepalika Bandara and Johannes von Lintig*

Department of Pharmacology, School of Medicine, Case Western Reserve University, Cleveland, OH, United States
*Corresponding author. e-mail address: johannes.vonlintig@case.edu

Contents

1. Introduction	320
2. Classes of retinoid metabolizing enzymes	322
3. Unlocking retinoids: the role of carotenoid cleavage dioxygenases in vitamin A conversion	322
4. Retinol and retinal dehydrogenases interconvert the oxidation states of vitamin A	324
5. Mastering retinoic acid levels: the role of cytochrome P450 enzymes	327
6. Lecithin: retinol acyltransferase: beyond vitamin A storage	328
7. Retinoid couriers: binding proteins navigating vitamin A across the body	329
8. Regulating vitamin A production: unveiling gut control mechanisms	330
9. Vitamin A express: delivering essential nutrients to the eyes	331
10. The visual cycle: the art of synthesis and recycling	334
11. Seeing in color: unveiling the cone visual pigment regeneration mechanism(s)	336
References	339

Abstract

Animals perceiving light through visual pigments have evolved pathways for absorbing, transporting, and metabolizing the precursors essential for synthesis of their retinylidene chromophores. Over the past decades, our understanding of this metabolism has grown significantly. Through genetic manipulation, researchers gained insights into the metabolic complexity of the pathways mediating the flow of chromophore precursors throughout the body, and their enrichment within the eyes. This exploration has identified transport proteins and metabolizing enzymes for these essential lipids and has revealed some of the fundamental regulatory mechanisms governing this process. What emerges is a complex framework at play that maintains ocular retinoid homeostasis and functions. This review summarizes the recent advancements and highlights future research directions that may deepen our understanding of this complex metabolism.

Abbreviations

BC	β-carotene
BCO1	β-carotene oxygenase 1
BCO2	β-carotene oxygenase 2

CCD	carotenoid cleavage dioxygenases
RPE65	retinal pigment epithelium specific 65 kDa protein
NinaB	neither inactivation nor afterpotential mutant B
RDH	retinol dehydrogenase
RALDH	retinal dehydrogenase
RBP	retinol binding protein
CRALBP	cellular retinal binding protein
LRAT	lecithin: retinol acyl transferase
DHRS3	dehydrogenase/reductase 3
SDR	short- chain dehydrogenase/reductase
RA	retinoic acid
NAD	nicotinamide adenine dinucleotide
NADP	nicotinamide adenine dinucleotide phosphate
at-RAL	all trans retinal
at-ROL	all trans retinol
at-RE	all trans retinyl ester
CRABP	cellular retinoic acid binding protein
IRBP	interphotoreceptor retinol binding protein
RPE	retinal pigment epithelium
ER	endoplasmic reticulum
DGAT1	diacylglycerol O-acyltransferase 1
STRA6	stimulated by retinoic acid 6
SCARB1	class B scavenger receptor type 1
ALDH	aldehyde dehydrogenase
ADH	alcohol dehydrogenase
StAR	steroidogenic acute regulatory protein
WT	wild type
CYP26A1	cytochrome P450 family 26 subfamily A member 1
OS	photoreceptor outer segment

1. Introduction

Vitamin A biology exemplifies the complex interplay between our bodies and foodstuff. The nutrient plays indispensable roles throughout the life cycle, contributing to a wide range of physiological processes from embryonic development to maintaining cell homeostasis and supporting vision.

In the diet, vitamin A exists in the form of certain carotenoids (provitamin A) found in produce and retinyl esters (REs) in dairy products and meats (Grune et al., 2010). These precursors are used for the synthesis of at least two critical metabolites: all-*trans*-retinoic acid (RA) and retinaldehyde (RAL). RA functions as a hormone in gene regulation in many cells (Chambon, 1996; Chawla et al., 2001) and RAL as chromophore of visual pigments in photoreceptors of the retina (Yau & Hardie, 2009).

The elucidation of the biochemical steps involved in vitamin A's role in the visual process led to significant discoveries, including the characterization of visual G protein signaling and the determination of the heptahelical transmembrane receptor rhodopsin's structure (Palczewski et al., 2000; Wald, 1968a). These transmembrane proteins, which activate heterotrimeric G proteins, play roles in various physiological processes throughout the body, responding to a wide range of chemical messengers such as hormones, neurotransmitters, odorants, and food ingredients (Weis & Kobilka, 2018).

Beyond its essential function in vision, vitamin A plays a crucial role in gene regulation through its derivative RA. RA binds to RA receptors (RARs) that belong to the nuclear receptor family (Giguere et al., 1987; Petkovich et al., 1987). Initially recognized for mediating steroid hormone signaling, nuclear receptors provide an important link between transcriptional regulation and physiology (Evans & Mangelsdorf, 2014). The human genome encodes 48 members of this transcription factor family. These include classic endocrine receptors that mediate the actions of steroid hormones, thyroid hormones, and the fat-soluble vitamins A and D, as well as nuclear receptors that act as lipid sensors and participate in numerous cellular processes. RARs form obligatory dimers with RXRs and control transcription by binding to conserved DNA motifs called RA response elements in the promoter or enhancer regions of numerous genes that vary from one cell type to another (Ghyselinck & Duester, 2019).

For vitamin A and its dietary precursors to exert their biological effects, they must undergo absorption in the intestine, metabolic conversion, and subsequent transportation to target tissues, particularly the eyes (Von Lintig et al., 2021). Mutations in the genes governing chromophore metabolism can lead to inherited retinal diseases such as retinitis pigmentosa (RP) and Leber congenital amaurosis (LCA) (Kiser & Palczewski, 2016) as well as complex inherited conditions such as the Matthew-Wood Syndrome (Blaner, 2007). Additionally, aberrant by-products of chromophore metabolism may contribute to pathologies like age-related macular degeneration (AMD) (Sears et al., 2017).

Thus, regulating the metabolic flux of dietary precursors is imperative to ensure the optimal production of biologically active retinoids. This regulation prevents deficiencies or excesses that could compromise vision or overall health. This review will introduce transporters, binding proteins, metabolizing enzymes of this pathway and describe their interplay with a particular focus on ocular retinoid homeostasis.

2. Classes of retinoid metabolizing enzymes

Vitamin A metabolizing enzymes play a crucial role in maintaining various physiological functions by catalyzing the conversion of vitamin A into its active forms, such as RA and RAL. These enzymes include carotenoid cleavage oxygenases (CCDs), retinol dehydrogenases (RDHs), retinaldehyde dehydrogenases (RALDHs), and cytochrome P450 enzymes, each responsible for different steps in the metabolic pathway (Fig. 1). Proper functioning of these enzymes is essential for vision, immune response, and cellular differentiation, highlighting their significance in overall health and disease prevention. Understanding the mechanisms and regulation of these enzymes offers insights into therapeutic strategies for vitamin A-related disorders.

3. Unlocking retinoids: the role of carotenoid cleavage dioxygenases in vitamin A conversion

All naturally occurring retinoids stem from carotenoid precursor molecules (Alvarez et al., 2014). The oxidative split of double bonds in carbon backbone of carotenoids produces apocarotenoids in most living organisms, including retinoids in animals (Fig. 1). Carotenoid cleavage dioxygenases (CCDs) are non-heme iron oxygenases and crucial in this transformation (Giuliano et al., 2003). Mammalian genomes encode three members of the CCD family: Two canonical CCDs, BCO1 (β-carotene oxygenase 1) and BCO2 (β-carotene oxygenase 2), and the retinyl ester metabolizing enzyme RPE65 (Retinal pigment epithelium specific 65 kDa protein) (Sui et al., 2013; Von Lintig et al., 2020). Cytosolic BCO1 cleaves pro-vitamin A carotenoids at the 15–15' position, yielding at-RAL, using substrates like β-carotene, α-carotene, and β-apo-carotenoids (Kelly et al., 2018; Lindqvist & Andersson, 2002). While β-carotene metabolism is well studied (Amengual et al., 2013), α-carotene metabolism needs more research to understand the fate of α-retinoids. Provitamins with hydroxylated ionone rings, such as β-cryptoxanthin, are seemingly converted in two steps: BCO2 first removes the hydroxylated ionone ring, followed by BCO1 converting the 10'-β-apocarotenal product (Kelly et al., 2018). Studies in mice also have demonstrated the use of long chain (> C20) β-apocarotenoids as vitamin A sources (Miller et al., 2023; Spiegler et al., 2018).

Mitochondrial BCO2 cleaves carotenoids and apocarotenoids at the 9–10 position, producing 10,10'-diapocarotene-dials and ionone products

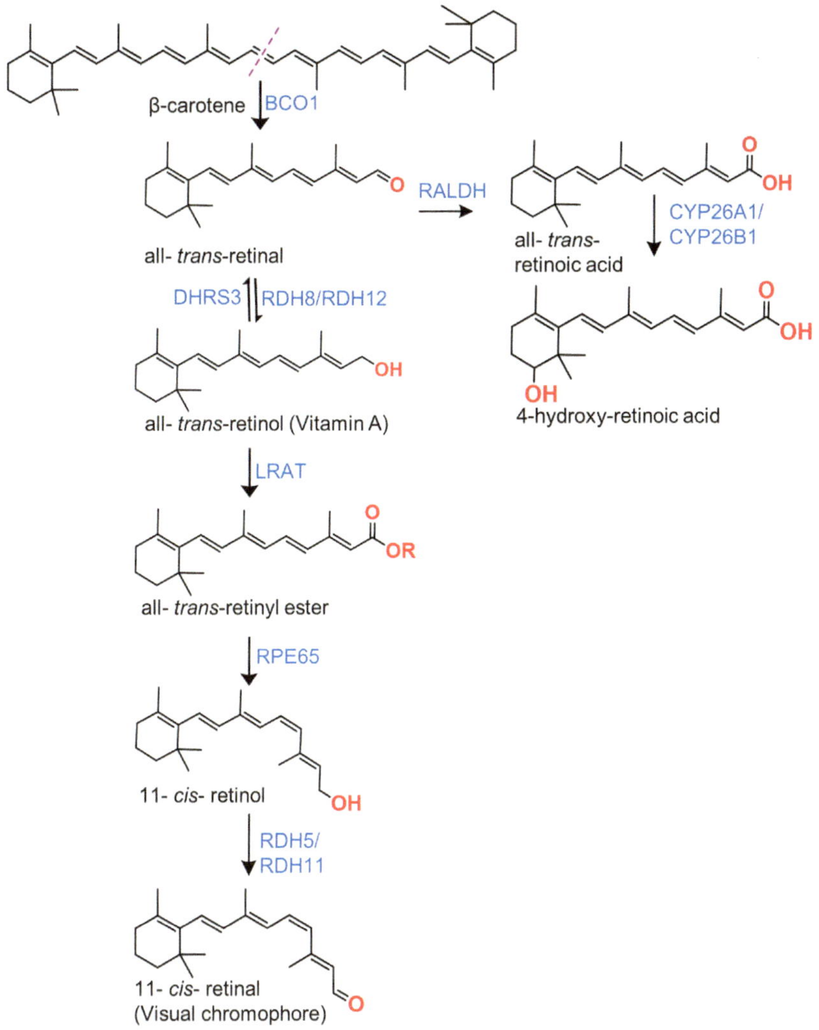

Fig. 1 Schematic pathway of retinoid processing in the human body. The involved enzymes catalyzing the different reactions are indicated in blue color.

(Amengual et al., 2011; Kiefer et al., 2001). While BCO2 shares a similar fold with BCO1, their substrate specificity differs. BCO2's bipartite substrate-binding cavity accommodates carotenoids with assorted ionone rings, cleaving them stepwise through a long-chain apocarotenoid intermediate (Bandara et al., 2021). In contrast, BCO1's narrow substrate tunnel excludes substituted ionone rings (Kelly et al., 2018). Studies in *Bco1* and *Bco2* knockout mice show their selectivity for specific carotenoids (Kelly et al., 2018). β-carotene accumulates in $Bco1^{-/-}$ mice similarly to $Bco1^{-/-}$ $Bco2^{-/-}$ double mutants, though recombinant BCO2 can convert β-carotene in vitro (Amengual et al., 2013). Recent studies indicate that Aster proteins, encoded by *Gramd1a, b*, and *c* genes, shuttle hydroxylated carotenoids but not β-carotene to mitochondria, making them accessible to BCO2 (Bandara et al., 2023).

RPE65 does not incorporate oxygen into carotenoids and apocarotenoids. Instead, it converts all *trans*-retinyl esters (at-REs) into 11-*cis*-ROL and palmitate through an ester cleavage and isomerase reactions, likely via a carbocation intermediate (Kiser et al., 2015; Redmond et al., 2005). Vertebrates use both BCO1 and RPE65 to synthesize the visual chromophore from carotenoids, while invertebrates combine these functions in a single enzyme, NinaB (Babino et al., 2016; Oberhauser et al., 2008). NinaB converts carotenoids into an 11-*cis* and at-RAL diastereomer and mutations in the NinaB gene causes chromophore deficiency and blindness (Von Lintig et al., 2001). Structural and biochemical comparisons of NinaB and RPE65 have revealed that the enzymes use distinct mechanisms for *cis*-to-*trans* isomerization of their substrates (Solano et al., 2024).

Metazoan CCDs features a common fold with a seven-bladed β-propeller scaffold, α-helices, and loops, relying on ferrous iron as a cofactor (Sui et al., 2013). The ferrous iron is accessible through a long tunnel that is aligned with aromatic amino acid side chains. NinaB's structure shows a surface region near the active site with lipophilic and cationic amino acid side chains for membrane interactions (Solano et al., 2024). This region starts with a conserved 'PDPC(+)' motif, where the cationic residue and a palmitoylated cysteine aid in membrane association. In RPE65, this motif extends into an amphipathic α-helix crucial for membrane targeting (Uppal et al., 2023). Whether this mode of membrane association is conserved in BCO1 and BCO2 requires further analysis.

4. Retinol and retinal dehydrogenases interconvert the oxidation states of vitamin A

The oxygen in retinoids exits in three oxidation states as ROLs, RALs, and RAs. ROL dehydrogenases (RDHs) and RAL dehydrogenases (RALDHs) catalyzes the interconversion of retinoids and belong to the oxidoreductase enzyme class (Fig. 2). RDHs, part of the SDR enzyme superfamily, play a significant role in vitamin A metabolism by catalyzing the redox reaction between RAL and ROL (Kedishvili, 2013). They use

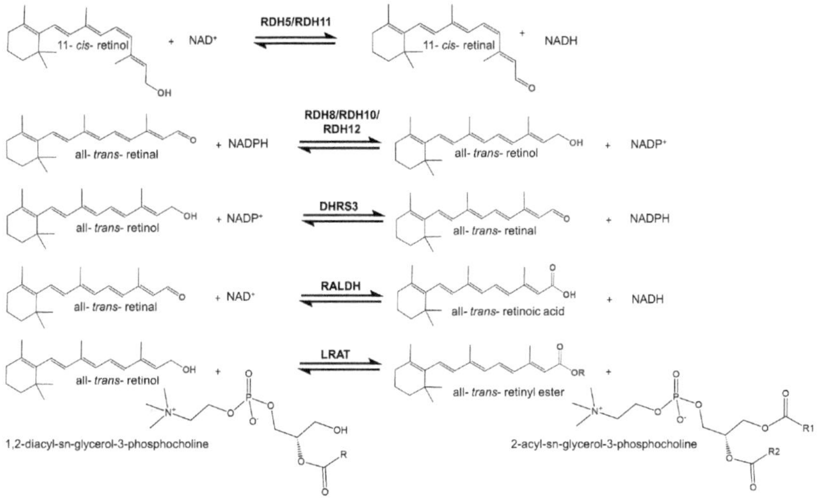

Fig. 2 Reactions catalyzed by retinoid processing enzymes. The functional groups involved in the catalytic reactions are highlighted with magenta color.

adenine dinucleotide cofactors NAD(H) and NADP(H) and bind the cofactors through a conserved Rossmann-fold motif, consisting of six to seven parallel β-strands flanked by three to four α-helices (Hofmann et al., 2016). Under physiological conditions, the NAD/NADH ratio in the cytoplasm is around 1000, while the NADP/NADPH ratio is 0.005. Thus, enzymes such as RDHs using NAD typically catalyze the oxidation, while those using NADP catalyze the reduction of retinoids.

Although RDHs can use both ROL and RAL as substrates, many of these enzymes also metabolize other alcohols, including sterols. Various RDH enzymes, including RDH5, RDH8, RDH10, RDH11, and RDH12, have been characterized for their tissue-specific expression and biochemical activities. RDH8 and RDH12 convert at-RAL to at-ROL, while RDH5 and RDH11 oxidize 11-*cis*-ROL into the visual chromophore, 11-*cis*-RAL; RDH10 oxidizes at-ROL to at-RAL.

The ocular localization of several RDH enzymes has been recently characterized: RDH8 is found in photoreceptor outer segments, while RDH11 and RDH12 are in photoreceptor inner segments, catalyzing the conversion of at-RAL to at-ROL (Maeda et al., 2009). RDH5 and RDH11 perform a core visual cycle reaction by converting 11-cis-ROL to 11-cis-RAL, supplementing the visual pigment (Kim et al., 2005). RDH12 exhibits a unique substrate preference, utilizing both *cis* and *trans* retinoids (Maeda et al., 2006, 2009).

Studies of knockout mice have provided significant insights into the specific functions of these enzymes in RA metabolism (Kedishvili, 2013). These studies identified RDH10 and DHRS3 as critical components for controlling RA homeostasis in the vertebrate embryo. DHRS3, a NADP-dependent enzyme related to RDH10, requires RDH10 to exhibit catalytic activity, catalyzing the reduction of at-RAL into at-ROL in the presence of a high NADPH/NADP ratio. Human DHRS3 and RDH10 activate each other reciprocally, independent of the catalytic activity of each protein. Adams et al. (2014) demonstrated DHRS3's high affinity for at-RAL but not for 11-*cis*-RAL, with a preference for NADPH as the cofactor. Additionally, DHRS3 is responsible for embryonic at-RAL reduction, playing a significant role in embryonic retinoid signaling.

Dhrs3$^{-/-}$ embryos show elevated levels of at-RA compared to wild-type embryos, resulting in decreased expression of genes encoding RA synthesizing enzymes *Raldh1–3* (*Aldh1a1–3*) and increased expression of the gene encoding the at-RA catabolizing enzyme *Cyp26a1* (Billings et al., 2013). This elevation in at-RA levels leads to late-gestational lethality and congenital defects, including cleft palate, heart malformations, and skeletal defects. Loss of DHRS3 in knockout mice also causes significantly reduced levels of ROL and REs, as well as embryonic RAL reductase and ROL dehydrogenase activities (Adams et al., 2014; Wang et al., 2018). On the other hand, RDH10 controls levels of RA synthesis, and embryos deficient in this enzyme display a severe lack of retinoid signaling, developing many malformations known from vitamin A deficient embryos (Sandell et al., 2007). RALDH enzymes play a crucial role in RA signaling pathways by oxidizing at-RAL into RA, a process that requires NADH as a cofactor (Wang et al., 1996) (Fig. 2). The primary RALDH enzymes, RALDH1 (ALDH1A1), RALDH2 (ALDH1A2), and RALDH3 (ALDH1A3), have been extensively studied in mouse models, each showing specific tissue expression patterns during embryogenesis (Niederreither et al., 2002). For instance, during fetal development, RALDH1 is expressed in the lungs and shows stage-specific expression in the stomach, intestinal epithelial, and mesenchymal layers, essential for localized RA synthesis during these stages. RALDH2 is predominantly found in the kidney nephrogenic zone, playing a vital role in kidney development and function. RALDH3 is expressed in the differentiating intestinal lamina propria, contributing to intestinal development. Genetic dissection in single and compound *Raldh1–3* knockout mice have significantly advanced our understanding of embryonic RA synthesis (Duester, 2008; Rhinn & Dollé, 2012). These studies have identified

RALDH2 as a major enzyme for RA production, evidenced by the early embryonic death of RALDH2-null mice, underscoring its critical role in early development (Niederreither et al., 1999). RALDH3 has specific expression in later stages of mouse development (Mic et al., 2000). Mice lacking RALDH3 exhibit eye defects and die at birth due to respiratory distress, highlighting its importance in later stages of development and specific organ formation (Dupe et al., 2003). In humans, mutations in the RALDH3 gene are associated with congenital microphthalmia, indicating its critical role in eye development (Yahyavi et al., 2013). In contrast, RALDH1 knockout mice develop normally and are fertile, suggesting that ALDH1A1 is not as critical for early development. However, these mice show several metabolic anomalies in adulthood, such as resistance to diet-induced obesity, highlighting the enzyme's role in metabolic regulation during adult life (Fan et al., 2003; Ziouzenkova et al., 2007).

5. Mastering retinoic acid levels: the role of cytochrome P450 enzymes

The regulation of RA levels in tissues is critical throughout the mammalian life cycle, primarily mediated by cytochrome P450-dependent hydroxylases CYP26A1, CYP26B1, and CYP26C1. These enzymes catalyze the production of 4-hydroxy and 4-oxo-RA that initiates degradation of RA essential for maintaining RA homeostasis. CYP26A1's expression is highly responsive to RA levels, indicating its significant role in acidic retinoid catabolism (Abu-Abed et al., 1998). This enzyme's activity helps prevent excessive RA accumulation by converting it into less active metabolites, exemplifying a feedback mechanism to ensure RA homeostasis. Such regulation is typical of many CYP enzymes involved in lipid metabolism, preventing toxic substrate build-ups.

The regulation and physiological roles of CYP26B1 and CYP26C1 are more complex and remain the focus of ongoing research (Isoherranen & Zhong, 2019). Unlike CYP26A1, these enzymes may have distinct and overlapping roles in different tissues and developmental stages. CYP26B1, for example, is crucial during embryonic development, particularly in the brain and skeletal system, where precise RA levels are necessary for proper differentiation and growth. CYP26C1 also plays a role in the brain and other specific tissues, though its full range of functions is still being elucidated.

The precise control of RA levels by these hydroxylases underscores their importance in preventing disorders associated with both RA deficiency and excess (Abu-Abed et al., 2001). Abnormal RA signaling can lead to developmental defects, immune dysfunctions, and other pathologies. Understanding the specific regulatory mechanisms and tissue-specific roles of CYP26B1 and CYP26C1 is crucial for developing therapeutic strategies for diseases related to RA imbalance.

6. Lecithin: retinol acyltransferase: beyond vitamin A storage

Retinoids resulting from carotenoid cleavage and preformed retinoids are primarily esterified in enterocytes by lecithin: acyltransferase (LRAT), with additional esterification by diacylglycerol O-acyltransferase 1 (DGAT1) enzymes (Wongsiriroj et al., 2008). LRAT, part of the ancestral N1pC/P60 thiol peptidase protein superfamily, is located in the endoplasmic reticulum and converts at-ROL into at-RE using phosphatidylcholine as the acyl donor (Golczak et al., 2015) (Fig. 2). Structurally, LRAT consists of a four-strand antiparallel β-sheet and three α-helices, forming a functional dimer for the esterification reaction. The C-terminal transmembrane helix anchors LRAT to the ER membrane, while His60, His72, and Cys161 form the catalytic triad in the LRAT active site, with His residues in one monomer and Cys in the other, the conserved Cys residue forming thioester bonds with acyl moieties.

Knockout mouse studies have identified LRAT as the major enzyme for retinoid storage (Liu & Gudas, 2005; O'Byrne et al., 2005). LRAT also is crucial for providing RE for chromophore synthesis via RPE65 in the retinal pigment epithelium (Batten et al., 2004). Furthermore, LRAT plays a significant role in RA homeostasis by directing the flux of ROL towards esterification, preventing excessive production of RA, the transcriptionally active form of vitamin A (Isken et al., 2007; Zolfaghari & Ross, 2002). This mechanism already is active in the enterocytes of the intestine, adjusting the concentration of retinoids by sequestering at-RE for distribution and storage (Ramkumar et al., 2021).

LRAT and DGAT1 belong to unrelated enzyme classes, with LRAT specifically targeting ROL esterification and DGAT1 having broad substrate specificity for esterification using acyl-CoA. Studies involving mice with intestinal DGAT1 deficiency do not indicate impaired ROL esterification (Ables et al., 2012), suggesting that LRAT likely plays the primary role in the synthesis of intestinal REs.

7. Retinoid couriers: binding proteins navigating vitamin A across the body

Retinoids are lipophilic compounds; therefore, they require protein-mediated transport mechanisms in aqueous mediums. To date, several retinoid binding proteins have been characterized with distinct structural and biochemical properties (Widjaja-Adhi & Golczak, 2019). These retinoid binding proteins include serum and cellular ROL binding proteins (RBPs), cellular retinal binding protein (CRALBP), and RA binding proteins (CRABPs).

RBPs play a vital role in transporting vitamin A. RBP4 is a serum ROL binding protein, and the major transport mode of vitamin A in the fasting circulation (Kanai et al., 1968; Quadro et al., 1999). The β-barrel structure of RBP4 consists of a hydrophobic binding cavity to bind a single molecule of ROL. Meanwhile, RBP1 and 2 bind to cellular ROL. RBP1 and 2 display distinct tissue distribution and expression in mice. RBP2 is expressed at high levels in enterocytes of the gut and facilitates ROL uptake from the diet (Blaner et al., 2020). RBP1 is expressed in hepatocytes and a critical role of the protein in liver retinoid homeostasis is indicated by studies in knockout mice that display reduced RE stores in hepatic stellate cells (Ghyselinck et al., 1999). Additionally, RBP1 is expressed in the eyes and $Rbp1^{-/-}$ mice show a delayed regeneration of chromophore after a bleach (Saari et al., 2002). Not only tissue distribution differs, but RBPs also show significant variation in affinity to retinoids. Even though RBP4 and RBP1 and 2 bind to the same ligands, the binding orientation differs for the ligand. In RBP4, the β-ionone ring binds to the middle of the binding cavity, and the hydrocarbon chain points towards the entrance of the cavity. However, in RBP1 and 2, ROL binds with the opposite orientation, where the hydrocarbon chain is in the center of the cavity, and the β-ionone ring is positioned at the entrance of the cavity (Franzoni et al., 2002; Perduca et al., 2018; Silvaroli et al., 2021).

Interphotoreceptor ROL binding protein, encoded by the RBP3 gene, is an eye-specific ROL binding protein that transports retinoids in RPE and rod outer segments. RBP3 is localized to the extracellular space between photoreceptors and RPE. RBP3 accelerates transport of at-ROL from bleached photoreceptors to the RPE, and 11-cis-RAL from RPE to photoreceptors (Parker et al., 2011). $Rbp3^{-/-}$ mice studies showed that its absence delayed the tranfer of the newly synthesized chromophore from RPE to photoreceptors (Parker et al., 2009).

There are two cellular RA binding proteins, CRABP1 and CRABP2, which only differ in tissue distribution. CRABP1 is expressed in many tissues whereas CRABP2 is expressed in tissues with high levels of RA synthesis (Wardlaw et al., 1997; Zheng & Ong, 1998). According to studies, CRABP2 expression is dependent on RA response elements. Both CRABPs have been implicated to facilitating the oxidative catabolism of RA by delivering to CYP enzymes, particularly CYP2B1 (Nelson et al., 2016). While CRABPs are structurally similar to RBPs, the functional residues that interact with ligands are not conserved within the two binding protein classes. This difference explains their specificity for either the binding of the alcoholic or acid forms of vitamin A (Zhang et al., 2012).

CRALBP (Cellular retinal binding protein), encoded by the *Rlbp1* gene, is an eye-specific retinoid-binding protein that binds to 11-*cis* and 9-*cis*-ROL and RAL. Binding to CRALBP protects the visual chromophore from thermal isomerization and other side reactions. CRALBP is expressed in the RPE and Müller glia cells of the retina (Bunt-Milam & Saari, 1983). CRALBP is a member of a family of carrier proteins containing the CRAL-TRIO domain. In terms of structure, CRALBP has two domains: an α-helical N-terminal domain and a C-terminal αβα sandwich domain. The C-terminal domain consists of the binding cavity for the ligand, where the β-ionone ring and the hydrocarbon chain are positioned through hydrophobic interactions, and the carbonyl carbon of the aldehyde group interacts through hydrogen bonding (He et al., 2009). Previously, it was proposed that CRALBP in RPE supports dim light vision mediated by rods and CRALBP in Müller glia cells daylight vision mediated by cone photoreceptors (Fleisch et al., 2008; Xue et al., 2015). Recent studies suggest that mainly CRALBP in the RPE is involved in chromophore metabolism. Knockout of *Rlbp1* in the RPE slows down chromophore synthesis, delays rod dark adaptation, and reduces protection against light-induced retinal degeneration (Bassetto et al., 2024; Schlegel et al., 2021). In a zebrafish model, RPE-specific deletion leads to the accumulation of 11-*cis*-RE that display as lipid droplets in the diseased RPE (Schlegel et al., 2021). A similar pathology has been reported in patients suffering from Bothnia dystrophy that is caused by homozygous mutations in the *RLBP1* gene.

8. Regulating vitamin A production: unveiling gut control mechanisms

Studies conducted in humans and animal models have demonstrated that vitamin A status markedly influences the bioconversion of carotenoids to vitamin A within the intestine (Lala & Reddy, 1970; Van Vliet et al., 1996). This regulation of vitamin A production in intestinal enterocytes represents an example of negative transcriptional feedback control. In conditions of vitamin A deficiency, the *Bco1* gene, encoding the enzyme responsible for vitamin A biosynthesis, is constitutively expressed at high levels. However, upon dietary intake of preformed vitamin A or its precursor β-carotene, enterocytes rapidly downregulate *Bco1* expression (Bachmann et al., 2002; Seino et al., 2008).

The master regulator governing vitamin A production has been identified as the intestine-specific homeobox transcription factor ISX (Lobo et al., 2010). This transcription factor binds to conserved motifs within the promoter region of the *Bco1* gene resulting in gene repression (Lobo et al., 2013). The small molecule that induces expression of the ISX repressor protein is RA (Lobo et al., 2010). Its concentrations in enterocytes are controlled by the LRAT enzyme, which directs the metabolic flow of newly synthesized retinoids towards retinyl ester formation and integration into chylomicrons (Ramkumar et al., 2021). Accordingly, $Lrat^{-/-}$ mice cannot balance intestinal RA levels, overexpress ISX, and suppress retinoid production in enterocytes. The resulting phenotype can be rescued by treatment with antagonists of RAR that reduce ISX protein and restore *Bco1* expression (Ramkumar et al., 2021). Physiologically, the regulation of BCO1 activity in enterocytes by ISX prevents excessive production of retinoids that can impair gut immunity and other processes (Widjaja-Adhi et al., 2017).

9. Vitamin A express: delivering essential nutrients to the eyes

Upon conversion to RE in enterocytes, dietary vitamin A is packaged along with other dietary lipids, into triacylglycerol-rich chylomicrons. Peripheral tissues, including the eyes can render RE from chylomicrons available in a lipoprotein lipase-dependent manner (Van Bennekum et al., 1999). The remainder in chylomicron remnants is deposited in the liver and stored in the form of RE in hepatic stellate cells. These stores are significant and allow mice to endure periods of prolonged vitamin A deficiency (for overview see Fig. 3).

Stored vitamin A is released from hepatocytes bound to RBP4 (D'Ambrosio et al., 2011; Quadro et al., 1999). Holo-RBP4 is the major

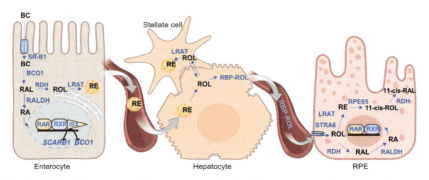

Fig. 3 Retinoid transport. The overview of retinoid transport and storage in humans includes several key processes. In the enterocytes, provitamin A absorption and its conversion to retinyl esters are shown. Retinoid storage and distribution occur in the hepatocytes and stellate cells. The steps involved in processing retinoids for visual chromophore generation are depicted in the retinal pigment epithelium (RPE). Additionally, retinoic acid signaling is highlighted in both enterocytes and the RPE. Proteins and genes involved in these processes are indicated in blue.

retinoid in the fasting circulation and forms a complex with the 55 kDa transthyretin (TTR) homotetramer at a 2:1 molar ratio (Episkopou et al., 1993). The levels of holo-RBP4 are homeostastic and only decline under conditions of severe vitamin A deficiency when hepatic stores are depleted.

The cellular uptake of ROL from holo-RBP4 is mediated by a receptor (Heller & Bok, 1976) which is encoded by the *Stra6* gene (Bouillet et al., 1997; Kawaguchi et al., 2007). Structural analysis revealed that STRA6 assembles as a dimer with 18 transmembrane helices (nine per protomer) and two long horizontal intramembrane helices interacting at the dimer core (Chen et al., 2016). The complex displays a lipophilic cleft to which holo-RBP4 binds with nanomolar affinity (Chen et al., 2016; Kawaguchi et al., 2007). Biochemical studies demonstrated that STRA6 facilitates the bidirectional flux of ROL between RBP4 and cells (Isken et al., 2008; Kawaguchi et al., 2011). The opening and closing of the channel are regulated by calcium and calmodulin that binds to the STRA6 homodimer (Chen et al., 2016; Zhong et al., 2020). Increased levels of calcium and calmodulin promotes ROL efflux of cells; however, the details of this regulation are yet not fully understood (Young et al., 2021).

The role of STRA6 and its RBP4 ligand in vitamin A transport has been studied in mice. At young ages, $Stra6^{-/-}$ mice display very low ocular retinoid concentration and impaired visual function with significantly reduced electroretinogram (ERG) responses (Amengual et al., 2014; Berry et al., 2013; Ruiz et al., 2012). Their rod photoreceptors are characterized by large amounts of unliganded opsin due to reduced 11-*cis*-RAL (Ramkumar et al., 2022). A comparable ocular phenotype has been described in RBP4-deficient mice (Montenegro et al., 2022; Quadro et al., 1999). In *Stra6*$^{-/-}$ mice, blood and other tissues, such as the lungs, fat, and liver, maintain normal vitamin A levels through chylomicrons when raised on vitamin A-rich chow (Amengual et al., 2014; Berry et al., 2013).

During adolescence of *Stra6*$^{-/-}$ mice, ocular retinoid concentrations increase when mice are nourished with vitamin A-rich chow (Amengual et al., 2014) but plateau with 6-month of age (Ramkumar et al., 2022). This increases restores rod photoreceptor ERG responses and increases the amount of rhodopsin in dark-adapted photoreceptors (Ramkumar et al., 2022). Despite this increase, cone photoreceptors of these mice consistently exhibit chromophore deficiency with absent or mislocalized cone opsins (Ramkumar et al., 2022; Ruiz et al., 2012).

The critical role of STRA6 for ocular retinoid homeostasis is mirrored in the regulation of its expression in different tissues and cell types (Kelly et al., 2016). STRA6 is highly expressed in the RPE and other epithelia forming blood-tissue barriers such as Sertoli cells of the testis, and ependymal cells of the choroid plexus (Amengual et al., 2014; Berry et al., 2013; Kelly et al., 2016). In the eyes, where the demand for vitamin A is highest in the body, *Stra6* is expressed constitutively (Kelly et al., 2016). This regulation maintains ocular vitamin A homeostasis as long as ROL is available from holo-RBP4 (Kelly et al., 2016). In the testis and brain, *Stra6* expression is under control of retinoid signaling (Kelly et al., 2016). This control mechanism reduces consumption of circulating ROL when supplies are low (Kelly et al., 2016).

By contrast, peripheral retinoid storing tissues such as the lungs exhibit very low STRA6 expression and rely on the receipt of post-meal dietary vitamin A through lipoproteins (Kelly et al., 2016; Shmarakov et al., 2023). Nevertheless, the lung and some tissues may serve as rapid removers of excess retinoids from the bloodstream since *Stra6* gene expression is responsive to retinoid signaling (Moon et al., 2022). Thus, STRA6 supports distribution of retinoids depending on tissue demands and vitamin A status.

In humans, missense and nonsense mutations in the *STRA6* gene cause a severe microphthalmic condition known as Matthew-Wood syndrome (Golzio et al., 2007; Blaner, 2007). The inherited disease is characterized by severe bilateral microphthalmia, often accompanied by pulmonary dysplasia, cardiac defects, diaphragmatic hernia, and other anomalies. The phenotypic manifestations of STRA6 mutations reflect the crucial role of retinoids in mammalian embryonic development. However, the severity of symptoms can vary even within the same family, ranging from isolated microphthalmia to severe systemic manifestations of the disease (Casey et al., 2011).

Compound missense mutations in the *RBP4* gene were initially described in two sisters who exhibited mild ocular coloboma and retinal dystrophy (Biesalski et al., 1999; Seeliger et al., 1999). Later, more severe congenital eye malformations, including bilateral anophthalmia, were reported in three families with *RBP4* mutations (Chou et al., 2015). Interestingly, some affected children were heterozygous for these mutations. It was suggested that in these cases the mutant RBP4 variant had lost the ability to bind ROL but displayed increased affinity to STRA6. This biochemical property blocked ROL delivery from the mother to the fetus and within fetal tissues. Thus, a dominant inheritance with incomplete penetrance was proposed for these RBP4 alleles. In contrast, reported mutations in STRA6 always demonstrate recessive inheritance.

Depending on genetic background, $Stra6^{-/-}$ and $Rbp4^{-/-}$ mice display choroidal colomboma as well as a persistent and hyperplasic primary vitreous body, demonstrating that STRA6 mutations also affect murine eye development (Amengual et al., 2014; Berry et al., 2013; Shen et al., 2016). The viability and absence of gross malformations in these mice demonstrates that mouse development can be maintained in the absence of STRA6 or RBP4 with vitamin A in serum lipoproteins. However, dietary vitamin A restriction leads to severe malformation in RBP4-deficient offspring (Quadro et al., 2005; Spiegler et al., 2012), suggesting that dietary vitamin A supplies are an important modulator of the phenotype of *STRA6* and *RBP4* mutations in patients.

10. The visual cycle: the art of synthesis and recycling

In mice, photoreceptors mature postnatally and this process is accompanied by chromophore synthesis and recycling through the visual cycle (Wald, 1968b) (Fig. 4). Once acquired by RPE from circulating precursors, ROL is rapidly esterified by LRAT. REs have a double function by retaining circulating ROL and by providing the substrate for the rate-limiting 11-*cis* isomerization reaction performed by RPE65 (Amengual et al., 2012). LRAT also helps to form oil droplet-like structures, named retinosomes that serve as cellular vitamin A reservoirs (Imanishi et al., 2004).

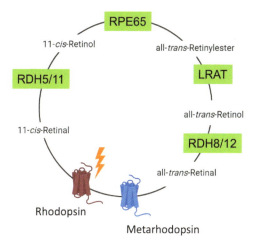

Fig. 4 The visual cycle for the generation and recycling of chromophore. The visual cycle takes place between rod and cone outer segments and involves distinct enzyme-catalyzed steps mediated by retinol dehydrogenases (RDHs), Lecithin: Retinol Acyltransferase (LRAT), and a retinoid isomerase encoded by the retina pigment epithelium protein of 65 kDa.

RPE65 catalyzes the transformation of at-RE to 11-*cis*-ROL and palmitate (Kiser, 2022). The next step in chromophore production is an oxidation of the alcohol that yields 11-*cis*-RAL mediated by the enzymes RDH5, RDH10 and RDH11 (Haeseleer et al., 2002; Sahu et al., 2015). There is evidence for the involvement of additional enzymes in this oxidation step (Maeda et al., 2007). CRALBP protects the newly synthesized 11-cis-RAL chromophore and facilitates its transport to the photoreceptor outer segments. Here the chromophore binds to opsins, via a Schiff base linkage thereby generating rhodopsin visual pigments in cone and rod photoreceptors (Saari & Bredberg, 1987).

Absorption of a photon by visual pigments results in a geometric isomerization of the rhodopsin-bound chromophore from 11-*cis*-RAL to at-RAL and the formation of metarhodopsin II that initiates phototransduction (Arshavsky et al., 2002). Hydrolysis of the Schiff base linkage releases the at-RAL photoproduct from metarhodopsin (Jastrzebska et al., 2011). At-RAL enters the disk lumen and transfers to the cytosol in an ATP-binding cassette transporter 4 (ABCA4) facilitated manner (Molday et al., 2000, 2009; Tsybovsky et al., 2010). For this transport, at-RAL forms conjugates with phosphatidylethanolamine, an abundant membrane lipid in disk membranes. ABCA4 catalyzes N-retinylidene-phosphatidylethanolamine import into the cytosol in an

ATP-dependent fashion (Quazi et al., 2012; Tsybovsky et al., 2013). This transporter also plays an important role in the control of chromophore homeostasis in the dark-adapted photoreceptors (Quazi & Molday, 2014).

In the cytoplasm, at-RAL is rapidly converted to at-ROL (Haeseleer et al., 1998; Rattner et al., 2000). Two enzymes, RDH8 and RDH12, are responsible for this reduction (Maeda et al., 2005). However, similar to chromophore production, redundancy exists for this redox reaction in photoreceptors of the mouse retina (Maeda et al., 2007).

The described sequence of reactions takes place between photoreceptors outer segments and RPE. Disruption of the cycle is associated with inherited blinding diseases that initially allowed researchers to dedicate specific genes to the different steps of the visual cycle (Thompson & Gal, 2003). The resulting phenotypes display in milder cases as retinitis pigmentosa and in more severe cases as Leber congenital amaurosis with slow but progressive rod photoreceptor cell degeneration (Woodruff et al., 2003), and rapid cone photoreceptor degeneration (Zhang et al., 2008).

11. Seeing in color: unveiling the cone visual pigment regeneration mechanism(s)

Cones are sensitive under bright light and rods under dim light conditions. Therefore, cone response kinetics differ significantly from rods (Perry & McNaughton, 1991). Rod photoresponses saturate at photoisomerization rates above 500 per second (Baylor et al., 1984) whereas cones remain responsive to light at far higher luminescence rates (Schnapf et al., 1990). These differences require mechanism(s) to maintain chromophore regeneration for cones when rod vision is saturated (Fig. 5).

A hallmark of retinas of day-active animals is the presence of 11-*cis*-RE (Mustafi et al., 2016). How this vitamin A metabolite is synthesized and whether it specifically supports cone vision was investigated in zebrafish (*Danio rerio*) eyes (Babino et al., 2015). The eyes of the zebrafish larva are amenable for genetic and pharmacological interventions and have been successfully used to study photoreceptor development and function (Fleisch & Neuhauss, 2010). These studies showed that the larval eyes express key components of a classical visual cycle, including *Stra6*, *Lrat*, *Rpe65*, and *Cralbp* (Fleisch et al., 2008; Isken et al., 2007, 2008; Schonthaler et al., 2007). Two-photon microscopy and biochemical analyses detected 11-*cis*-RE and at-RE mainly in the RPE of dark-adapted fish larvae (Babino et al., 2015).

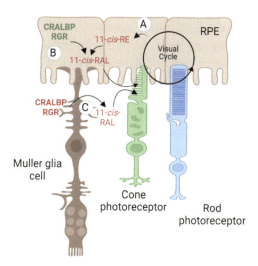

Fig. 5 **Cone specific pathways for regeneration of chromophore.** The visual cycle contributes to both cone and rod visual pigment regeneration. As indicated under A, a dark-generated pool of 11-*cis*-RE supports cone vision. As indicated under B, a light-dependent pathway involving RGR and CRALBP exists in the RPE and supports cone vision. Additionally, RGR and CRALBP in Muller glia cells, as indicated under C, contribute to cone visual pigment regeneration.

When larvae were exposed to bright light, chromophore concentrations initially decreased but then reached a steady state concentration. This steady state was maintained by a consumption of the 11-*cis*-REs stores under bright light illumination. The specificity of this pathway for cones was confirmed in a cone-only zebrafish mutant. Interestingly, evidence exists that a light dependent hydrolysis of 11-*cis*-RE also occurs in homogenates of human retinal epithelial cells (Blaner et al., 1987), suggesting that a similar pathway exists in the human eyes.

Moreover, pharmacological inhibition of RPE65 with retinylamine showed that 11-*cis*-REs synthesis depends in the classical retinoid cycle. 11-*cis*-RE pools were generated in the dark and served cone vision under bright light luminance (Babino et al., 2015) when the relatively slow RPE65 catalyzed isomerization reaction becomes rate limiting (Lyubarsky et al., 2005; Wenzel et al., 2005). The mobilization of this 11-*cis*-RE reservoir depends on CRALBP and in its absence, REs accumulate in the RPE of the fish (Schlegel et al., 2021).

Additionally, researchers reported light-dependent regeneration mechanisms for chromophores in the retina. Blue light illumination of a bleached mouse retina significantly increased the levels of 9-*cis*-RAL and 11-*cis*-RAL, providing evidence for light-dependent re-isomerization of chromophores in photoreceptors (Ramkumar et al., 2024). Moreover, it has been shown that the at-N-retinylidene-PE conjugates are converted to the corresponding 11-*cis*-conjugates by blue light (450 nm). The quantum efficiency of this enzyme-independent photo-conversion is similar to that of rhodopsin, and blue light (450 nm) induces visual pigment regeneration in the living mouse retina (Kaylor et al., 2017).

Moreover, the retina contains a photoisomerase-dependent pathway, involving retinal G-protein-coupled receptor (RGR) and CRALBP (Yang & Fong, 2002). Initial evidence showed that RGR contributes to chromophore regeneration in $Rpe65^{-/-}$ mice, but not in $Rpe65^{-/-}Rgr^{-/-}$ compound knockout mice (Van Hooser et al., 2002). However, these mice displayed a minimal rate of 11-*cis*-RAL synthesis (Hao & Fong, 1999). Recent biochemical studies have revived the idea of an RGR-dependent visual chromophore regeneration pathway (Zhang et al., 2019). The Palczewski laboratory reported high-level production of 11-*cis*-RAL in RPE membranes upon narrow-band green light illumination (Zhang et al., 2019). The activity increased with protein concentration and displayed substrate saturation, indicating that it is enzyme catalyzed. CRALBP presence stimulated this enzyme activity in the assays. Using specific inhibitors for RPE65, the researchers demonstrated that the light dependent isomerase activity was associated with RGR. Accordingly, recombinant RGR exhibited the same activity in cell-based and cell-free enzyme assays. Site-directed mutagenesis and biochemical testing revealed that the isomerase reaction involves a Schiff base linkage of the at-RAL substrate at Lys255 of RGR. Single-cell RNA-Seq analysis of the retina and RPE tissue showed that RGR is expressed in human and bovine RPE and Müller glia cells, whereas mouse RGR is expressed in RPE but at much lower levels in Müller glial cells (Tworak et al., 2023). In keeping with this observation, recent studies in mice indicate that, similar to zebrafish, CRALBP in the RPE but not in Müller cells contributes to chromophore regeneration. This contrasts with other studies in mice indicating that RGR, in concert with a retinol dehydrogenase enzyme (Xue et al., 2017), contributes to cone visual pigment regeneration in Müller glia cells (Morshedian et al., 2019). Thus, further research is required to elucidate all details of chromophore regeneration for cone photoreceptor function.

References

Ables, G. P., Yang, K. J., Vogel, S., Hernandez-Ono, A., Yu, S., Yuen, J. J., ... Ginsberg, H. N. (2012). Intestinal DGAT1 deficiency reduces postprandial triglyceride and retinyl ester excursions by inhibiting chylomicron secretion and delaying gastric emptying. *Journal of Lipid Research, 53*, 2364–2379.

Abu-Abed, S., Dolle, P., Metzger, D., Beckett, B., Chambon, P., & Petkovich, M. (2001). The retinoic acid-metabolizing enzyme, CYP26A1, is essential for normal hindbrain patterning, vertebral identity, and development of posterior structures. *Genes & Development, 15*, 226–240.

Abu-Abed, S. S., Beckett, B. R., Chiba, H., Chithalen, J. V., Jones, G., Metzger, D., ... Petkovich, M. (1998). Mouse P450RAI (CYP26) expression and retinoic acid-inducible retinoic acid metabolism in F9 cells are regulated by retinoic acid receptor gamma and retinoid X receptor alpha. *The Journal of Biological Chemistry, 273*, 2409–2415.

Adams, M. K., Belyaeva, O. V., Wu, L., & Kedishvili, N. Y. (2014). The retinaldehyde reductase activity of DHRS3 is reciprocally activated by retinol dehydrogenase 10 to control retinoid homeostasis. *The Journal of Biological Chemistry, 289*, 14868–14880.

Alvarez, R., Vaz, B., Gronemeyer, H., & De Lera, A. R. (2014). Functions, therapeutic applications, and synthesis of retinoids and carotenoids. *Chemical Reviews, 114*, 1–125.

Amengual, J., Golczak, M., Palczewski, K., & von Lintig, J. (2012). Lecithin:Retinol acyltransferase is critical for cellular uptake of vitamin A from serum retinol-binding protein. *The Journal of Biological Chemistry, 287*, 24216–24227.

Amengual, J., Lobo, G. P., Golczak, M., Li, H. N., Klimova, T., Hoppel, C. L., ... von Lintig, J. (2011). A mitochondrial enzyme degrades carotenoids and protects against oxidative stress. *The FASEB Journal, 25*, 948–959.

Amengual, J., Widjaja-Adhi, M. A., Rodriguez-Santiago, S., Hessel, S., Golczak, M., Palczewski, K., & von Lintig, J. (2013). Two carotenoid oxygenases contribute to mammalian provitamin A metabolism. *The Journal of Biological Chemistry, 288*, 34081–34096.

Amengual, J., Zhang, N., Kemerer, M., Maeda, T., Palczewski, K., & Von Lintig, J. (2014). STRA6 is critical for cellular vitamin A uptake and homeostasis. *Human Molecular Genetics, 23*, 5402–5417.

Arshavsky, V. Y., Lamb, T. D., & Pugh, E. N., Jr. (2002). G proteins and phototransduction. *Annual Review of Physiology, 64*, 153–187.

Babino, D., Golczak, M., Kiser, P. D., Wyss, A., Palczewski, K., & Von Lintig, J. (2016). The biochemical basis of vitamin A3 production in arthropod vision. *ACS Chemical Biology, 11*, 1049–1057.

Babino, D., Perkins, B. D., Kindermann, A., Oberhauser, V., & Von Lintig, J. (2015). The role of 11-cis-retinyl esters in vertebrate cone vision. *The FASEB Journal, 29*, 216–226.

Bachmann, H., Desbarats, A., Pattison, P., Sedgewick, M., Riss, G., Wyss, A., ... Grolier, P. (2002). Feedback regulation of beta,beta-carotene 15,15'-monooxygenase by retinoic acid in rats and chickens. *The Journal of Nutrition, 132*, 3616–3622.

Bandara, S., Moon, J., Ramkumar, S., & von Lintig, J. (2023). ASTER-B regulates mitochondrial carotenoid transport and homeostasis. *Journal of Lipid Research*, 100369.

Bandara, S., Thomas, L. D., Ramkumar, S., Khadka, N., Kiser, P. D., Golczak, M., & von Lintig, J. (2021). The structural and biochemical basis of apocarotenoid processing by beta-carotene oxygenase-2. *ACS Chemical Biology, 16*, 480–490.

Bassetto, M., Kolesnikov, A. V., Lewandowski, D., Kiser, J. Z., Halabi, M., Einstein, D. E., ... Kiser, P. D. (2024). Dominant role for pigment epithelial CRALBP in supplying visual chromophore to photoreceptors. *Cell Reports, 43*, 114143.

Batten, M. L., Imanishi, Y., Maeda, T., Tu, D. C., Moise, A. R., Bronson, D., ... Palczewski, K. (2004). Lecithin-retinol acyltransferase is essential for accumulation of all-trans-retinyl esters in the eye and in the liver. *The Journal of Biological Chemistry, 279*, 10422–10432.

Baylor, D. A., Nunn, B. J., & Schnapf, J. L. (1984). The photocurrent, noise and spectral sensitivity of rods of the monkey Macaca fascicularis. *The Journal of Physiology, 357*, 575–607.
Berry, D. C., Jacobs, H., Marwarha, G., Gely-Pernot, A., O'Byrne, S. M., Desantis, D., ... Ghyselinck, N. B. (2013). The STRA6 receptor is essential for retinol-binding protein-induced insulin resistance but not for maintaining vitamin A homeostasis in tissues other than the eye. *The Journal of Biological Chemistry, 288*, 24528–24539.
Biesalski, H. K., Frank, J., Beck, S. C., Heinrich, F., Illek, B., Reifen, R., ... Zrenner, E. (1999). Biochemical but not clinical vitamin A deficiency results from mutations in the gene for retinol binding protein. *The American Journal of Clinical Nutrition, 69*, 931–936.
Billings, S. E., Pierzchalski, K., Butler Tjaden, N. E., Pang, X. Y., Trainor, P. A., Kane, M. A., & Moise, A. R. (2013). The retinaldehyde reductase DHRS3 is essential for preventing the formation of excess retinoic acid during embryonic development. *The FASEB Journal, 27*, 4877–4889.
Blaner, W. S. (2007). STRA6, a cell-surface receptor for retinol-binding protein: The plot thickens. *Cell Metabolism, 5*, 164–166.
Blaner, W. S., Brun, P. J., Calderon, R. M., & Golczak, M. (2020). Retinol-binding protein 2 (RBP2): Biology and pathobiology. *Critical Reviews in Biochemistry and Molecular Biology, 55*, 197–218.
Blaner, W. S., Das, S. R., Gouras, P., & Flood, M. T. (1987). Hydrolysis of 11-cis- and all-trans-retinyl palmitate by homogenates of human retinal epithelial cells. *The Journal of Biological Chemistry, 262*, 53–58.
Bouillet, P., Sapin, V., Chazaud, C., Messaddeq, N., Decimo, D., Dolle, P., & Chambon, P. (1997). Developmental expression pattern of Stra6, a retinoic acid-responsive gene encoding a new type of membrane protein. *Mechanisms of Development, 63*, 173–186.
Bunt-Milam, A. H., & Saari, J. C. (1983). Immunocytochemical localization of two retinoid-binding proteins in vertebrate retina. *The Journal of Cell Biology, 97*, 703–712.
Casey, J., Kawaguchi, R., Morrissey, M., Sun, H., McGettigan, P., Nielsen, J. E., ... Ennis, S. (2011). First implication of STRA6 mutations in isolated anophthalmia, microphthalmia, and coloboma: a new dimension to the STRA6 phenotype. *Human Mutation, 32*, 1417–1426.
Chambon, P. (1996). A decade of molecular biology of retinoic acid receptors. *The FASEB Journal, 10*, 940–954.
Chawla, A., Repa, J. J., Evans, R. M., & Mangelsdorf, D. J. (2001). Nuclear receptors and lipid physiology: Opening the X-files. *Science (New York, N. Y.), 294*, 1866–1870.
Chen, Y., Clarke, O. B., Kim, J., Stowe, S., Kim, Y. K., Assur, Z., ... Mancia, F. (2016). Structure of the STRA6 receptor for retinol uptake. *Science (New York, N. Y.), 353*.
Chou, C. M., Nelson, C., Tarle, S. A., Pribila, J. T., Bardakjian, T., Woods, S., ... Glaser, T. (2015). Biochemical basis for dominant inheritance, variable penetrance, and maternal effects in RBP4 congenital eye disease. *Cell, 161*, 634–646.
D'Ambrosio, D. N., Clugston, R. D., & Blaner, W. S. (2011). Vitamin A metabolism: An update. *Nutrients, 3*, 63–103.
Duester, G. (2008). Retinoic acid synthesis and signaling during early organogenesis. *Cell, 134*, 921–931.
Dupe, V., Matt, N., Garnier, J. M., Chambon, P., Mark, M., & Ghyselinck, N. B. (2003). A newborn lethal defect due to inactivation of retinaldehyde dehydrogenase type 3 is prevented by maternal retinoic acid treatment. *Proceedings of the National Academy of Sciences of the United States of America, 100*, 14036–14041.
Episkopou, V., Maeda, S., Nishiguchi, S., Shimada, K., Gaitanaris, G. A., Gottesman, M. E., & Robertson, E. J. (1993). Disruption of the transthyretin gene results in mice with depressed levels of plasma retinol and thyroid hormone. *Proceedings of the National Academy of Sciences of the United States of America, 90*, 2375–2379.

Evans, R. M., & Mangelsdorf, D. J. (2014). Nuclear receptors, RXR, and the big bang. *Cell, 157*, 255–266.

Fan, X., Molotkov, A., Manabe, S., Donmoyer, C. M., Deltour, L., Foglio, M. H., ... Duester, G. (2003). Targeted disruption of Aldh1a1 (Raldh1) provides evidence for a complex mechanism of retinoic acid synthesis in the developing retina. *Molecular and Cellular Biology, 23*, 4637–4648.

Fleisch, V. C., & Neuhauss, S. C. (2010). Parallel visual cycles in the zebrafish retina. *Progress in Retinal and Eye Research, 29*, 476–486.

Fleisch, V. C., Schonthaler, H. B., Von Lintig, J., & Neuhauss, S. C. (2008). Subfunctionalization of a retinoid-binding protein provides evidence for two parallel visual cycles in the cone-dominant zebrafish retina. *The Journal of Neuroscience, 28*, 8208–8216.

Franzoni, L., Lucke, C., Perez, C., Cavazzini, D., Rademacher, M., Ludwig, C., ... Ruterjans, H. (2002). Structure and backbone dynamics of Apo- and holo-cellular retinol-binding protein in solution. *The Journal of Biological Chemistry, 277*, 21983–21997.

Ghyselinck, N. B., Bavik, C., Sapin, V., Mark, M., Bonnier, D., Hindelang, C., ... Chambon, P. (1999). Cellular retinol-binding protein I is essential for vitamin A homeostasis. *The EMBO Journal, 18*, 4903–4914.

Ghyselinck, N. B., & Duester, G. (2019). Retinoic acid signaling pathways. *Development (Cambridge, England), 146*.

Giguere, V., Ong, E. S., Segui, P., & Evans, R. M. (1987). Identification of a receptor for the morphogen retinoic acid. *Nature, 330*, 624–629.

Giuliano, G., Al-Babili, S., & von Lintig, J. (2003). Carotenoid oxygenases: Cleave it or leave it. *Trends in Plant Science, 8*, 145–149.

Golczak, M., Sears, A. E., Kiser, P. D., & Palczewski, K. (2015). LRAT-specific domain facilitates vitamin A metabolism by domain swapping in HRASLS3. *Nature Chemical Biology, 11*, 26–32.

Golzio, C., Martinovic-Bouriel, J., Thomas, S., Mougou-Zrelli, S., Grattagliano-Bessieres, B., Bonniere, M., ... Etchevers, H. C. (2007). Matthew-Wood syndrome is caused by truncating mutations in the retinol-binding protein receptor gene STRA6. *American Journal of Human Genetics, 80*, 1179–1187.

Grune, T., Lietz, G., Palou, A., Ross, A. C., Stahl, W., Tang, G., ... Biesalski, H. K. (2010). Beta-carotene is an important vitamin A source for humans. *The Journal of Nutrition, 140*, 2268S–2285S.

Haeseleer, F., Huang, J., Lebioda, L., Saari, J. C., & Palczewski, K. (1998). Molecular characterization of a novel short-chain dehydrogenase/reductase that reduces all-trans-retinal. *The Journal of Biological Chemistry, 273*, 21790–21799.

Haeseleer, F., Jang, G. F., Imanishi, Y., Driessen, C. A., Matsumura, M., Nelson, P. S., & Palczewski, K. (2002). Dual-substrate specificity short chain retinol dehydrogenases from the vertebrate retina. *The Journal of Biological Chemistry, 277*, 45537–45546.

Hao, W., & Fong, H. K. (1999). The endogenous chromophore of retinal G protein-coupled receptor opsin from the pigment epithelium. *The Journal of Biological Chemistry, 274*, 6085–6090.

He, X., Lobsiger, J., & Stocker, A. (2009). Bothnia dystrophy is caused by domino-like rearrangements in cellular retinaldehyde-binding protein mutant R234W. *Proceedings of the National Academy of Sciences of the United States of America, 106*, 18545–18550.

Heller, M., & Bok, D. (1976). A specific receptor for retinol binding protein as detected by the binding of human and bovine retinol binding protein to pigment epithelial cells. *American Journal of Ophthalmology, 81*, 93–97.

Hofmann, L., Tsybovsky, Y., Alexander, N. S., Babino, D., Leung, N. Y., Montell, C., ... Palczewski, K. (2016). Structural Insights into the Drosophila melanogaster Retinol Dehydrogenase, a Member of the Short-Chain Dehydrogenase/Reductase Family. *Biochemistry, 55*, 6545–6557.

Imanishi, Y., Gerke, V., & Palczewski, K. (2004). Retinosomes: New insights into intracellular managing of hydrophobic substances in lipid bodies. *The Journal of Cell Biology, 166*, 447–453.

Isken, A., Golczak, M., Oberhauser, V., Hunzelmann, S., Driever, W., Imanishi, Y., ... von Lintig, J. (2008). RBP4 disrupts vitamin A uptake homeostasis in a STRA6-deficient animal model for Matthew-Wood syndrome. *Cell Metabolism, 7*, 258–268.

Isken, A., Holzschuh, J., Lampert, J. M., Fischer, L., Oberhauser, V., Palczewski, K., & von Lintig, J. (2007). Sequestration of retinyl esters is essential for retinoid signaling in the zebrafish embryo. *The Journal of Biological Chemistry, 282*, 1144–1151.

Isoherranen, N., & Zhong, G. (2019). Biochemical and physiological importance of the CYP26 retinoic acid hydroxylases. *Pharmacology & Therapeutics, 204*, 107400.

Jastrzebska, B., Palczewski, K., & Golczak, M. (2011). Role of bulk water in hydrolysis of the rhodopsin chromophore. *The Journal of Biological Chemistry, 286*, 18930–18937.

Kanai, M., Raz, A., & Goodman, D. S. (1968). Retinol-binding protein: The transport protein for vitamin A in human plasma. *The Journal of Clinical Investigation, 47*, 2025–2044.

Kawaguchi, R., Yu, J., Honda, J., Hu, J., Whitelegge, J., Ping, P., ... Sun, H. (2007). A membrane receptor for retinol binding protein mediates cellular uptake of vitamin a. *Science (New York, N. Y.), 315*, 820–825.

Kawaguchi, R., Yu, J., Ter-Stepanian, M., Zhong, M., Cheng, G., Yuan, Q., ... Sun, H. (2011). Receptor-mediated cellular uptake mechanism that couples to intracellular storage. *ACS Chemical Biology, 6*, 1041–1045.

Kaylor, J. J., Xu, T., Ingram, N. T., Tsan, A., Hakobyan, H., Fain, G. L., & Travis, G. H. (2017). Blue light regenerates functional visual pigments in mammals through a retinyl-phospholipid intermediate. *Nature Communications, 8*, 16.

Kedishvili, N. Y. (2013). Enzymology of retinoic acid biosynthesis and degradation. *Journal of Lipid Research, 54*, 1744–1760.

Kelly, M., Widjaja-Adhi, M. A., Palczewski, G., & Von Lintig, J. (2016). Transport of vitamin A across blood-tissue barriers is facilitated by STRA6. *The FASEB Journal, 30*, 2985–2995.

Kelly, M. E., Ramkumar, S., Sun, W., Colon Ortiz, C., Kiser, P. D., Golczak, M., & von Lintig, J. (2018). The Biochemical Basis of Vitamin A Production from the Asymmetric Carotenoid beta-Cryptoxanthin. *ACS Chemical Biology, 13*, 2121–2129.

Kiefer, C., Hessel, S., Lampert, J. M., Vogt, K., Lederer, M. O., Breithaupt, D. E., & von Lintig, J. (2001). Identification and characterization of a mammalian enzyme catalyzing the asymmetric oxidative cleavage of provitamin A. *The Journal of Biological Chemistry, 276*, 14110–14116.

Kim, T. S., Maeda, A., Maeda, T., Heinlein, C., Kedishvili, N., Palczewski, K., & Nelson, P. S. (2005). Delayed dark adaptation in 11-cis-retinol dehydrogenase-deficient mice: A role of RDH11 in visual processes in vivo. *The Journal of Biological Chemistry, 280*, 8694–8704.

Kiser, P. D. (2022). Retinal pigment epithelium 65 kDa protein (RPE65): An update. *Progress in Retinal and Eye Research, 88*, 101013.

Kiser, P. D., & Palczewski, K. (2016). Retinoids and retinal diseases. *Annual Review of Vision Science, 2*, 197–234.

Kiser, P. D., Zhang, J., Badiee, M., Li, Q., Shi, W., Sui, X., ... Palczewski, K. (2015). Catalytic mechanism of a retinoid isomerase essential for vertebrate vision. *Nature Chemical Biology, 11*, 409–415.

Lala, V. R., & Reddy, V. (1970). Absorption of beta-carotene from green leafy vegetables in undernourished children. *The American Journal of Clinical Nutrition, 23*, 110–113.

Lindqvist, A., & Andersson, S. (2002). Biochemical properties of purified recombinant human beta-carotene 15,15'-monooxygenase. *The Journal of Biological Chemistry, 277*, 23942–23948.

Liu, L., & Gudas, L. J. (2005). Disruption of the lecithin:retinol acyltransferase gene makes mice more susceptible to vitamin A deficiency. *The Journal of Biological Chemistry, 280,* 40226–40234.

Lobo, G. P., Amengual, J., Baus, D., Shivdasani, R. A., Taylor, D., & von Lintig, J. (2013). Genetics and diet regulate vitamin A production via the homeobox transcription factor ISX. *The Journal of Biological Chemistry, 288,* 9017–9027.

Lobo, G. P., Hessel, S., Eichinger, A., Noy, N., Moise, A. R., Wyss, A., ... von Lintig, J. (2010). ISX is a retinoic acid-sensitive gatekeeper that controls intestinal beta,beta-carotene absorption and vitamin A production. *The FASEB Journal, 24,* 1656–1666.

Lyubarsky, A. L., Savchenko, A. B., Morocco, S. B., Daniele, L. L., Redmond, T. M., & Pugh, E. N., Jr. (2005). Mole quantity of RPE65 and its productivity in the generation of 11-cis-retinal from retinyl esters in the living mouse eye. *Biochemistry, 44,* 9880–9888.

Maeda, A., Golczak, M., Maeda, T., & Palczewski, K. (2009). Limited roles of Rdh8, Rdh12, and Abca4 in all-trans-retinal clearance in mouse retina. *Investigative Ophthalmology & Visual Science, 50,* 5435–5443.

Maeda, A., Maeda, T., Imanishi, Y., Kuksa, V., Alekseev, A., Bronson, J. D., ... Palczewski, K. (2005). Role of photoreceptor-specific retinol dehydrogenase in the retinoid cycle in vivo. *The Journal of Biological Chemistry, 280,* 18822–18832.

Maeda, A., Maeda, T., Imanishi, Y., Sun, W., Jastrzebska, B., Hatala, D. A., ... Palczewski, K. (2006). Retinol dehydrogenase (RDH12) protects photoreceptors from light-induced degeneration in mice. *The Journal of Biological Chemistry, 281,* 37697–37704.

Maeda, A., Maeda, T., Sun, W., Zhang, H., Baehr, W., & Palczewski, K. (2007). Redundant and unique roles of retinol dehydrogenases in the mouse retina. *Proceedings of the National Academy of Sciences of the United States of America, 104,* 19565–19570.

Mic, F. A., Molotkov, A., Fan, X., Cuenca, A. E., & Duester, G. (2000). RALDH3, a retinaldehyde dehydrogenase that generates retinoic acid, is expressed in the ventral retina, otic vesicle and olfactory pit during mouse development. *Mechanisms of Development, 97,* 227–230.

Miller, A. P., Hornero-Mendez, D., Bandara, S., Parra-Rivero, O., Limon, M. C., von Lintig, J., ... Amengual, J. (2023). Bioavailability and provitamin A activity of neurosporaxanthin in mice. *Communications Biology, 6,* 1068.

Molday, L. L., Rabin, A. R., & Molday, R. S. (2000). ABCR expression in foveal cone photoreceptors and its role in Stargardt macular dystrophy. *Nature Genetics, 25,* 257–258.

Molday, R. S., Zhong, M., & Quazi, F. (2009). The role of the photoreceptor ABC transporter ABCA4 in lipid transport and Stargardt macular degeneration. *Biochimica et Biophysica Acta, 1791,* 573–583.

Montenegro, D., Zhao, J., Kim, H. J., Shmarakov, I. O., Blaner, W. S., & Sparrow, J. R. (2022). Products of the visual cycle are detected in mice lacking retinol binding protein 4, the only known vitamin A carrier in plasma. *The Journal of Biological Chemistry, 298,* 102722.

Moon, J., Ramkumar, S., & von Lintig, J. (2022). Genetic dissection in mice reveals a dynamic crosstalk between the delivery pathways of vitamin A. *Journal of Lipid Research, 63,* 100215.

Morshedian, A., Kaylor, J. J., Ng, S. Y., Tsan, A., Frederiksen, R., Xu, T., ... Travis, G. H. (2019). Light-driven regeneration of cone visual pigments through a mechanism involving RGR opsin in muller glial cells. *Neuron.*

Mustafi, D., Kevany, B. M., Bai, X., Golczak, M., Adams, M. D., Wynshaw-Boris, A., & Palczewski, K. (2016). Transcriptome analysis reveals rod/cone photoreceptor specific signatures across mammalian retinas. *Human Molecular Genetics, 25,* 4376–4388.

Nelson, C. H., Peng, C. C., Lutz, J. D., Yeung, C. K., Zelter, A., & Isoherranen, N. (2016). Direct protein-protein interactions and substrate channeling between cellular retinoic acid binding proteins and CYP26B1. *FEBS Letters, 590,* 2527–2535.

Niederreither, K., Fraulob, V., Garnier, J. M., Chambon, P., & Dolle, P. (2002). Differential expression of retinoic acid-synthesizing (RALDH) enzymes during fetal development and organ differentiation in the mouse. *Mechanisms of Development, 110*, 165–171.

Niederreither, K., Subbarayan, V., Dolle, P., & Chambon, P. (1999). Embryonic retinoic acid synthesis is essential for early mouse post-implantation development. *Nature Genetics, 21*, 444–448.

O'Byrne, S. M., Wongsiriroj, N., Libien, J., Vogel, S., Goldberg, I. J., Baehr, W., ... Blaner, W. S. (2005). Retinoid absorption and storage is impaired in mice lacking lecithin:retinol acyltransferase (LRAT). *The Journal of Biological Chemistry, 280*, 35647–35657.

Oberhauser, V., Voolstra, O., Bangert, A., von Lintig, J., & Vogt, K. (2008). NinaB combines carotenoid oxygenase and retinoid isomerase activity in a single polypeptide. *Proceedings of the National Academy of Sciences of the United States of America, 105*, 19000–19005.

Palczewski, K., Kumasaka, T., Hori, T., Behnke, C. A., Motoshima, H., Fox, B. A., ... Miyano, M. (2000). Crystal structure of rhodopsin: A G protein-coupled receptor. *Science (New York, N. Y.), 289*, 739–745.

Parker, R., Wang, J. S., Kefalov, V. J., & Crouch, R. K. (2011). Interphotoreceptor retinoid-binding protein as the physiologically relevant carrier of 11-cis-retinol in the cone visual cycle. *The Journal of Neuroscience, 31*, 4714–4719.

Parker, R. O., Fan, J., Nickerson, J. M., Liou, G. I., & Crouch, R. K. (2009). Normal cone function requires the interphotoreceptor retinoid binding protein. *The Journal of Neuroscience, 29*, 4616–4621.

Perduca, M., Nicolis, S., Mannucci, B., Galliano, M., & Monaco, H. L. (2018). Human plasma retinol-binding protein (RBP4) is also a fatty acid-binding protein. *Biochimica et Biophysica Acta (BBA)—Molecular and Cell Biology of Lipids, 1863*, 458–466.

Perry, R. J., & McNaughton, P. A. (1991). Response properties of cones from the retina of the tiger salamander. *The Journal of Physiology, 433*, 561–587.

Petkovich, M., Brand, N. J., Krust, A., & Chambon, P. (1987). A human retinoic acid receptor which belongs to the family of nuclear receptors. *Nature, 330*, 444–450.

Quadro, L., Blaner, W. S., Salchow, D. J., Vogel, S., Piantedosi, R., Gouras, P., ... Gottesman, M. E. (1999). Impaired retinal function and vitamin A availability in mice lacking retinol-binding protein. *The EMBO Journal, 18*, 4633–4644.

Quadro, L., Hamberger, L., Gottesman, M. E., Wang, F., Colantuoni, V., Blaner, W. S., & Mendelsohn, C. L. (2005). Pathways of vitamin A delivery to the embryo: insights from a new tunable model of embryonic vitamin A deficiency. *Endocrinology, 146*, 4479–4490.

Quazi, F., Lenevich, S., & Molday, R. S. (2012). ABCA4 is an N-retinylidene-phosphatidylethanolamine and phosphatidylethanolamine importer. *Nature Communications, 3*, 925.

Quazi, F., & Molday, R. S. (2014). ATP-binding cassette transporter ABCA4 and chemical isomerization protect photoreceptor cells from the toxic accumulation of excess 11-cis-retinal. *Proceedings of the National Academy of Sciences of the United States of America, 111*, 5024–5029.

Ramkumar, S., Jastrzebska, B., Montenegro, D., Sparrow, J. R., & Von Lintig, J. (2024). Unraveling the mystery of ocular retinoid turnover: Insights from albino mice and the role of STRA6. *The Journal of Biological Chemistry, 300*, 105781.

Ramkumar, S., Moon, J., Golczak, M., & Von Lintig, J. (2021). LRAT coordinates the negative-feedback regulation of intestinal retinoid biosynthesis from beta-carotene. *Journal of Lipid Research, 62*, 100055.

Ramkumar, S., Parmar, V. M., Samuels, I., Berger, N. A., Jastrzebska, B., & von Lintig, J. (2022). The vitamin A transporter STRA6 adjusts the stoichiometry of chromophore and opsins in visual pigment synthesis and recycling. *Human Molecular Genetics, 31*, 548–560.

Rattner, A., Smallwood, P. M., & Nathans, J. (2000). Identification and characterization of all-trans-retinol dehydrogenase from photoreceptor outer segments, the visual cycle enzyme that reduces all-trans-retinal to all-trans-retinol. *The Journal of Biological Chemistry, 275*, 11034–11043.

Redmond, T. M., Poliakov, E., Yu, S., Tsai, J. Y., Lu, Z., & Gentleman, S. (2005). Mutation of key residues of RPE65 abolishes its enzymatic role as isomerohydrolase in the visual cycle. *Proceedings of the National Academy of Sciences of the United States of America, 102*, 13658–13663.

Rhinn, M., & Dollé, P. (2012). Retinoic acid signalling during development. *Development (Cambridge, England), 139*, 843–858.

Ruiz, A., Mark, M., Jacobs, H., Klopfenstein, M., Hu, J., Lloyd, M., ... Bok, D. (2012). Retinoid content, visual responses, and ocular morphology are compromised in the retinas of mice lacking the retinol-binding protein receptor, STRA6. *Investigative Ophthalmology & Visual Science, 53*, 3027–3039.

Saari, J. C., & Bredberg, D. L. (1987). Photochemistry and stereoselectivity of cellular retinaldehyde-binding protein from bovine retina. *The Journal of Biological Chemistry, 262*, 7618–7622.

Saari, J. C., Nawrot, M., Garwin, G. G., Kennedy, M. J., Hurley, J. B., Ghyselinck, N. B., & Chambon, P. (2002). Analysis of the visual cycle in cellular retinol-binding protein type I (CRBPI) knockout mice. *Investigative Ophthalmology & Visual Science, 43*, 1730–1735.

Sahu, B., Sun, W., Perusek, L., Parmar, V., Le, Y. Z., Griswold, M. D., ... Maeda, A. (2015). Conditional ablation of retinol dehydrogenase 10 in the retinal pigmented epithelium causes delayed dark adaption in mice. *The Journal of Biological Chemistry, 290*, 27239–27247.

Sandell, L. L., Sanderson, B. W., Moiseyev, G., Johnson, T., Mushegian, A., Young, K., ... Trainor, P. A. (2007). RDH10 is essential for synthesis of embryonic retinoic acid and is required for limb, craniofacial, and organ development. *Genes & Development, 21*, 1113–1124.

Schlegel, D. K., Ramkumar, S., von Lintig, J., & Neuhauss, S. C. (2021). Disturbed retinoid metabolism upon loss of rlbp1a impairs cone function and leads to subretinal lipid deposits and photoreceptor degeneration in the zebrafish retina. *eLife, 10*.

Schnapf, J. L., Nunn, B. J., Meister, M., & Baylor, D. A. (1990). Visual transduction in cones of the monkey Macaca fascicularis. *The Journal of Physiology, 427*, 681–713.

Schonthaler, H. B., Lampert, J. M., Isken, A., Rinner, O., Mader, A., Gesemann, M., ... Von Lintig, J. (2007). Evidence for RPE65-independent vision in the cone-dominated zebrafish retina. *The European Journal of Neuroscience, 26*, 1940–1949.

Sears, A. E., Bernstein, P. S., Cideciyan, A. V., Hoyng, C., Charbel Issa, P., Palczewski, K., ... Scholl, H. P. N. (2017). Towards treatment of stargardt disease: Workshop organized and sponsored by the foundation fighting blindness. *Translational Vision Science & Technology, 6*, 6.

Seeliger, M. W., Biesalski, H. K., Wissinger, B., Gollnick, H., Gielen, S., Frank, J., ... Zrenner, E. (1999). Phenotype in retinol deficiency due to a hereditary defect in retinol binding protein synthesis. *Investigative Ophthalmology & Visual Science, 40*, 3–11.

Seino, Y., Miki, T., Kiyonari, H., Abe, T., Fujimoto, W., Kimura, K., ... Seino, S. (2008). Isx participates in the maintenance of vitamin A metabolism by regulation of beta-carotene 15,15'-monooxygenase (Bcmo1) expression. *The Journal of Biological Chemistry, 283*, 4905–4911.

Shen, J., Shi, D., Suzuki, T., Xia, Z., Zhang, H., Araki, K., ... Li, Z. (2016). Severe ocular phenotypes in Rbp4-deficient mice in the C57BL/6 genetic background. *Laboratory Investigation; A Journal of Technical Methods and Pathology, 96*, 680–691.

Shmarakov, I. O., Gusarova, G. A., Islam, M. N., Marhuenda-Munoz, M., Bhattacharya, J., & Blaner, W. S. (2023). Retinoids stored locally in the lung are required to attenuate the severity of acute lung injury in male mice. *Nature Communications, 14*, 851.

Silvaroli, J. A., Plau, J., Adams, C. H., Banerjee, S., Widjaja-Adhi, M. A. K., Blaner, W. S., & Golczak, M. (2021). Molecular basis for the interaction of cellular retinol binding protein 2 (CRBP2) with nonretinoid ligands. *Journal of Lipid Research, 62*, 100054.

Solano, Y. J., Everett, M. P., Dang, K. S., Abueg, J., & Kiser, P. D. (2024). Carotenoid cleavage enzymes evolved convergently to generate the visual chromophore. *Nature Chemical Biology*.

Spiegler, E., Kim, Y. K., Hoyos, B., Narayanasamy, S., Jiang, H., Savio, N., ... Quadro, L. (2018). beta-apo-10'-carotenoids support normal embryonic development during vitamin A deficiency. *Scientific Reports, 8*, 8834.

Spiegler, E., Kim, Y. K., Wassef, L., Shete, V., & Quadro, L. (2012). Maternal-fetal transfer and metabolism of vitamin A and its precursor beta-carotene in the developing tissues. *Biochimica et Biophysica Acta, 1821*, 88–98.

Sui, X., Kiser, P. D., Lintig, J., & Palczewski, K. (2013). Structural basis of carotenoid cleavage: From bacteria to mammals. *Archives of Biochemistry and Biophysics, 539*, 203–213.

Thompson, D. A., & Gal, A. (2003). Vitamin A metabolism in the retinal pigment epithelium: Genes, mutations, and diseases. *Progress in Retinal and Eye Research, 22*, 683–703.

Tsybovsky, Y., Molday, R. S., & Palczewski, K. (2010). The ATP-binding cassette transporter ABCA4: Structural and functional properties and role in retinal disease. *Advances in Experimental Medicine and Biology, 703*, 105–125.

Tsybovsky, Y., Orban, T., Molday, R. S., Taylor, D., & Palczewski, K. (2013). Molecular organization and ATP-induced conformational changes of ABCA4, the photoreceptor-specific ABC transporter. *Structure (London, England: 1993), 21*, 854–860.

Tworak, A., Kolesnikov, A. V., Hong, J. D., Choi, E. H., Luu, J. C., Palczewska, G., ... Palczewski, K. (2023). Rapid RGR-dependent visual pigment recycling is mediated by the RPE and specialized Muller glia. *Cell Reports, 42*, 112982.

Uppal, S., Liu, T., Galvan, E., Gomez, F., Tittley, T., Poliakov, E., ... Redmond, T. M. (2023). An inducible amphipathic alpha-helix mediates subcellular targeting and membrane binding of RPE65. *Life Science Alliance, 6*.

Van Bennekum, A. M., Kako, Y., Weinstock, P. H., Harrison, E. H., Deckelbaum, R. J., Goldberg, I. J., & Blaner, W. S. (1999). Lipoprotein lipase expression level influences tissue clearance of chylomicron retinyl ester. *Journal of Lipid Research, 40*, 565–574.

Van Hooser, J. P., Liang, Y., Maeda, T., Kuksa, V., Jang, G. F., He, Y. G., ... Palczewski, K. (2002). Recovery of visual functions in a mouse model of Leber congenital amaurosis. *The Journal of Biological Chemistry, 277*, 19173–19182.

Van Vliet, T., Van Vlissingen, M. F., Van Schaik, F., & Van Den Berg, H. (1996). beta-Carotene absorption and cleavage in rats is affected by the vitamin A concentration of the diet. *The Journal of Nutrition, 126*, 499–508.

Von Lintig, J., Dreher, A., Kiefer, C., Wernet, M. F., & Vogt, K. (2001). Analysis of the blind Drosophila mutant ninaB identifies the gene encoding the key enzyme for vitamin A formation invivo. *Proceedings of the National Academy of Sciences of the United States of America, 98*, 1130–1135.

Von Lintig, J., Moon, J., & Babino, D. (2021). Molecular components affecting ocular carotenoid and retinoid homeostasis. *Progress in Retinal and Eye Research, 80*, 100864.

Von Lintig, J., Moon, J., Lee, J., & Ramkumar, S. (2020). Carotenoid metabolism at the intestinal barrier. *Biochimica et Biophysica Acta (BBA)—Molecular and Cell Biology of Lipids, 1865*, 158580.

Wald, G. (1968a). The molecular basis of visual excitation. *Nature, 219*, 800–807.

Wald, G. (1968b). Molecular basis of visual excitation. *Science (New York, N. Y.)*, *162*, 230–239.
Wang, S., Yu, J., Jones, J. W., Pierzchalski, K., Kane, M. A., Trainor, P. A., ... Moise, A. R. (2018). Retinoic acid signaling promotes the cytoskeletal rearrangement of embryonic epicardial cells. *The FASEB Journal*, *32*, 3765–3781.
Wang, X., Penzes, P., & Napoli, J. L. (1996). Cloning of a cDNA encoding an aldehyde dehydrogenase and its expression in Escherichia coli. Recognition of retinal as substrate. *The Journal of Biological Chemistry*, *271*, 16288–16293.
Wardlaw, S. A., Bucco, R. A., Zheng, W. L., & Ong, D. E. (1997). Variable expression of cellular retinol- and cellular retinoic acid-binding proteins in the rat uterus and ovary during the estrous cycle. *Biology of Reproduction*, *56*, 125–132.
Weis, W. I., & Kobilka, B. K. (2018). The molecular basis of G protein-coupled receptor activation. *Annual Review of Biochemistry*, *87*, 897–919.
Wenzel, A., Oberhauser, V., Pugh, E. N., Jr., Lamb, T. D., Grimm, C., Samardzija, M., ... von Lintig, J. (2005). The retinal G protein-coupled receptor (RGR) enhances isomerohydrolase activity independent of light. *The Journal of Biological Chemistry*, *280*, 29874–29884.
Widjaja-Adhi, M. A. K., & Golczak, M. (2019). The molecular aspects of absorption and metabolism of carotenoids and retinoids in vertebrates. *Biochimica et Biophysica Acta (BBA)—Molecular and Cell Biology of Lipids*.
Widjaja-Adhi, M. A. K., Palczewski, G., Dale, K., Knauss, E. A., Kelly, M. E., Golczak, M., ... von Lintig, J. (2017). Transcription factor ISX mediates the cross talk between diet and immunity. *Proceedings of the National Academy of Sciences of the United States of America*, *114*, 11530–11535.
Wongsiriroj, N., Piantedosi, R., Palczewski, K., Goldberg, I. J., Johnston, T. P., Li, E., & Blaner, W. S. (2008). The molecular basis of retinoid absorption: A genetic dissection. *The Journal of Biological Chemistry*, *283*, 13510–13519.
Woodruff, M. L., Wang, Z., Chung, H. Y., Redmond, T. M., Fain, G. L., & Lem, J. (2003). Spontaneous activity of opsin apoprotein is a cause of Leber congenital amaurosis. *Nature Genetics*, *35*, 158–164.
Xue, Y., Sato, S., Razafsky, D., Sahu, B., Shen, S. Q., Potter, C., ... Kefalov, V. J. (2017). The role of retinol dehydrogenase 10 in the cone visual cycle. *Scientific Reports*, *7*, 2390.
Xue, Y., Shen, S. Q., Jui, J., Rupp, A. C., Byrne, L. C., Hattar, S., ... Kefalov, V. J. (2015). CRALBP supports the mammalian retinal visual cycle and cone vision. *The Journal of Clinical Investigation*, *125*, 727–738.
Yahyavi, M., Abouzeid, H., Gawdat, G., de Preux, A. S., Xiao, T., Bardakjian, T., ... Slavotinek, A. M. (2013). ALDH1A3 loss of function causes bilateral anophthalmia/microphthalmia and hypoplasia of the optic nerve and optic chiasm. *Human Molecular Genetics*, *22*, 3250–3258.
Yang, M., & Fong, H. K. (2002). Synthesis of the all-trans-retinal chromophore of retinal G protein-coupled receptor opsin in cultured pigment epithelial cells. *The Journal of Biological Chemistry*, *277*, 3318–3324.
Yau, K. W., & Hardie, R. C. (2009). Phototransduction motifs and variations. *Cell*, *139*, 246–264.
Young, B. D., Varney, K. M., Wilder, P. T., Costabile, B. K., Pozharski, E., Cook, M. E., ... Weber, D. J. (2021). Physiologically relevant free Ca(2+) ion concentrations regulate STRA6-calmodulin complex formation via the BP2 region of STRA6. *Journal of Molecular Biology*, *433*, 167272.
Zhang, H., Fan, J., Li, S., Karan, S., Rohrer, B., Palczewski, K., ... Baehr, W. (2008). Trafficking of membrane-associated proteins to cone photoreceptor outer segments requires the chromophore 11-cis-retinal. *The Journal of Neuroscience*, *28*, 4008–4014.

Zhang, J., Choi, E. H., Tworak, A., Salom, D., Leinonen, H., Sander, C. L., ... Palczewski, K. (2019). Photic generation of 11-cis-retinal in bovine retinal pigment epithelium. *The Journal of Biological Chemistry, 294*, 19137–19154.

Zhang, Y. R., Zhao, Y. Q., & Huang, J. F. (2012). Retinoid-binding proteins: Similar protein architectures bind similar ligands via completely different ways. *PLoS One, 7*, e36772.

Zheng, W. L., & Ong, D. E. (1998). Spatial and temporal patterns of expression of cellular retinol-binding protein and cellular retinoic acid-binding proteins in rat uterus during early pregnancy. *Biology of Reproduction, 58*, 963–970.

Zhong, M., Kawaguchi, R., Costabile, B., Tang, Y., Hu, J., Cheng, G., ... Sun, H. (2020). Regulatory mechanism for the transmembrane receptor that mediates bidirectional vitamin A transport. *Proceedings of the National Academy of Sciences of the United States of America, 117*, 9857–9864.

Ziouzenkova, O., Orasanu, G., Sharlach, M., Akiyama, T. E., Berger, J. P., Viereck, J., ... Plutzky, J. (2007). Retinaldehyde represses adipogenesis and diet-induced obesity. *Nature Medicine, 13*, 695–702.

Zolfaghari, R., & Ross, A. C. (2002). Lecithin:retinol acyltransferase expression is regulated by dietary vitamin A and exogenous retinoic acid in the lung of adult rats. *The Journal of Nutrition, 132*, 1160–1164.